SDGs and Textiles

Series Editor

Hafeezullah Memon, International Institute of Silk, Zhejiang Sci-Tech University, Hangzhou, China

The book series "SDGs and Textiles" addresses the strategies to achieve sustainable development goals (SDGs) in the present, past, and future. It presents books about the present and future policies of textile ministries of different countries, and books related to sustainability education around different parts of the world in the textile sector. Moreover, it would welcome the conference proceeding related to SDGs and Textiles. The series would cover books comparing the sustainability and SDGs of different institutions and countries. The individual book volumes in the series are thematic. The goal of each book is to give readers a comprehensive overview of a different area of sustainability in the textile sector. As a collection, the series provides valuable resources to a broad audience in academia, the research community, industry, and anyone looking to expand their knowledge of SDGs and Textiles.

Textiles and life are together – life cannot be separated from textiles as it is the most important need for human beings after food. In 2015, the United Nations General Assembly proposed 17 interlinked global goals to be achieved by 2030. Since then, academia and industry have paid much attention to achieving these goals. Textile found its close relation with almost all of these 17 goals.

SDG 1 - No Poverty: Poverty would never be overcome by a charity only; it is essential to develop people's skills to have a better and wealthy life. Thus, the textile can be considered an excellent discipline to achieve this goal by creating jobs and small and medium businesses.

SDG 2 - Zero Hunger: Through the effective utilization of advanced application of Agrotech Textiles, it is possible to have higher crop yields and save crops from rough weather, unexpected rains, floods, insects, etc.; thus, geotextiles play an essential in achieving this goal of sustainable development.

SDG 3 - Good Health & well-being: There has been much health consciousness after Covid19, and medical textiles assist in getting good health and well-being.

SDG 4 - Learning & Education: Textile or fashion has remained a significant discipline for societies for ages, and there has always remained much to explore in this field. Textile-related universities may play a vital role by offering free access to their education resources, training and spreading information among the locals.

SDG 5 - Gender Equality: The textile sector is one of the industrial sectors that accepted gender equality long ago; in particular, the garment sector has more females than males. Thus, the textile sector has been doing gender equality. Moreover, there has been a recent trend for Gender Neutral Clothing, which need worth studying and may further assist gender equality.

SDG 6 - Clean Water & Sanitation: Textiles could be achieved through filtration, and of course, textile is one of the critical materials for filtration.

SDG 7 - Affordable & Clean Energy: With the recent advancement in material science and engineering, the textile sector has come on the front for, not only by using this clean energy during textile production but also by assisting the production of this clean energy, either in the form of wind turbines blades made of textile composites or by energy harvesting from T-Shirts, etc.

SDG 8 - Decent Work: Recently, there has been much attention that the textile workers are not paid well, labor rights are not cared about, etc.

SDG 9 - Industry and innovation: Textile Industry always follows innovation; the textile companies that do not chase innovation cannot survive in the market.

SDG 10 - Reduced Inequalities: Getting better life and well-being would help reduce inequalities in the textile industry.

SDG 11 - Sustainable Cities: Sustainable Textile Cities through Buildtech and transport textiles.

SDG 12 - Consumption and Production: Textile and garment consumption and production all come under.

SDG 13 - Climate Action: Oekotech or Ecotech Textile, waste management of textiles are upfront to achieve this goal of sustainable development.

SDG 14 - Life Below Water: Mitigating microfiber waste in rivers and oceans may come under the context of it. There has been much attention on this subject after passing the bill at the parliament level of the UK.

SDG 15 - Life on Land: Geotech or Geotextiles studies life on land.

SDG 16 - Peace, Justice, and Strong Institutions: Protective textiles are doing their best to achieve peace, justice, and strong institutions.

SDG 17 - Partnerships for the Goals: The application of textiles to achieve sustainable development goals is only an example. In all textiles sectors, combined efforts of all the goals are essential to achieve true sustainability.

Martin Wynn · Tina Wiegand
Editors

Sustainability, the Circular Economy and Digitalisation in the European Textile and Clothing Industry

How Digital Technologies are Enabling the Circular Economy

Springer

Editors
Martin Wynn
School of Business, Computing and Social Sciences
University of Gloucestershire
Cheltenham, Gloucestershire, UK

Tina Wiegand
Hof University of Applied Sciences
Hof, Germany

ISSN 2948-1236 ISSN 2948-1244 (electronic)
SDGs and Textiles
ISBN 978-981-97-9115-6 ISBN 978-981-97-9116-3 (eBook)
https://doi.org/10.1007/978-981-97-9116-3

This Springer imprint is published by the registered company Springer Nature Singapore Pte Ltd.
The registered company address is: 152 Beach Road, #21-01/04 Gateway East, Singapore 189721, Singapore

Foreword

The textile and clothing sector was the cradle of the first industrial revolution in Europe and it has historically been shaped by structural changes driven by evolving external conditions. Nowadays the textile sector is one of the least sustainable industries worldwide. It is resource intensive and wasteful. A recent European Union report argued that "the textile industry is among the top three pressures on water and land use and the top five for raw materials use and greenhouse gas emissions worldwide".[1] The report emphasised the importance of taking action in order to address alarming figures that show that the production of textiles around the world doubled in the years 2000–2015, and that it is set to "more than double by 2030".

Due to its significant impact on the environment, health, and economic development across many parts of the world, the transformation of the textile and clothing industry towards sustainability and the circular economy is a crucial component in achieving the United Nations' Sustainable Development Goals. To this end, the European Commission published an *EU Strategy for sustainable and circular textiles,* which proposes strategic interventions that aim to tackle the lifecycle of textile products with a view of implementing actions that impact the production and consumption of textiles. The strategy was launched in 2020 and has gone through various revisions and adaptations. Although its goals and proposed actions are clearly laid out[2], their impact is still to be demonstrated.

Today, the European textile and clothing industry stands at a critical crossroads, influenced by sustainability and digital transformation. This is notable in some of the recent initiatives that have been introduced in order to ensure that consumers are

---

[1] European Commision. *How is the EU making fashion sustainable?.* Available at: https://environment.ec.europa.eu/topics/circular-economy/reset-trend/how-eu-making-fashion-sustainable_en. Accessed on: August 23, 2024.

[2] European Commission. *EU strategy for sustainable and circular textiles.* Available at: https://environment.ec.europa.eu/strategy/textiles-strategy_en#timeline. Accessed on: August 23, 2024.

protected from greenwashing[3] and that the new Ecodesign for Sustainable Products Regulations (ESPR) are understood and implemented.[4]

If the European textile and clothing industries embrace digital transformation, and if they succeed in reducing their overall environmental and climate impact on the EU market, then they have the potential to usher in the 5th Industrial Revolution in Europe by embracing and shaping the digital shift from linear mass production and outsourcing to more sustainable, collaborative practices and new consumption patterns, once again becoming the engine of economic and social development in Europe.

This book contains chapters on nine different European countries. A combination of qualitative and quantitative research methods is used across these chapters to answer three core research questions concerning sustainability strategies, circular economy practices and digital technology deployment in the textile and clothing industry. Martin Wynn and Tina Wiegand have provided us with unique insights into the intersection of sustainability, the circular economy, and digitalisation within the industry, and the carefully selected expert contributors offer new perspectives on how digital technologies can revolutionise sustainability and circular practices in manufacturing, supply chain management, and product lifecycle strategies.

The findings and case studies will be of interest to researchers and practitioners and for anyone who wants to support the transition to a sustainable and circular future in the European textile and clothing industry. For researchers, it offers broad approaches for further interdisciplinary research that combines environmental science, digital technology, economics and social sciences to address complex challenges related to industrial and social transformation. For policy makers and industry leaders, it offers approaches, strategies and ideas to shape sustainable and digital transformation in a way that could make the industry a beacon of international competitiveness again.

We are each implicated in this debate; and we ought to be invested in the successful transition of the European textile and clothing industry towards more sustainable models of production. Ultimately, however, it is our consumer behaviour that will ensure that these industries shift towards more responsible approaches.

Prof. Cathia Jenainati
Head of School of Business,
Computing and Social Sciences
University of Gloucestershire
Cheltenham, UK
CJenainati@glos.ac.uk

[3] European Commission. *The Road to Green 12: How is the EU Combatting Greenwashing*. Available at: https://environment.ec.europa.eu/news/video-how-eu-combating-greenwashing-2024-06-26_en. Accessed on: December 24, 2024.

[4] European Commission. *What's changing with new Ecodesign for Sustainable Products Regulation?*. Available at: https://environment.ec.europa.eu/news/video-ecodesign-sustainable-products-regulation-explained-2024-07-19_en. Accessed on: July 19, 2024.

Preface

The European textile and clothing (T&C) industry is at a critical crossroads, facing unprecedented challenges and opportunities driven by the twin forces of sustainability and digitalisation. The integration of these elements into the industry is not merely a trend but a fundamental shift that will determine the future of this sector. The journey towards a sustainable and circular textile economy, alongside the adoption of digital technologies, presents both immense potential and significant obstacles for companies of all sizes across Europe.

This book, entitled *Sustainability, the Circular Economy and Digitalisation in the European Textile and Clothing Industry—How Digital Technologies are Enabling the Circular Economy* seeks to explore this transition in depth. It aims to provide a comprehensive understanding of how the industry is navigating complex changes within various national and regional contexts. The chapters contained herein analyse the strategies, challenges, and success factors of various companies within the textile value chain, from small and medium-sized enterprises (SMEs) to multinational companies. This exploration is further enriched by the perspectives of various stakeholders, including suppliers, customers, consumers, and recovery actors, all of whom play a crucial role in shaping the industry's path forward.

The European Union's legislative framework, increasingly focused on sustainability, circularity, and digital transformation, provides the backdrop for this investigation. Various laws and regulations, such as the Extended Producer Responsibility (EPR) scheme and the Eco-design for Sustainable Products Regulation (ESPR), are pushing companies toward more sustainable practices. However, as the chapters in this book reveal, the pace and extent of this transformation vary significantly across countries and companies, depending on local market conditions, the availability of human and financial resources, and corporate culture.

This book covers nine countries, each presented in a dedicated chapter, with the findings eventually synthesised into an overarching conclusion. A brief overview of the main topics of each chapter is provided below.

The **UK chapter** focuses on five major textile manufacturing companies and five leading clothes retailers in the UK, predominantly multinational companies with global supply chains. The **Romania chapter** also examines large companies but pays particular attention to the strategies and practices of international brands operating in the country.

In contrast, the **Denmark chapter** focuses on SMEs with domestic production and recovery actors, intentionally excluding large international brands and retailers from the analysis.

The **Italy chapter** examines both small businesses within a traditional wool-processing industrial district and multinational luxury brands, both with a focus on innovative and collaborative approaches. The analysis of primary data also provides deep insights into company practices and allows conclusions to be drawn about industry-wide development paths.

The **chapter on France** limits its analysis also to SMEs involved in national sustainability initiatives, while the **chapters on Bulgaria, Türkiye, and Portugal** analyse randomly selected industry representatives from their respective countries. Finally, the **Germany chapter** builds upon a previous study to examine various manufacturing SMEs across different stages of the textile value chain. Drawing on primary data from interviews with industry representatives, the chapter provides new insights into the sustainability strategies, CE practices, and the state of digitalisation within these companies.

The extent to which the portrayal of each country, as presented in these studies, is comprehensive or influenced by the authors' disciplinary perspectives and the specific companies analysed, is difficult to determine. Nevertheless, the involvement of numerous authors, case studies, and analytical focuses collectively provides a broad overview of a highly heterogeneous European T&C industry. Despite significant national differences and challenges, the industry shares an awareness of its significant impact on sustainable development and is increasingly committed to developing innovative solutions to meet this responsibility.

The insights derived from these studies offer indications of general trends and innovative solutions within the European T&C industry, which are synthesised in the **final conclusion** into a maturity matrix which may usefully serve as a basis for further research and the development of innovative approaches and policy concepts to shape a sustainable and circular future for the industry in Europe.

This book provides a comprehensive overview of the diverse ways in which the European T&C industry is responding to the demands of sustainability and digitalisation. These responses not only reflect the industry's adaptability but also highlight the critical role that innovation, collaboration, and strategic foresight will play in determining its future. It is our hope that this book will serve as both a resource and an inspiration for those seeking to understand and contribute to the ongoing transformation of this vital industry.

Martin Wynn
School of Business, Computing
and Social Sciences
University of Gloucestershire
Cheltenham, Gloucestershire, UK
mwynn@glos.ac.uk

Tina Wiegand
Hof University of Applied Sciences
Hof, Germany
tina.wiegand@hof-university.de

Contents

Editors and Contributors

About the Editors

Martin Wynn is Associate Professor in Information Technology in the School of Business, Computing and Social Sciences at the University of Gloucestershire, UK, and holds a Ph.D. from Nottingham Trent University. He was appointed Research Fellow at East London University, and he then spent 20 years in industry at Glaxo Pharmaceuticals and HP Bulmer Drinks. His research interests include digitalisation, information systems, sustainability, project management, and urban planning. His latest book, Handbook of Research on Digital Transformation, Industry Use Cases, and the Impact of Disruptive Technologies, was published in 2022. e-mail: mwynn@glos.ac.uk

Tina Wiegand is a lecturer at Hof University of Applied Sciences, Germany. She completed a dual degree in Business Administration with a focus on supply chain management and a Master's degree with a focus on project management. She worked in the textile industry for over ten years in project and process management, with managerial responsibilities in operational and strategic purchasing for different corporate brands. Additionally, she gained experience in significant systems projects. Currently, she is pursuing a Ph.D. degree at the University of Gloucestershire, UK, researching digitalisation and circularity in the textile supply chain. e-mail: tina.wiegand@hof-university.de

Contributors

Sevil Acar Bogazici University, Istanbul, Türkiye

Teresa Raquel Barata Fiber Materials and Environmental Technologies (FibEnTech-UBI), Universidade da Beira Interior, Covilhã, Portugal

Nuno Belino Fiber Materials and Environmental Technologies (FibEnTech-UBI), Universidade da Beira Interior, Covilhã, Portugal

Erica Bicchi CERADE, ESAIP La Salle, Saint-Barthélemy-d'Anjou, France; LPG (Laboratoire Planétologie et Géosciences), University of Angers, Angers, France

Marta Bicho Fiber Materials and Environmental Technologies (FibEnTech-UBI), Universidade da Beira Interior, Covilhã, Portugal

Gianmarco Bressanelli University of Brescia, Brescia, Italy

Guilherme Hörner Bussolo University of Florence, Florence, Italy

Voicu D. Dragomir Bucharest University of Economic Studies, Bucharest, Romania

Mădălina Dumitru Bucharest University of Economic Studies, Bucharest, Romania

Sandra Ferreira Fiber Materials and Environmental Technologies (FibEnTech-UBI), Universidade da Beira Interior, Covilhã, Portugal

Cathryn Anneka Hall Kolding School of Design, Kolding, Denmark

Karen Marie Hasling Kolding School of Design, Kolding, Denmark

Peter Jones University of Gloucestershire, Cheltenham, UK

Kerli Kant Hvass Aaborg University, Copenhagen, Denmark; Revaluate, Hellerup, Denmark

Nikola Kiørboe Revaluate, Hellerup, Denmark

Assunção Mesquita CITEVE—Technological Centre for the Textile and Clothing Industry, Vila Nova de Famalicão, Portugal

Rui Miguel Fiber Materials and Environmental Technologies (FibEnTech-UBI), Universidade da Beira Interior, Covilhã, Portugal; CIAUD—University of Lisbon, Lisbon, Portugal

Semra Ocak Bogazici University, Istanbul, Türkiye

Madalena Pereira Fiber Materials and Environmental Technologies (FibEnTech-UBI), Universidade da Beira Interior, Covilhã, Portugal; UNIDCOM—IADE/UE, Lisbon, Portugal

Radoslav Radev University of Economics, Varna, Bulgaria

Nicola Saccani University of Brescia, Brescia, Italy

Ana Santiago Fiber Materials and Environmental Technologies (FibEnTech-UBI), Universidade da Beira Interior, Covilhã, Portugal

Mehri Shabani CERADE, ESAIP La Salle, Saint-Barthélemy-d'Anjou, France; Group Analysis and Processes (GA&P), University of Angers, Angers, France

Carla Joana Silva CITEVE—Technological Centre for the Textile and Clothing Industry, Vila Nova de Famalicão, Portugal

Rachel Vieira University of Gloucestershire, Cheltenham, UK

Helena Vilaça CITEVE—Technological Centre for the Textile and Clothing Industry, Vila Nova de Famalicão, Portugal

Filippo Visintin University of Florence, Florence, Italy

Tina Wiegand Hof University of Applied Sciences, Hof, Germany

Martin Wynn University of Gloucestershire, Cheltenham, UK

Chapter 1
Introduction

Tina Wiegand and Martin Wynn

1.1 Background

The production and consumption of industrial-scale products result in substantial natural resource depletion, with a projected demand that could exceed the capacity of almost three Earths by 2050 to sustain current lifestyles (United Nations, 2022). Within the same timescale, annual waste generation is expected to increase by 70 percent to over 3 billion tonnes (Kaza, 2018). If unchecked, this combination of resources depletion and waste increase will have catastrophic environmental impacts, including climate-damaging greenhouse gas emissions, biodiversity loss, and air, soil and water pollution (OECD, 2019).

The textiles and clothing (T&C) industry exemplifies wasteful resource utilisation. In recent decades, T&C production and consumption have undergone a significant escalation (Gözet & Wilts, 2022). Goods are discarded after an increasingly short use period, with only a low percentage being recycled into new textile fibres, resulting in significant environmental harm and resource depletion. The T&C industry is now the world's fourth-largest consumer of primary raw materials and water, and the fifth-largest emitter of greenhouse gases (European Environment Agency, 2019). Given the escalating detrimental impacts on both people and the environment, the prevailing linear "cradle-to-grave" approach to production and consumption, and the accompanying waste generation, are facing growing scrutiny.

The circular economy (CE) is viewed as a way to decouple future economic growth from resource consumption (European Commission, 2022a). The CE aims to achieve

T. Wiegand (✉)
Hof University of Applied Sciences, Hof, Germany
e-mail: tina.wiegand@hof-university.de

M. Wynn
University of Gloucestershire, Cheltenham, UK
e-mail: mwynn@glos.ac.uk

M. Wynn and T. Wiegand (eds.), *Sustainability, the Circular Economy and Digitalisation in the European Textile and Clothing Industry*, SDGs and Textiles,
https://doi.org/10.1007/978-981-97-9116-3_1

"zero waste and pollution throughout materials lifecycles" (Nobre & Tavares, 2021, p. 10) by replacing "the end-of-life concept with reducing, alternatively reusing, recycling, and recovering of materials throughout the supply chain" (Kirchherr et al., 2023, p. 4). In this context, the CE primarily pertains to the microeconomic level, focusing on businesses and their strategies to reduce resource consumption, mitigate environmental impacts like waste and emissions, and promote closed material loops.

While the concept of the CE is not new, it has grown in significance in recent years, becoming a focal point in various governmental and supranational sustainability agendas, such as the United Nations Sustainable Development Goals (United Nations, 2015) and the European Union's Green Deal (European Commission, 2019, 2020, 2022b). Digital technology deployment is seen as a crucial catalyst for orchestrating the comprehensive transition from a linear to a more circular economy, serving as a key enabler for implementing circular strategies and practices (European Commission, 2022a).

This transition poses significant challenges for companies, as decision-makers must accurately assess and reshape products, processes, relationships, systems and even business models, while considering competitive advantages and stakeholder interests (Battesini Teixeira et al., 2023; Franco, 2017; Jia et al., 2020).

1.2 The European Textile and Clothing Industry

The terms "textiles" and "clothing" are often used interchangeably, but there are significant differences. "Textiles" refer to structures made of natural or synthetic fibres, and the textile industry comprises companies involved in textile production or processing across four key areas: fibre preparation, fibre processing, yarn processing (e.g., knitting, weaving), and finishing (e.g., dyeing, printing) (Neugebauer & Schewe, 2015). In contrast, the "clothing" industry further processes textile products into clothing, encompassing activities such as design, creation and distribution of clothing on a national and international scale (Gesamtverband textil+mode, 2023). Figure 1.1 illustrates the primary products of the T&C industry.

With about 197,000 companies, more than 1.3 million employees and a turnover of around €170 billion in the T&C industry in 2023, the EU plays a pivotal role as a significant importer, exporter, and producer in a highly globalised T&C value chain (EURATEX, 2024). More than 90 percent of the companies are small and medium-sized enterprises with less than 250 employees. Italy leads the EU, with more than 40% of the total turnover of the EU's T&C industry, followed by Germany, France, and Spain (see Fig. 1.2). Within the EU, approximately 7 kg of textile products are produced per person per year, while per capita consumption is nearly 26 kg. The EU is, therefore, a net importer of textiles, these being mainly clothing products from Asia (European Environment Agency, 2019). Furthermore, in the EU, approximately 11 kg of textiles per person are discarded annually, with a very low collecting and recycling rate (EURATEX, 2022), with the majority of collected used clothing and textile waste being exported to Eastern European countries, Asia, and Africa, or

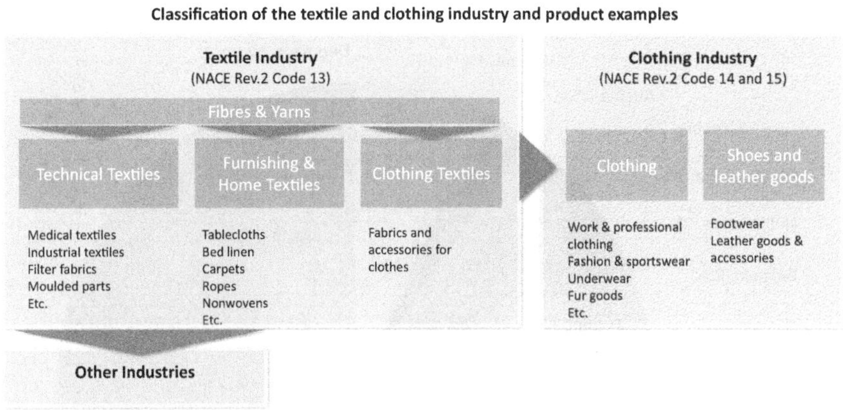

Fig. 1.1 Overview of the T&C industry. Based on Neugebauer and Schewe (2015) and EURATEX (2024)

ending up in landfills or incineration plants (European Environment Agency, 2019, 2023).

1.3 Problem Discussion and Research Objectives

According to the European Environment Agency (2019), clothing, shoes, and household textiles constitute the fourth-largest burden category (after food, housing, and transportation), as regards consumption of primary resources in the supply chain in the EU. This, coupled with the high environmental impacts and a low recycling rate, underscores the linear model of the textile supply chain and justifies its status as a focus industry within EU's Circular Economy Action Plan (European Commission, 2020), which aims to make textiles more durable, repairable, reusable, and recyclable, and promote innovation in the industry. New legal measures require T&C companies to have or provide new eco-design specifications, clearer information, a digital product passport, and specify carbon and environmental footprint reduction targets, among many other regulations (European Commission, 2022b). To achieve a fully circular economy, circularity and sustainability must be integrated into all stages of the value chain: from design and production to the use phase and waste management. However, the empirical exploration of what changes and collaborations are necessary and of value to implement sustainable and circular production and consumption patterns remains largely unexplored (Franco, 2017; Rinaldi et al., 2022). In particular, how digital technologies, such as Blockchain, Artificial Intelligence, the Internet of Things and others (Furferi et al., 2022; Rinaldi et al., 2022) can be effectively harnessed to support this transition has received relatively little focus in the research literature.

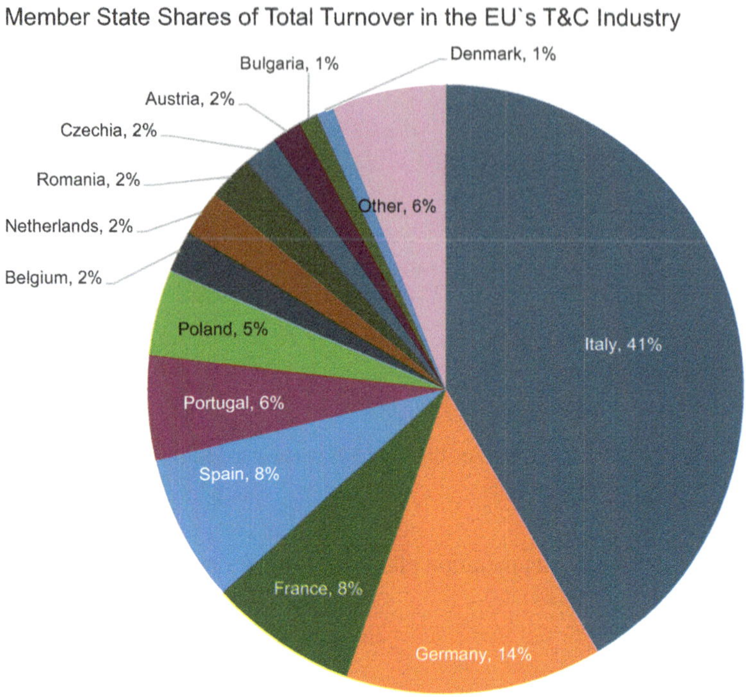

Fig. 1.2 Member state shares of total turnover in the EU's T&C industry (incl. NACE Rev. 2 Codes 13, 14, 15). *Source* Eurostat (2024)

Recently published research (Wiegand & Wynn, 2023) suggests that there are four conceptual and operational areas that must be progressed to successfully transition the T&C industry to a new model and sustainable future. First, the adoption of sustainable business strategies and activities that align with triple bottom line accounting and the United Nations Sustainable Development Goals; second, the integration of CE practices into the full range of business processes, from procurement and product design to production and customer engagement across the extended supply chain; third, the deployment and exploitation of digital technologies to enable and leverage CE practices and benefits; and fourth, an integrated management approach to holistic change that involves all relevant stakeholders—suppliers, manufacturers, retailers and consumers. The authors also suggested that this transition must encompass coordinated change in three main interrelated areas—processes, products and the organisation (Fig. 1.3).

The aim of this book is, therefore, to examine the nature of the European T&C industry, focusing on the relationship between sustainability strategies, CE practices, and digital technology deployment. The book provides a profile of the industry in several European countries, enabling comparisons and contrasts, as well as the development of recommendations for future changes and developments by:

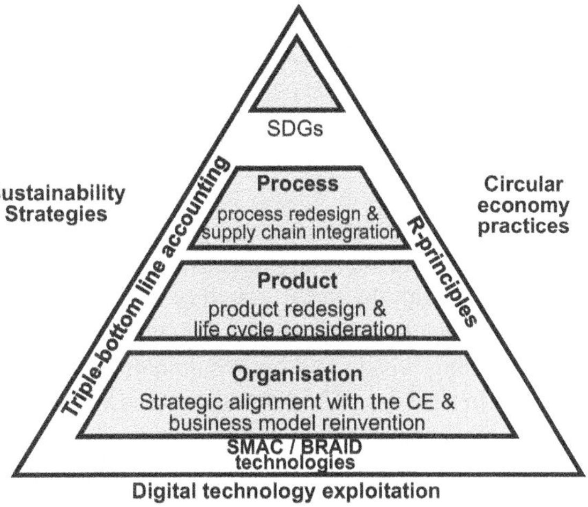

Fig. 1.3 A conceptual model for transitioning to the circular economy. *Source* Wiegand and Wynn (2023)

(1) exploring how the T&C industry is addressing sustainability in corporate strategies and activities
(2) assessing what strategies and activities relating to the CE are being pursued and
(3) examining the role of digital technologies in the transition to sustainability and the CE.

1.4 Research Methodology

The research methodology for this book is based upon a pragmatic case study approach, which emphasises the importance of practical implications and real-world applications and acknowledges the complex nature of the T&C industry. This is case study research in the sense that nine European countries are seen as case studies of how sustainability, the CE and digitalisation is impacting the T&C industry in each national context. However, within each case study, a number of different approaches are adopted to address the three main research objectives noted above. In the main, the approach is qualitative, with some chapters including primary data from interviews, for example, whilst other chapters rely mainly on secondary data sources. Overall, however, the chapters share the common ground of focusing on the three themes noted above, albeit the approach and methods vary somewhat from country to country (and from research team to research team).

According to Babbie (2010), the purpose of research is either exploratory, descriptive, or explanatory. Theory development in the field of sustainability, the CE and digitalisation is still in its infancy, and thus this research can be classified as mainly exploratory, with elements of descriptive research design which attempt to describe different characteristics of a phenomenon. Case studies are empirical inquiries which collect detailed information using a variety of data collection procedures over a sustained period of time (Stake, 2010). There is also a strong epistemological justification for using the case study methodology, since the aim is to achieve an understanding of a modern-day phenomenon that is both multifaceted and dynamic (Lewin & Johnston, 1997). In this context, the study aims to generate a comprehensive understanding of the status of the transition towards sustainability and circularity in the European T&C industry and the role digital technologies are playing in this transition.

A case study research strategy can be adopted to examine complex issues, allowing a full investigation and comprehension of emergent themes (Yin, 2017). It empowers a researcher with the tools necessary to inspect the information within a particular setting (Bernard, 2012) and it is a powerful research strategy that provides an all-encompassing, inside and out, examination of a particular phenomenon (Thomas, 2015). Yin (2017, p. 13) defines the case study research strategy as "an empirical inquiry that investigates a contemporary phenomenon within its real-life context, especially when the boundaries between phenomenon and context are not clearly evident; and in which multiple sources of evidence are used". Simons (2009, p. 21) also defines case study as "an in-depth exploration from multiple perspectives on the complexity and uniqueness of a particular project, policy, institution, programme or system in a real-life context".

The case study is a widely used research method within business and academic research. Bryman and Bell (2011) argue that the case study is particularly appropriate when used in conjunction with qualitative research methods, allowing detailed and intensive research activity, usually in combination with an inductive approach as regards the relationship between theory, research and practice. This is of particular relevance to the study of the T&C industry in a range of European countries. Saunders et al. (2019) argue that case studies are of particular value for explanatory or exploratory investigation, such as that pursued in this research. While case study research usually involves working with one or more related cases, it often leads to the generation and collection of a lot of data (Gomm et al., 2000), and some researchers have indicated that conducting case study research can sometimes be rigorously long and hard to direct (Noor, 2008). Here, the case studies under investigation focus on the T&C industry in a range of European countries. The book reports on the findings from nine different countries and also draws upon available literature in a number of other European countries where the T&C industry has been the subject of recent research.

The generalisation of findings to a wider population is a significant issue. With the case study strategy, being able to generalise from one case study is quite complex (Thomas, 2015), and this is often seen as a limitation of the case study strategy. Tight (2017) suggests that the use of multiple case studies offers an obvious acceptable strategy to combat this limitation, and this study makes use of multiple case studies in order to support generalisation and validate findings. The use of multiple case studies allows the researcher to seek similarities or differences between cases (Baxter & Jack, 2008), and although information from one case study can provide insight into a phenomenon, multiple case studies allow for the aggregation of knowledge from multiple units (Noor, 2008). With multiple case studies, data from wider perspectives can be gathered and analysed rigorously with a mind to generalise the findings beyond

Table 1.1 T&C industry (incl. NACE 13, 14, 15) parameters in countries featured in the research as of 2020

Country	Turnover (€ million)	Number of companies	Number of employees
Italy	67,708	352,928	451,623
Germany	22,746	9,495	136,231
Türkiye	14,287[a]	91,788	1,131,497
France	13,010	27,433	121,476
UK[b]	10,695	8,987	95,136
Portugal	9,089	14,799	174,148
Romania	4,018	8,194	159,216
Bulgaria	1,753	4,893	89,898
Denmark	1,421	715	4,934

Source Eurostat (2024) and Turkish Statistical Institute (2023)
[a] Turnover in Euro (converted with exchange rate 1 TL = 0,034 Euro as of Oct 2023)
[b] Data for the UK as of 2018 (last year of EU Membership)

just a single unit but across industries, countries or other particular interests. Although they are generally considered time consuming, expensive and complex to carry out (Gustafsson, 2017), the results from these studies are generally accepted in academia to produce more reliable results when compared with single case studies (Baxter & Jack, 2008; Blaxter, 2010). Yin (2017) suggests that the use of six to ten case studies is sufficient to provide compelling support for an initial set of propositions, and this aligns with the approach adopted here.

The cases within the EU countries studied here include the top three in terms of revenue (Italy, Germany and France), the fifth and ninth (Portugal and Romania), and two of the smaller nations (Bulgaria and Denmark). Outside the EU, the UK has a turnover in excess of €10 billion in the industry, at roughly the same level as Portugal. Table 1.1 indicates turnover in 2020[1] for the countries studied, and in addition shows the number of companies and employees in the industry in each country. This reveals significant differences in the size of companies, with France averaging four employees per enterprise, whilst Germany has almost 15, Bulgaria more than 18 and Romania over 19. Turnover per employee also varies significantly with Denmark, Italy, Germany, and France being the most revenue generative, and Romania and Bulgaria the least. This no doubt reflects the differing nature of the industry in these countries, which is explored in more detail in the core chapters of the book that follow.

[1] Data from Eurostat for annual detailed enterprise statistics for industry (NACE Rev. 2, B-E) are only available for the years 2005–2020.

References

Babbie, E. R. (2010). *The practice of social research*. Wadsworth Inc Fulfillment.

Battesini Teixeira, T. G., de Medeiros, J. F., Kolling, C., Duarte Ribeiro, J. L., & Morea, D. (2023). Redesign in the textile industry: Proposal of a methodology for the insertion of circular thinking in product development processes. *Journal of Cleaner Production, 397*, 136588. https://doi.org/10.1016/j.jclepro.2023.136588

Baxter, P., & Jack, S. (2008). Qualitative case study methodology: Study design and implementation for novice researchers. *The Qualitative Report, 13*(4), 544–559. https://doi.org/10.46743/2160-3715/2008.1573

Bernard, H. R. (2012). *Social research methods: Qualitative and quantitative approaches* (2nd ed.). SAGE.

Blaxter, L. (2010). *How to research* (4th ed.). *UK higher education OUP humanities and social sciences study skills Ser*. McGraw-Hill/Open University Press. http://site.ebrary.com/lib/academiccompletetitles/home.action

Bryman, A., & Bell, E. (2011). *Business research methods* (3rd ed.). Oxford University.

EURATEX. (2022). *Facts & key figures 2022 of the European textile and clothing industry*. https://euratex.eu/wp-content/uploads/EURATEX_FactsKey_Figures_2022rev-1.pdf

EURATEX. (2024). *Facts & key figures 2024 of the European textile and clothing industry*. https://euratex.eu/wp-content/uploads/EURATEX-Facts-Key-Figures-2024.pdf

European Commission. (2019). *The European green deal: COM(2019) 640 final*. https://eur-lex.europa.eu/resource.html?uri=cellar:b828d165-1c22-11ea-8c1f-01aa75ed71a1.0002.02/DOC_1&format=PDF

European Commission. (2020). *COM(2020) 98 final: A new circular economy action plan: For a cleaner and more competitive Europe*. https://eur-lex.europa.eu/resource.html?uri=cellar:9903b325-6388-11ea-b735-01aa75ed71a1.0017.02/DOC_1&format=PDF

European Commission. (2022a). *Circular economy*. https://joinup.ec.europa.eu/collection/rolling-plan-ict-standardisation/circular-economy-0

European Commission. (2022b). *EU strategy for sustainable and circular textiles: COM(2022) 141 final*. https://environment.ec.europa.eu/strategy/textiles-strategy_en

European Environment Agency. (2019). *Textiles and the environment in a circular economy: Eionet report—ETC/WMGE 2019/6*. https://www.eea.europa.eu/publications/textiles-and-the-environment-the

European Environment Agency. (2023). *EU exports of used textiles in Europe's circular economy. EEA briefing: no. 1/2023*. Publications Office of the European Union.

Eurostat. (2024). *Annual detailed enterprise statistics for industry (NACE Rev. 2, B-E)*. https://ec.europa.eu/eurostat/databrowser/view/SBS_NA_IND_R2__custom_7709083/default/table

Franco, M. A. (2017). Circular economy at the micro level: A dynamic view of incumbents' struggles and challenges in the textile industry. *Journal of Cleaner Production, 168*, 833–845. https://doi.org/10.1016/j.jclepro.2017.09.056

Furferi, R., Volpe, Y., & Mantellassi, F. (2022). Circular economy guidelines for the textile industry. *Sustainability, 14*(17), 11111. https://doi.org/10.3390/su141711111

Gesamtverband textil+mode. (2023). *Branchen* [In English: Industry sectors]. https://textil-mode.de/de/verband/branchen/

Gomm, R., Hammersley, M., & Foster, P. (2000). Case study and generalization. In R. Gomm, M. Hammersley, & P. Foster (Eds.), *Case study method: Key issues, key texts* (Reprinted., pp. 98–115). SAGE.

Gözet, B., & Wilts, H. (2022). *The circular economy as a new narrative for the textile industry: An analysis of the textile value chain with a focus on Germany's transformation to a circular economy*. Zukunftsimpuls (No. 23). https://wupperinst.org/en/a/wi/a/s/ad/7733

Gustafsson, J. (2017). *Single case studies vs. multiple case studies: A comparative study*. http://www.diva-portal.org/smash/get/diva2:1064378/FULLTEXT01.pdf

Jia, F., Yin, S., Chen, L., & Chen, X. (2020). The circular economy in the textile and apparel industry: A systematic literature review. *Journal of Cleaner Production, 259*, 120728. https://doi.org/10.1016/j.jclepro.2020.120728

Kaza, S. (2018). *What a waste 2.0: A global snapshot of solid waste management to 2050. Urban development*. The World Bank. https://doi.org/10.1596/978-1-4648-1329-0

Kirchherr, J., Yang, N.-H.N., Schulze-Spüntrup, F., Heerink, M. J., & Hartley, K. (2023). Conceptualizing the circular economy (revisited): An analysis of 221 definitions. *Resources, Conservation and Recycling, 194*, 107001. https://doi.org/10.1016/j.resconrec.2023.107001

Lewin, J. E., & Johnston, W. J. (1997). Relationship marketing theory in practice: A case study. *Journal of Business Research, 39*(1), 23–31. https://doi.org/10.1016/S0148-2963(96)00152-X

Neugebauer, C., & Schewe, G. (2015). Wirtschaftsmacht Modeindustrie – Alles bleibt anders [In English: Economic power of the fashion industry—Everything remains different]. *Aus Politik Und Zeitgeschichte, 65*(1–3), 31–41. https://www.bpb.de/shop/zeitschriften/apuz/198384/wirtschaftsmacht-modeindustrie-alles-bleibt-anders/#footnote-target-54

Nobre, G. C., & Tavares, E. (2021). The quest for a circular economy final definition: A scientific perspective. *Journal of Cleaner Production, 314*, 127973. https://doi.org/10.1016/j.jclepro.2021.127973

Noor, K. B. M. (2008). Case study: A strategic research methodology. *American Journal of Applied Sciences, 5*(11), 1602–1604. https://doi.org/10.3844/ajassp.2008.1602.1604

OECD. (2019). *Global material resources Outlook to 2060*. https://www.oecd-ilibrary.org/content/publication/9789264307452-en10.1787/9789264307452-en

Rinaldi, F. R., Di Bernardino, C., Cram-Martos, V., & Pisani, M. T. (2022). Traceability and transparency: Enhancing sustainability and circularity in garment and footwear. *Sustainability: Science, Practice and Policy, 18*(1), 132–141. https://doi.org/10.1080/15487733.2022.2028454

Saunders, M., Lewis, P., & Thornhill, A. (2019). *Research methods for business students* (8th ed.). Pearson.

Simons, H. (2009). *Case study research in practice* (1. publ). SAGE.

Stake, R. E. (2010). *The art of case study research*. SAGE.

Thomas, G. (2015). *How to do your case study* (2nd ed.). SAGE.

Tight, M. (2017). *Understanding case study research: Small-scale research with meaning*. SAGE.

Turkish Statistical Institute. (2023). *Annual industry and service statistics (NACE Rev.2)*. https://data.tuik.gov.tr/Kategori/GetKategori?p=sanayi-114&dil=2

United Nations. (2015). *Transforming our world: The 2030 agenda for sustainable development*. https://sustainabledevelopment.un.org/post2015/transformingourworld

United Nations. (2022). *Goal 12: Ensure sustainable consumption and production patterns*. https://www.un.org/sustainabledevelopment/sustainable-consumption-production/

Wiegand, T., & Wynn, M. (2023). Sustainability, the circular economy and digitalisation in the German textile and clothing industry. *Sustainability, 15*(11), 9111. https://doi.org/10.3390/su15119111

Yin, R. K. (2017). *Case study research and applications: Design and methods* (6th ed.). SAGE.

Tina Wiegand (tina.wiegand@hof-university.de) is a lecturer at Hof University of Applied Sciences, Germany. She completed a dual degree in Business Administration with a focus on supply chain management and a Master's degree with a focus on project management. She worked in the textile industry for over ten years in project and process management, with managerial responsibilities in operational and strategic purchasing for different corporate brands. Additionally, she gained experience in significant systems projects. Currently, she is pursuing a Ph.D. degree at the University of Gloucestershire, UK, researching digitalisation and circularity in the textile supply chain.

Martin Wynn (mwynn@glos.ac.uk) is Associate Professor in Information Technology in the School of Business, Computing and Social Sciences at the University of Gloucestershire, UK and holds a Ph.D. from Nottingham Trent University. He was appointed Research Fellow at East London University, and then spent 20 years in industry at Glaxo Pharmaceuticals and HP Bulmer Drinks. His research interests include digitalisation, information systems, sustainability, project management, and urban planning. His latest book, Handbook of Research on Digital Transformation, Industry Use Cases, and the Impact of Disruptive Technologies, was published in 2022.

Chapter 2
United Kingdom

Rachel Vieira and Peter Jones

Abstract The UK textile and clothing (T&C) industry has a global reach and embraces a range of elements, including raw material production, treatment and manufacturing, and all of these elements pose testing sustainability challenges. With this in mind, this chapter looks to explore how some of the major players within the UK's T&C industry are publicly addressing sustainability. More specifically, it focuses on three research questions. Firstly, how are the leading companies within the UK's T&C industry publicly communicating their transition to sustainability? Secondly, in what ways are the UK T&C companies harnessing the CE to support this transition? Thirdly, what role are digital technologies playing in facilitating sustainability and the CE in UK T&C companies? The findings reveal that while the majority of the selected companies have publicly addressed sustainability, albeit in varying measure, the companies have provided much less material in the public realm on the role of the CE and the digital technologies in facilitating corporate approaches to sustainability. More generally, the authors offer some critical reflections on the selected companies' approaches to sustainability, the CE, and the digital technologies.

Keywords Sustainability · Circular economy · CE · Digital technologies · UK textile and clothing industry

2.1 Introduction and Background

The UK textile and clothing (T&C) industry has a global reach and embraces a range of elements, including the production of the raw materials, a variety of treatment and manufacturing processes, retailing and consumption, and all these elements face

R. Vieira (✉) · P. Jones
University of Gloucestershire, Cheltenham, UK
e-mail: RVieira@glos.ac.uk

P. Jones
e-mail: PJones@glos.ac.uk

M. Wynn and T. Wiegand (eds.), *Sustainability, the Circular Economy and Digitalisation in the European Textile and Clothing Industry*, SDGs and Textiles,
https://doi.org/10.1007/978-981-97-9116-3_2

a variety of environmental and social challenges. The production of the raw materials, spinning them into fibres, weaving these fibres into fabrics and dyeing, for example, require large volumes of water, as well as chemicals, notably pesticides and insecticides. Widespread concerns have been expressed about working conditions, notably forced labour and modern slavery, in T&C manufacturing factories in retailers' supply chains. Here consumer demand for cheap clothes and fast fashion is seen to be driving the clothing retailers' business model, which is based on cheap labour and low prices. At the same time, western consumers' clothing shopping behaviour generates the emission of greenhouse gases, associated shopping journeys, and online shopping deliveries, and is more generally seen to be part of unsustainable patterns of consumption. In short, the transition to sustainable patterns of development within the T&C industry faces a variety of pressing public challenges, and according to DeSL (2022), a digital transformation solutions consultancy, "sustainability is no longer optional" in the "retail, fashion, apparel and footwear industries" (para. 14). A variety of approaches have been put forward to tackle the industry's sustainability challenges, including adopting a more circular business model and harnessing the digital technologies.

The T&C industry is complex and can be seen to include a number of interlinked elements, namely the production of materials, including cotton fibres, leather, fur and silk that form the primary units in cloth making the manufacture of men's, women's and children's apparel; and retailing. Within the UK, the retail element is arguably the most prominent. The retailing of clothes embraces a wide range of routes to market, including department stores, speciality stores, boutiques, discount stores, outlet stores, supermarkets, retail warehouses, online sales, and second-hand shops.

The T&C industry was a driving force in the UK's industrial revolution in the seventeenth and eighteenth centuries and a leader in world exports. It developed as a result of the demand for clothing for the country's rapid growing population. Before the eighteenth century, the T&C industry was essentially a domestic industry and had been undertaken by individual workers in their own homes. Initially the clothes produced from wool and flax were functional, but by the middle of the eighteenth century, increasingly skilled craftsmen and craftswomen were using cotton and linen to make more fashionable clothes, demanded by the more affluent sections of the population.

During the eighteenth century, a number of inventions and innovations, including the flying shuttle, the water frame, the spinning jenny, and the power loom, along with the harnessing of steam power, saw the textile industry transformed from a craft-based cottage industry to a highly mechanised one, centred on factories (or mills, as they were widely known) in Manchester and South Lancashire. Textile manufacture in the UK was at its height at the turn of the twentieth century, but declined during that century in the face of increasing competition from cheaper overseas producers and the growing importance of synthetic fibres, including rayon and nylon. In 2020, there were 8,245 manufacturers operating in the UK fashion and textile sector (ukft, 2021).

In this context, this chapter explores the T&C industry in the UK, and assesses how some of the industry's major players are publicly addressing sustainability. In doing so, it considers if, and how, the circular economy (CE) and digital technologies are playing a role in the transition to sustainability within the industry in the UK. More specifically it addresses the following inter-related research questions (RQs):

RQ1 How are the leading companies within the UK's T&C industry publicly communicating their transition to sustainability?
RQ2 In what ways are the UK T&C companies harnessing the CE to support this transition?
RQ3 What role are digital technologies playing in facilitating sustainability and the CE in UK T&C companies?

This chapter comprises five main sections. Following this introduction, the research method is briefly discussed in Sect. 2.2, based on a literature analysis of the available material provided by eight selected companies in the T&C industry in the UK. Section 2.3 then addresses the RQs, followed by a discussion Sect. 2.4, in which some of the key emergent themes are set out and examined. Finally, the concluding Sect. 2.5 provides a summary of the chapter and offers some thoughts on possible areas for future research.

2.2 Research Method

The research method employed in this study is inductive and qualitative, being founded on an interpretivist philosophy. Gill and Johnson (2002) concluded that such an approach is well-suited to exploratory research which aims to develop an explanation of the phenomenon being studied. There were four main phases to the research process (Fig. 2.1). First, a scoping literature review was undertaken to research key themes and develop the research questions noted above in Sect. 2.1. As observed by Hanneke et al. (2017), scoping reviews "are best employed when there is limited literature to inform the research question of interest" (p. 5), as was the case in this study. Various academic databases, including Google Scholar and Science Direct, were searched for relevant literature.

In phase 2 of the research, and in order to explore the RQs, leading companies in the T&C sector were identified for more detailed scrutiny. The top five textile manufacturing companies, namely Coats, MAF, Quantum Clothing Group, Karazeen Textile and Jimmy Choo, as listed by ZoomInfo (2023), and the top five clothing retailers, namely Next, J.D. Sports, Primark, Marks and Spencer (M&S), and Asda, as listed by Retail Economics (2023), were selected for investigation. The authors then chose a simple method for further enquiry, based on three sets of Internet searches, conducted on the Google search engine, in October 2023. These searches used the key terms "sustainability", "CE", and "digital technologies" in conjunction with the names of the top five textile manufacturing companies, and the top five clothing retailers.

The search revealed that of the selected ten companies, five companies (Coats, Next, JD Sports, Primark, and Asda) posted dedicated sustainability reports (or

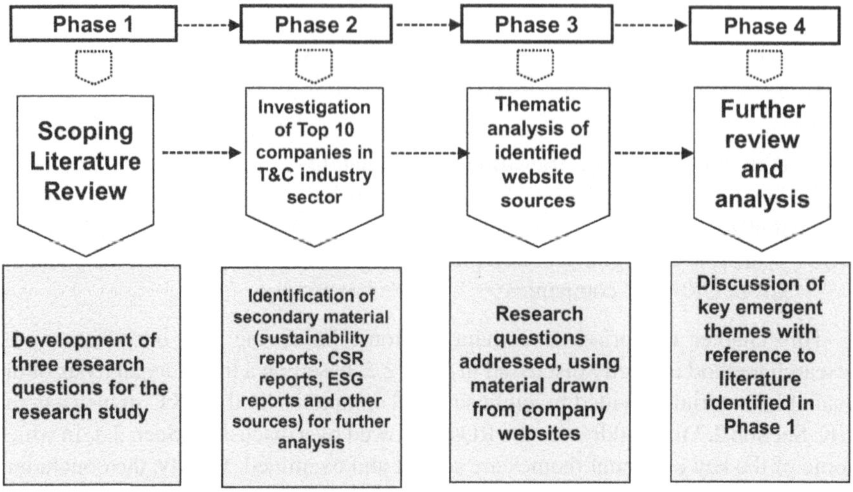

Fig. 2.1 The four phases in the research process

Corporate Social Responsibility reports or Environmental, Social and Governance reports, or Integrated Annual Reports which embraced sustainability), and that a further three companies (Quantum Clothing Company, Jimmy Choo, and M&S) posted information on their approach to sustainability on their websites. The investigation therefore focused on these eight companies. Further, while only Coats had posted specific information on its approach to the CE, the sustainability reports posted by Jimmy Choo, Next, JD Sports, Primark, and M&S included some material on circularity, and while Asda's sustainability report did not mention the CE explicitly, it did include material on recycling. Only three of the selected companies posted material on how the digital technologies were harnessed to promote sustainability. Phase 3 of the research analysed this material to directly address the three RQs set out in Phase 1. Then, in Phase 4, further analysis identified additional perspectives for review and discussion. The empirical material on which this chapter is based was drawn from the corporate reports and information generated by these searches. Pen pictures of the selected companies are provided below.

Coats is a UK based multinational thread manufacturing company, and was founded in Paisley, Scotland in 1755, has operations across 50 countries and employs 17,000 people. The company works with 30,000 apparel and footwear manufacturers and 4,000 retailers and its products are sold in over 100 countries. The Quantum Clothing Company manufactures and supplies clothing and apparel, particularly women's lingerie, sports and shapewear products, to a number of European retail companies. The company, founded in 2000, has its Head Office in the UK and owns manufacturing sites in Sri Lanka and Cambodia, employing 5,000 employees. Jimmy Choo, founded in 1996 but acquired by Capri Holdings in 2006, is a luxury UK fashion company.

Next is a multinational UK clothing and footwear retailer, originally founded in 1864. The company trades from some 700 stores, of which 500 are in the UK, where it is the country's largest clothing retailer, with a further 200 in mainland Europe, Asia and the Middle East. JD Sports is a UK sports omnichannel fashion retailer, founded in 1981. The company trades from 3,400 stores principally in the UK, but also in mainland Europe, North America, and the Asia Pacific region. Primark is a leading clothing international retailer, founded by Associated British Foods in 1969, and has over 400 outlets across Europe and the US. The retailer's products include menswear, womenswear, and children's wear, and the company has some 70,000 employees. Marks and Spencer is a UK multinational retailer, established in 1894. The company's product range includes, clothing, food, and household goods. Asda is a UK supermarket retailer founded in Yorkshire in 1965, it trades from over 600 locations within the UK, and the company's product range includes food, general merchandise and clothing. The George at Asda brand is one of the UKs leading clothing brands, and is available within the company's stores and in some locations in free standing outlets.

The chapter draws extensively on quotations from the selected T&C companies' websites, and the aim here was to explore how the selected companies publicly expressed, and looked to evidence, their approaches to sustainability, the CE, and the digital technologies. Here, the authors were of the opinion that an important way of capturing such approaches was to cite the companies' own words, not least in that such citations could convey corporate authenticity, and offer greater depth of understanding.

2.3 Findings

2.3.1 How Are the Leading Companies Within the UK's T&C Industry Publicly Communicating Their Transition to Sustainability? (RQ1)

The sustainability reports and information posted by the selected UK T&C companies varied in their detail and content, but rather than looking to describe each company's approach to sustainability in detail, the authors looked to identify, and draw out, a number of interlinked general themes to provide a narrative account to capture how the companies were addressing sustainability. These are: strategic corporate commitment; climate change and greenhouse gas emissions; energy efficiency; materials; water; biodiversity; waste; responsible sourcing and supply chains; human rights and modern slavery; health and safety at work; diversity, equity and inclusion; links with communities; and the United Nations Sustainable Development Goals (SDGs).

Strategic corporate commitment was articulated in a variety of ways. In his introduction to Primark's "Sustainability and Ethics Report 2021/22" (Primark, 2023), Paul Marchant, the company's Chief Executive Officer, claimed:

> at Primark, we have always been about quality clothing at affordable prices. Now we want to help people buy better and more sustainably. This is what our customers tell us they want, and now more than ever, they want to be able to do this without breaking the bank. Our Primark Cares strategy is deliberately set out to challenge the premise that affordable fashion and sustainability are incompatible. We want to show that we can use our size and scale to drive meaningful change to make more sustainable fashion affordable for all, to reduce our impact on the planet and to better support the livelihoods of the people who make our clothes. We understand that we have a critical role to play along with the whole industry to deliver this change (p. 2).

Jimmy Choo (2024a, para. 1) introduced its "Sustainability Manifesto" with the claim that "at Jimmy Choo we are glamorous optimists, designing products that create beauty and deliver joy, holding ourselves to the highest standards of excellence and craftsmanship. We are now deepening that promise by embarking on a journey towards a sustainable future." In a similar vein, in his introduction to Coats' 2022 sustainability report, Rajiv Sharma, the Group Chief Executive, emphasised that "sustainability lies at the heart of our business strategy" (Coats, 2023, p. 6), while Regis Schultz, Chef Executive Officer at JD Sports (2023, p. 23), claimed "we are very passionate at JD about sustainability".

All the selected T&C companies identified climate change is a major, perhaps *the* major, challenge to sustainability. Asda (2023, p. 8), for example, recognised that "climate change is the defining crisis of our time, it's happening much more quickly than previously anticipated, and will have a major impact on our business". Marks and Spencer (2023, para. 7) emphasised "the impact of climate change affects all of us and the scientific evidence is clear—urgent action is needed by everyone". More specifically, a few of the selected T&C companies posted details of their targets to reduce greenhouse gas emissions. Next, for example, reported its commitment to reduce Scope 1 emissions (which occur within a company and are within its control) and Scope 2 emissions (which occur when energy is produced and supplied from outside the company) by 55% by 2030, against a 2016/2017 baseline, while Marks and Spencer (2023) reported its commitment to a similar reduction.

Notwithstanding the role of reducing and changing energy use in tackling climate change, several of the selected companies reported on their plans to increase energy efficiency. Next (2023, p. 22) emphasised that one of its priorities was "to work to improve energy efficiency", and the company reported that it actively tracked its energy performance in an attempt to "ensure our properties are operating efficiently". More specifically, next reported that during 2022, it had optimised the heating and cooling systems within its stores to reduce the amount of electricity used, and that it fitted additional solar panels in its warehouses. More generally next claimed that it worked closely with its energy provider to actively identify new energy efficiency opportunities. In addressing "distribution efficiency", Next (2023, p. 22) reported that it was involved its testing electric vehicles and investigating the possibility of using hydrogen as an alternative fuel.

The consumption of raw materials is widely seen as a challenging issue within the T&C industry. Coats (2023, p. 33), for example, reported that "during 2022, we consumed 121 thousand tonnes of direct raw materials, the majority of which comprised the fibres and filaments", that "84% of our products are made from oil-derived plastic materials, principally polyester", and that the company is looking "to deliver a transition to 60% sustainable raw materials by 2026". Next (2023, p. 15) reported that "cotton is our most significant raw material, both in terms of volume used and the associated environmental and social issues linked with its production", and the company reported employing a number of measures, including looking to reduce the dependency on pesticides and synthetic fertilisers, to reduce water use, to improve the social conditions for farmers, to improve the economic sustainability of farmers, and to reduce water use.

Indeed, the T&C industry uses large volumes of water, especially in the dyeing process. Coats (2023) claimed:

> we recognise our responsibility to use water as efficiently as possible to ensure that our water extraction activities create no negative impact on the local communities and biodiversity in the areas in which we operate. We are committed to minimising freshwater abstraction, and commit to reducing water consumption especially in areas of high-water stress, and we strive to return the water we have used to the environment in a fit state for use by others. (p. 41)

More specifically, Coats (2023) reported on its investment in Twine, a start-up company, which is developing digital dyeing technology for yarns. Primark (2023) acknowledged:

> water is a critical natural resource within the fashion industry – from the irrigation of the cotton fields to the dyeing and finishing of our fabrics and materials. It is vital that we manage it effectively. Ensuring factories of our suppliers manage their water use effectively and control their wastewater responsibly is crucial for helping to reduce our environmental impact as a business (p. 31).

Many of the damaging impacts of the T&C industry are generated at three stages of the industry's value chain, namely material production; material preparation and processing; and product use and disposal. Next (2023, p. 12) reported that it looked to "reduce our overall impact on biodiversity and natural ecosystems in our supply chain", while JD Sports (2023, p. 78) claimed to ensure that "licensed cotton farms and farmers adopt management plans to conserve and enhance biodiversity". More expansively, Primark (2023, p. 29) reported commissioning a biodiversity consultancy to conduct a biodiversity risk assessment of its operations, which helped shape its biodiversity strategy, and claimed that "given that cotton is our most widely used raw material, our biodiversity strategy has cotton at its core". Here, the focus is on regenerative farming, but the company acknowledged that given that its suppliers source cotton from different geographical locations, this effectively means that no single approach is appropriate, and perhaps more importantly that improving biodiversity will be a long-term project. However, Primark claimed to have been working closely with farmers and to be developing a monitoring framework to assess biodiversity.

Waste is a major problem across the T&C industry and many of the selected companies have developed programmes to reduce its impact. Coats (2023), for example, recognised two types of waste generated within its overall operations, namely, product and packaging waste generated principally by its manufacturing activity, and liquid effluent and sludge, generated by wet processing activities. Further, Coats (2023, p. 46) argued that "reducing materials usage and minimising the impact of the materials that we consume is a very high priority for us", and claimed that adopting the European Union's Waste Framework Directive in 2019 gave the company increased transparency across the business, and assisted the company in the identification of opportunities for reduction, and where possible, prevention of waste generation streams across its business. In addressing effluent waste, Coats (2023, p. 48) claimed to recognise its responsibilities "as temporary custodians of the water we use, and are committed to ensure that this sometimes-scarce resource is returned to the local communities in which we operate in a fit state". The Quantum Clothing Group (n.d., para. 3) simply claimed to be "minimising the use of resources and the creation of waste wherever possible and practical".

Next (2023, p. 14) emphasised that "we are committed to responsible sourcing and ensuring traceability back to source", and acknowledged that "we use raw materials such as cotton, polyester, man-made cellulosics, wool, timber and leather. These raw materials can have a wide-ranging environmental and social impact during their production and extraction. If not responsibly sourced, this can contribute to critical issues such as deforestation, water scarcity and climate change". Further, the company claimed that its "responsible sourcing strategy sets out our ambition to source 100% of the key raw materials we use through known, responsible or certified routes by 2025" (Next, 2023, p. 14). Next also reported on how it looked to secure the responsible sourcing of polyester, man-made cellulosic fibres, microfibres and leather. Marks and Spencer (2023) reported that 100% of the cotton used in its clothing products is responsibly sourced and that the company was empowering farmers to use water more efficiently, to care for the soil and to respect biodiversity.

Responsible sourcing also covers a series of major economic and social challenges, not least in supply chains. Under the banner "Improving People's Lives", Primark (2023, p. 35) acknowledged its commitment to "pursuing a living wage for workers in our supply chains" and reported that "our approach builds on what is already a well-established process to monitor the payment of workers' wages", and that "this is a key aspect of social audits conducted in our suppliers' factories through our Ethical Trade programme". In a similar vein, JD Sports (2022, p. 94) reported that its ethical code of practice was reviewed by "our ESG Committee and is shared with all suppliers. The Code follows the International Labour Organization minimum standards". In addition "living wages are paid in line with local laws and for a standard working week, overtime must be paid at premium rate", that "the organisation shall respect the right of personnel to a living wage and ensure that wages for a normal work week, not including overtime, shall always meet at least legal or industry minimum standards, or collective bargaining agreements", and that "wages shall be sufficient to meet the basic needs of personnel and to provide some discretionary income" (p. 73). JD Sport's ethical code of practice also emphasises freedom of association and the right to collective bargaining, that working hours must not be excessive and must be voluntary, and that regular employment be provided. Here again, the company reported that factories in its supply chain are independently audited.

Protecting workers in supply chains is certainly a major challenge for T&C companies, and a few of them have explicitly addressed human rights, and more specifically, modern slavery. Next (2023, p. 7), for example, reported "the violation of human rights anywhere in our operations is unacceptable and we deal firmly with any infringements identified in our supply chain". Coats emphasised its full support for the United Nations Guiding Principles on Business and Human Rights in its operations, and that it upheld the United Nations Declaration of Human Rights. At the same time, Coats (2023, p. 58) reported that "our Group Internal Audit team include a series of 30 human resource audit areas in their audit templates", that most of these are related to compliance "with our employment policies or directly with human rights issues", and that "during 2022 they completed 8 audits (compared to 11 in 2021) and identified 10 minor issues requiring remediation within a number of people-related process areas, compared to 13 in 2021".

Some, but not all, of the selected T&C companies, reported on their policies on modern slavery, as part of their approach to human rights and protecting workers. Next (2023, p. 6) emphasised "we will not tolerate any instance of modern slavery in our business or in our supply chain. Next products should be made by workers who are treated honestly and fairly for the work they undertake and whose human rights and wellbeing are respected". In looking to evidence this commitment, Next (2023, p. 6) reported that it had "worked with 13 factories to successfully remediate modern slavery issues", and "an additional four sites are being supported through an agreed remediation process", and that it had "disengaged seven factories where remediation of modern slavery issues had not been achieved to an acceptable level." Primark (2023, p. 11) reported that "we ban all forms of modern slavery, including child labour, forced labour and human trafficking", and that "we publish our Modern Slavery Statement annually". These statements are accessible via a hyperlink from the company's sustainability report.

Health and safety at work are also important elements in protecting workers. Under the banner "Health and Safety", Coats (2023), for example, outlined its approach to "occupational safety", which is seen to "be at the heart of our business strategy", and "health and wellbeing" (p. 56). In focussing upon its approach to health and wellbeing, the company emphasised its commitment "to the wellbeing of all employees" and acknowledged "workplace mental health as an important issue for the business and its employees". More specifically, they reported that its "Energy4Performance strategy" looked to provide a health promoting workplace", which embraced "health-related opportunities and risks being identified and regularly assessed"; "incentives being provided to local management to offer health-promoting programs"; "medical services in plants based on local law and regulations"; and "increased employee awareness about the benefits of a healthy lifestyle" (Coats, 2023, p. 56). Jimmy Choo (2024a, para. 1) emphasised it commitment "to ensuring that our employees enjoy a healthy working life, both within our business and across our supply chain", that "we prioritize the safety and wellbeing of our Jimmy Choo community, our benefits are designed to meet the healthcare, financial, work, life and mental wellbeing needs of eligible employees", and that "we are committed to strengthening the labour protections of our artisans by routinely auditing suppliers around health and safety".

A commitment to diversity, equity and inclusion feature in the sustainability reports of many of the selected companies. Next (2023, p. 27), emphasised "we are an equal opportunities employer and we offer career opportunities without discrimination", "we treat all employees fairly regardless of gender, sexual orientation, marital status, race, colour, nationality, religion, ethnic or national origin, age, disability or union membership status", and "we believe that having a diverse and inclusive working environment is vital to supporting the health and wellbeing of our employees." Under the banner "Championing Diversity and Inclusion", Primark (2023, p. 43) claimed "it's important that all our people are treated with dignity and respect, which means listening to our colleagues and learning what's important to each person", that "our workforce is a diverse collection of individuals, all with their own stories, viewpoints, ideas, and insights. It makes us a stronger, more adaptive and resilient business", and that "we are also careful to understand local variations and comply with relevant legislation in each market".

Primark (2023) also reported its commitment to "promote equal opportunities for women." In addressing this commitment, Primark emphasised that "women account for the majority of its global supplier factory workforce", and that the company's goal is to strengthen the position of women in the garment industry "through skills development and addressing barriers to progression by 2030". These barriers were seen to "include the risks of gender-based violence and harassment in the workplace and unpaid care work, which often forces women out of the workplace" (Primark, 2023, p. 37). More specifically, the company outlined its largest skills programme, which supports factories to introduce an efficient in-house training system for both new and existing sewing operators, most of whom are women. They receive quality technical training which improves their productivity and can often open-up opportunities for promotion. JD Sports (2022, p. 88) reported on its "Behind the Label" project, which supports women and girls around the world, with specific focus on women working within the textile manufacture supply chain, and this project includes a commitment to fund initiatives to support women in those countries seen to be most at risk of gender-based harassment and violence in the workplace.

A few of the selected apparel companies reported on their links with local communities and on their charitable donations, as part of their sustainability commitments. Coats (2023), for example, reported on its contribution to flood alleviation in Pakistan, a Save the Children campaign in Mexico, a book donation and tree planting programme for school children in Bangladesh, tree planting programmes in Mexico and Romania, and beach clear ups in Sri Lanka and Bangladesh. In addressing "Supporting Our Communities", Next (2023) emphasised its commitment "to support causes that make a real difference", noting "we particularly focus on supporting charities and organisations that have an impact in the countries and communities we source from and operate in". The company reported on its approach which looked to "provide donations that are of the most benefit—this can be in the form of financial or product donations, expertise, knowledge or time", and to support charities for a number of years with a specified annual donation, as this commitment helps them plan their work with confidence and "allows us to become strategic partners" (Next, 2023, p. 29). More specifically, Next reported that its charitable giving policy focused on the advancement and promotion of health and supporting both end of life and emergency care services; the advancement of education, life and work skills and the development of youth amateur sport; the advancement of environmental protection or improvement; and reducing inequality, by supporting the promotion of diversity, inclusion, human rights and by the prevention or relief of poverty.

Some of the selected T&C companies outlined how their approach to sustainability was aligned to the United Nations SDGs, which were agreed at a United Nations General Assembly in 2015 and designed to shift the world on to a sustainable and resilient path. In reporting its alignment to eight of the SDGs, Coats (2023), outlined why each of them was relevant to the company, identified the company's opportunities and responsibilities, its priorities, planned actions, and desired goals. In addressing SDG13, climate action, for example, Coats recognised that their activities contributed to global warming, and that it had a responsibility to reduce its greenhouse gas emissions and those of its value chain, in line with minimising the damage from climate change, and that the principal opportunity to achieve this objective was via a transition to renewable electricity and via energy conservation schemes.

In a similar vein, Next (2023) reported its alignment to nine of the SDGs, covering poverty, health and well-being, gender equality, clean water and sanitation, decent work and economic growth, responsible consumption and production, climate action, life on land, and partnerships. In outlining its focus on life on land, Next claimed to be, "committed to the protection of natural resources and will source all land based raw materials responsibly", (Next, 2023, p. 5), and that partnerships underpin all the company's business activities. In contrast, JD Sports (2022, p. 62) reported that its sustainability strategy was aligned to the three SDGs it considered "most relevant" to the T&C industry, namely the promotion of inclusive and sustainable economic growth, sustainable production and consumption, and acting to combat climate change and its impacts.

2.3.2 *In What Ways Are the UK T&C Companies Harnessing the CE to Support This Transition? (RQ2)*

Only two of the selected T&C companies, namely Coats and Jimmy Choo, posted specific information on the CE on their websites, but some of the other companies included details of their approach to circularity in their sustainability reports. Jimmy Choo specify that this stance is due to the fact that

> "our products are designed to last and come with a five-year repair service.... we aim to give our products a second life and be active in the resale market", and "we are committed to ongoing research and innovation in circular product design, as well as collaborating with others to contribute to a redesign of the systems required to create a truly circular fashion industry" (Jimmy Choo, 2024b, para. 3).

In January 2021, Coats announced that it had joined the Ellen MacArthur Foundation Network, which looks to develop and promotes the concept of the CE. More specifically Coats (2023, p. 34) showcased its "Ecocycle" innovation, which the company claimed will "propel us towards a fashion landscape where circularity is not just a concept but a reality". The company claimed that "Ecocycle" introduced "the ground-breaking concept of water dissolvable threads", which dissolves the seams at 95 degrees Centigrade during a 30-min wash cycle, and enables garments to be pulled apart into recyclable components, with zips, buttons, and labels being easily removed, thus paving the way for end-of-lifecycle recycling and championing the ideals of a CE. The company claims that Ecocycle can be used in a range of garments, including knitwear, dresses, underwear, children's apparel and workwear. Coats also provided details of Epic EcoVerde, a 100% recycled premium polyester sewing thread which offers durable, attractive seams across a wide range of recycled fabrics and materials.

For some of the selected companies, the CE seemed to be predominantly linked to waste management, while others looked to embrace circularity across their value chain. Under the banner "Moving Towards Circularity", Next (2023), for example, reported it was "supporting the transition to a more circular economy by designing, producing and selling products that limit pollution and waste and help to keep materials in use for longer" (p. 17). In looking to evidence its commitment to circularity next outlined that it was working to reduce packaging, using customers' returned packaging to create new packaging, and to develop takeback schemes to ensure valuable resources were kept in circulation.

Marks and Spencer (2024a) claimed "as a business we are committed to driving circularity and sending zero waste to landfill from our own operations and construction activities in the UK and Republic of Ireland". Further, the company emphasised that it was "working across our value chain to drive circularity and reduce waste in key areas such as used clothing, food waste, packaging, and carrier bags", that "we've started our transition to new circular business models, by improving the recyclability of our packaging and goods, increasing the durability and use of recycled materials in our products, and introducing refill and reuse options for customers", and that "we support the transition to a sustainable circular economy and will prioritise business model innovation, working with our partners to put circular ways of working into practice" (Marks & Spencer, 2024a, paras. 1–3).

More specifically focussed on clothing, Marks and Spencer (2024a) outlined the company's "Circular Partnerships", namely its "M&S Hire Street", "Shwopping" initiatives, and reported that it had joined the "Dotte collective" (paras. 11–12). The M&S Hire Street initiative, founded in 2021 and run by the Zoa Group, enables customers to rent up to seven items, while the Shwopping initiative, introduced in 2008, allows customers to recycle their clothes, which are donated to Oxfam, by placing them in Shwop boxes in the company's UK stores. The company reported that in 2021/22, the initiative generated £227,00 for Oxfam. In addition, they described the Dotte collective as "the UK's first fully circular kidswear peer-to-peer marketplace where parents can buy, sell, donate and recycle kidswear" (Marks & Spencer, 2024b, para. 1).

More expansively, Asda (2023) argued that "circularity is a key pillar of our George clothing sustainability strategy", suggested that its "customers were more likely than the average UK customer to participate in circular initiatives, such as repairing, donating, buying second-hand and selling used items", and claimed that "as we develop our ranges, we are designing with longevity in mind, aiming to ensure our products last longer, can have a second life and at the end of life be recycled or disposed of in a responsible and low-impact way" (p. 16). JD Sports (2022) reported developing its "supply chain to extend material and product life at every opportunity", and while the company acknowledged that "this is not circularity by definition", it argued that "extending product life represents an investment in the same principles that support the CE" (p. 86).

2.3.3 *What Role Are Digital Technologies Playing in Facilitating Sustainability and the CE in the UK T&C Companies? (RQ3)*

Only two of the selected companies, namely Coats and Primark, had posted information on how the digital technologies were contributing to sustainability and/or the CE, though two other companies, Marks and Spencer and JD Sports, referred to such contributions in their sustainability reports. Coats (2023) claimed to offer "business critical garment software and fashion industry expertise, powering an agile, resilient, transparent and sustainable supply chain", and that "our fashion supply chain software harnesses industry best practices and the latest technology to improve speed to market, agility, efficiency and sustainability". The company argues that digital technologies enabled it to "simplify complexity", and that "in an increasingly volatile, uncertain and complex world, in which sped, efficiency and sustainability are terms of trade, we enable fashion businesses to meet these challenges by focusing on simplifying and optimising key business processes which have disproportionate impact on performance" (para. 2). Although Coats looked to showcase a number of its digital technology innovations on its corporate website, the focus was on improving productivity and business efficiency, rather than on sustainability and the CE. Primark (2023) reported on using a digital platform, "Tracebale", to track cotton from farmers in its supply chain.

Marks and Spencer (2023) reported on a variety of ways in which the digital technologies were contributing on its approach to sustainability, including integrating data from environmental performance in the company's supply chain into its clothing footprint, improving the quality and granularity of packaging data, tracking greenhouse gas emissions, complementing supplier audit processes, and providing equal opportunities for women. JD Sports (2023, p. 75) reported on using QR code identifiers on digital passports, to enable private label products to be traced through the supply chain, from mill to distribution centre, thus enhancing supply chain transparency.

2.4 Discussion

The findings revealed that selected T&C companies publicly addressed a wide range of themes in reporting on their approach to sustainability, though they were much less fulsome in reporting on the role that the CE and the digital technologies were playing in their overall transition to sustainability. That said, a few wider issues, relating to sustainability, the CE and the digital technologies, merit reflection and discussion. At the outset, it is important to recognise that that sustainability is a contested concept, which "means different things to different people" (Aras & Crowther, 2008, p. 436), and while some definitions are rooted in ecological principles, others include social and economic, as well as environmental, goals, and look to embrace equity in meeting human needs. Typical of the first set is Sutton's (2004) definition of environmental sustainability as "the ability to maintain things or qualities that are valued in the physical environment" (p. i), while the second set is reflected in McCann-Erickson's (2007) definition that sustainability "is a collective term for everything to do with the world in which we live", and "is about consuming differently and consuming efficiently" (p. 6).

In some ways, the selected T&C companies' approach to sustainability can be interpreted as being built around economic efficiencies and thus to reflect the demands of their businesses, as much as a fundamental concern for the maintenance and enhancement of natural resources and ecosystems, or the welfare of their employees, or of the communities in which they operate. While several of the selected T&C companies' environmental policies are designed to reduce energy consumption, and thus to reduce greenhouse gas emissions, they also produce cost savings. At the same time, commitments to diversity in the workplace, to health and safety initiatives, and to investment in local communities, can be seen to promote stability and loyalty amongst the workforce.

At the same time, these approaches and commitments to sustainability can be seen as aspirational, in that they are often couched in terms of what the companies hope and plan to achieve in the future. However, such approaches could essentially be seen as public relations, or marketing exercises, rather than genuine commitments to sustainable development. This could be seen as "greenwashing", designed to appeal to consumers and investors, who are increasingly concerned about the environmental and social impact of the T&C industry's operations, while effectively ignoring fundamental and social concerns.

Adamkiewicz et al. (2022) looked to demonstrate the extent to which greenwashing could jeopardise the fashion industry. The authors identified a number of challenges related to the implementation of a more sustainable CE, suggesting that these included a lack of accurate information, a lack of transparency and doubts about sharing information, and argued that education programmes and the active participation of all stakeholders in the circular value chain were important in addressing such challenges. At the same time, the independent external assurance of the sustainability reports published by some of the selected companies was, at best, limited, and this could be seen to undermine the credibility of the companies claims about their sustainability achievements. More generally, Williams (2024) argued that the processes of obscuring social and ecological relations via greenwashing are central to the (dis) functioning of contemporary capitalism.

While a few of the selected T&C companies emphasised their commitment to responsible sourcing, not least from their supply chains, such commitments present several challenges. Bhandari et al. (2022), for example, argued that a transition to sustainable sourcing was essential if the fashion industry was to address some of the environmental and social challenges facing it, but suggested that a transition from conventional methods to a new sustainable business model posed a variety of challenges for companies and for their supply chain partners. More specifically, Bhandari et al. (2022) identified a range of barriers to the implementation of sustainable sourcing in the fashion industry but suggested that "insufficient commitment from top management" (p. 225) was the most important barrier to implementing sustainable sourcing in the fashion industry. This finding emphasises the critical importance of securing full executive support for sustainability strategies, and for ensuring that sustainability is reflected throughout the business.

Working conditions, and more specifically, concerns about human rights and modern slavery in the selected T&C companies' supply chains, have attracted attention from consumers, pressure groups and governments, not least in that the Global Slavery Index 2018 (Walk Free Foundation, 2018) listed the textiles and fashion industry as one of the largest global perpetrators of modern slavery. Here, the approach adopted by some of the selected companies can be seen to be expectational, at least one step removed from their direct corporate control, in that they put the emphasis on the suppliers to ensure compliance with their policies on human rights and modern slavery. In an attempt to verify that suppliers are complying with their supply chain policies, a number of the selected companies reported employing a variety of audit processes. Here, it is important to recognise that though widely used in industries with complex and geographically diverse supply chains, the role of auditing to verify compliance has attracted considerable criticism. LeBaron et al. (2017, p. 958), for example, argued that "the growing adoption of auditing as a governance tool is a puzzling trend, given two decades of evidence that audit programs generally fail to detect or correct labour and environmental problems in global supply chains".

In many ways, fast fashion epitomises the sustainability challenges facing the T&C industry. In focusing on the environmental costs of fast fashion, the United Nations Environment Programme (2022) argued that "sustainable fashion and circularity in the textiles value chain are possible, yet this century the world's consumers are buying more clothes and wearing them for less time than ever before, discarding garments as fast as trends shift" (para. 3). More specifically, the Geneva Environment Network (2024, para. 5) has emphasised that "fast fashion also has a human cost: textile workers, primarily women in developing countries, are often paid derisory wages and forced to work long hours in appalling conditions. In many places, these conditions create infringements on human rights", while the "use of chemicals in clothes production also raise serious health concerns, both for the workers in the industry and consumers". Looking more positively towards the future, Chen (2024) looks to explore the complex interplay between financial imperatives and environmental responsibilities in the fast fashion industry, and concludes that digital transformation could be an effective solution to this problem for the fast fashion industry.

More generally, there are, arguably fundamental, tensions between sustainability and economic growth, and the issues of sustainable consumption and the current structure of market economies loom large. Sustainable consumption was described by Cohen (2005, p. 25) as "the most obdurate challenge for the sustainable development agenda", and indeed in many ways the T&C industry, and more specifically, the fashion industry, epitomise unsustainable consumption. The selected T&C companies' business models are based on continuing growth and aggressive marketing strategies, essentially to promote consumption, but such growth is dependent on the continuing depletion of the earth's finite resources, and is incompatible with sustainability. At the same time, many consumers may be unwilling to reduce their levels of consumption, as any such reduction may be seen as a retrograde step, in societies where consumption is prized and where consumption has become part of many people's identity.

Here, Ropke (2009, p. 2492), for example, argued "trying to persuade individuals that curbing consumption would contribute to environmental improvements and increased wellbeing, has little chance of success". On the other hand, major changes in the levels of consumption also pose challenges for the current structure of market economies. In looking to address such challenges Jackson (2018, p. 18) argued "it is entirely fanciful to suppose that deep emission and resource cuts can be achieved without confronting the structure of market economies", while Mansfield (2009, p. 37) claimed that mainstream approaches to sustainable development fail to recognise "the political nature of sustainability".

While a transition to a more circular economy has been proposed as a way of tackling the sustainability challenges facing the T&C industry, as with sustainability, the concept of the CE can have different meanings. Korhonen et al. (2018, p. 545), for example, saw "the circular economy as an essentially contested concept", and "in terms of scientific research, it appears to be vague". Arguably more critically, Gregson et al. (2015, p. 218) suggested the idea of the CE is "more often celebrated than critically interrogated" and that "its actual enactment is limited and fragile". Corvellec et al. (2020, p. 97) acknowledged that the CE "allows for a whole range of interpretations and approaches to be bundled together". As such, this effectively allows companies to define the CE to mean what they want it to mean.

The transition to a circular business model within the T&C industry would certainly constitute a dramatic change in the ways in which consumers approach consumption and could see the emergence of a "new consumption culture" (Korhonen et al., 2018, p. 41). Indeed, the emergence of such a new culture of consumption where the focus would be on co-operative endeavour, is seen to be central to the CE, with "user groups and communities sharing the use of the function, service and value of physical products" (Korhonen et al. 2018, p. 548). Such a transition towards a CE would surely be seen to challenge the current social value which consumers often ascribe to many of the products and services they buy, which may in turn, make it difficult for large numbers of consumers to buy into second hand, or reusable, patterns of consumption. While Marks and Spencer, for example, do facilitate a clothes recycling scheme, as noted earlier, at the present time, the company seems unlikely to radically change their current business model to fully accommodate reuse and recycling at the expense of their current retail operations.

While harnessing the digital technologies has been suggested as offering a way forward in tackling the sustainability challenges in the T&C industry, the increasing use of the digital technologies also brings new sets of responsibilities and generates new environmental and social challenges. In some ways, the increasing adoption of digital technologies, which effectively enable retail customers, for example, to buy a wide range of clothes instantly, regardless of their location or the time of day, at the touch of a button or key, might be seen to be the antithesis of sustainability. Not least because the data centres which drive the digital technologies, and on which they ultimately depend, are major energy users and, as such, contribute to greenhouse gas emissions and to global warming.

There are also thorny social issues about whose interests are best served by the digital technologies. Lobschat et al. (2021, p. 879), for example, argued "the multi-sided natures of many markets for digital products and services makes the assessment of beneficence for all involved stakeholders so complex". More worryingly, van der Merwe and Al Achkar (2022) suggested that even when companies claim to have recognised their digital responsibilities, this can offer corporations an opportunity to build a cover for unethical behaviours and practices, and the authors emphasised the need for "a comprehensive approach to ensure data are managed responsibly and protect customers" (para. 2).

2.5 Conclusion

This chapter has provided an exploratory narrative on how some of the leading players in the T&C industry in the UK are publicly addressing sustainability, and the extent to which the CE and the digital technologies are contributing to such corporate policies and strategies. The findings reveal that while the majority of the selected companies have publicly addressed sustainability, albeit in varying measure, the companies have provided much less material in the public realm on the role of the CE and the digital technologies in facilitating corporate approaches to sustainability. Such approaches to sustainability embraced a wide range of environmental and social issues, but many of the selected companies' approaches to sustainability can be seen to be aspirational and expectational. A minority of the selected companies emphasised their corporate commitment to the CE, and while some companies seemed to focus their approach to circularity principally in terms of waste management, some of the selected retail companies outlined their initiatives on recycling and reuse. The role of the digital technologies in promoting sustainability received little public attention from the selected companies.

More generally, the authors offered some wider critical reflections on the selected companies' approaches to sustainability, the CE and the digital technologies. Reflections on sustainability focused on the different meanings of sustainability and the relative importance of business drivers vis-à-vis environmental and social sustainability drivers, the aspirational nature of many of the selected companies' approaches to sustainability, and the spectre of greenwashing. In addition, the challenges of securing sustainability, and more specifically of safeguarding human rights and tackling modern slavery, within supply chains were discussed, as were the tensions between sustainability and continuing business growth, as regards, for example, sustainable consumption and the structure of market economies. All these issues have important implications for the T&C industry as it looks to pursue the transition to a more sustainable future.

In reflecting on the role of the CE in contributing to the selected companies' approaches to sustainability, the authors drew attention to the varied meanings of the concept, which effectively enables companies to define it to suit their own purposes, without radically changing their traditional business models, and perhaps more tellingly, argued that corporate moves to adopt a comprehensive circular approach would have major implications for consumers. Finally, the authors also argued that harnessing the digital technologies to contribute to sustainability had important environmental and social implications for corporate approaches to sustainability.

The chapter has several limitations, not least in that it is drawn exclusively from Internet sources, in that it involves the authors' selection from these sources, and that in offering a narrative of how the selected T&C companies publicly addressed sustainability and how the CE and the digital technologies are contributing to those companies' approaches to sustainability, it involves considerable generalisation. Nevertheless, the authors believe that the chapter not only offers a valuable exploratory picture of sustainability challenges and corporate strategies and measures to address these challenges within the UK's T&C industry, but that it may also provide a platform for future research. Future research agendas might, for example, include detailed empirical investigations into how major companies within the UK's T&C industry are looking to develop their sustainability strategies, and into if, how, and why, they are looking to adopt more circular business models and utilise digital technologies within their approach to sustainability. At the same time, research into UK consumers' perceptions of the nature and importance of sustainability within the T&C industry, and into the extent to which consumers may be willing to change their patterns of T&C buying behaviour given the increasing awareness of the varied environmental and social challenges facing T&C companies, may also pay dividends.

References

Adamkiewicz, J., Kochańska, E., Adamkiewicz, I., & Lukasik, R. M. (2022). Greenwashing and sustainable fashion industry. *Current Opinion in Green and Sustainable Chemistry, 38*. https://www.sciencedirect.com/science/article/pii/S2452223622001225

Aras, G., & Crowther, D. (2008). Governance and sustainability: An investigation into the relationship between corporate governance and corporate sustainability. *Management Decision, 43*(6), 433–448. https://doi.org/10.1108/00251740810863870

Asda. (2023). *Environmental, social and governance report*. https://asdagroceries.scene7.com/is/content/asdagroceries/Asda.com/7.%20Sites/Creating%20Change%20for%20Better/Asda-ESG-Report-2023.pdf

Bhandari, N., Garza-Reyes, J. A., Rocha-Lona, L., Kumar, A., Naz, F., & Joshi, R. (2022). Barriers to sustainable sourcing in the apparel and fashion luxury industry. *Sustainable Production and Consumption, 31*, 220–235. https://www.sciencedirect.com/science/article/pii/S2352550922000410

Chen, Z. (2024). Transforming fast fashion: The digital pathway to sustainable supply chains. In *Proceedings of the 3rd International Conference on Education, Language and Art (ICELA 2023): Advances in Social Science, Education and Humanities Research* (pp. 537–545). https://www.atlantis-press.com/proceedings/icela-23/125998592

Coats. (2023). *Sustainability report*. https://assets.lsegissuerservices.com/original-5a713518-a3db-499a-bb1c-9e21378463cc.pdf

Cohen, M. J. (2005). Sustainable consumption in national context; an introduction to the special issue. *Sustainability: Science, Practice and Policy, 1*(1), 22–28. https://www.researchgate.net/publication/26404569_Sustainable_consumption_in_national_context_An_introduction_to_the_symposium

Corvellec, H., Bohm, S., Stowell, A., & Valenzuela, F. (2020). Introduction to the special issue on the contested realities of the CE. *Culture and Organization, 26*(2), 97–102.

DeSL. (2022). *Sustainability*. https://www.desl.net/apparel-sustainability/

Geneva Environmental Network. (2024). *Environmental sustainability in the fashion industry*. https://www.genevaenvironmentnetwork.org/resources/updates/sustainable-fashion/

Gill, J., & Johnson, P. (2002). *Research methods for managers* (3rd ed.). London.

Gregson, N., Crang, M., Fuller, S., & Holmes, H. (2015). Interrogating the CE: The moral economy of resource recovery in the EU. *Economy and Society, 44*(2), 218–243. https://doi.org/10.1080/03085147.2015.1013353

Hanneke, R., Asada., Y., Lieberman, L., Neubauer, L., & Fagan, M. (2017). *The scoping review method: Mapping the literature in structural change public health interventions* (Vol. 94, pp. 1–14). Department of Public Health Scholarship and Creative Works. SAGE Research Methods Cases Part 2. https://doi.org/10.4135/9781473999008

Jackson, T. (2018). Governance for prosperity. *Review of Economic Philosophy, 11*(1), 11–33. https://www.cairn.info/revue-de-philosophie-economique-2010-1-page-11.htm&wt.src=pdf

JD Sports. (2022). *Annual report and accounts 2022*. https://www.annualreports.com/HostedData/AnnualReportArchive/J/LSE_JD_2022.pdf

JD Sports. (2023). *Annual Report and Accounts 2023*. https://www.annualreports.com/HostedData/AnnualReports/PDF/LSE_JD_2023.pdf

Jimmy Choo. (2024a). *Our sustainability manifesto*. https://www.jimmychoo.com/en/sustainability-manifesto.html

Jimmy Choo. (2024b). *Sustainability: CE*. https://www.jimmychoo.com/en/circular-economy.html

Korhonen, J., Honkasalo, A., & Seppala, J. (2018). CE: The concept and its limitations. *Ecological Economics, 143*(1), 37–46. https://econpapers.repec.org/article/eeeecolec/v_3a143_3ay_3a2018_3ai_3ac_3ap_3a37-46.htm

LeBaron, G., Lister, J., & Dauvergne, P. (2017). Governing global supply chain sustainability through the ethical audit regime. *Globalizations, 14*(6), 958–975. https://doi.org/10.1080/14747731.2017.1304008

Lobschat, L., Mueller, B., Eggers, F., Brandmarte, L., Diefenbach, S., Kroschke, M., & Wirtz, J. (2021). Corporate digital responsibility. *Journal of Business Research, 122*, 875–888.

Mansfield, B. (2009). Sustainability. In N. Castree, D. Demeriff, D. Liverman, & B. Rhoads (Eds.), *A companion to environmental geography* (pp. 37–49). Wiley.

Marks and Spencer. (2023). *Reshaping M&S; sustainability report 2023.* https://corporate.marksandspencer.com/sites/marksandspencer/files/sustainability-report-2023.pdf

Marks and Spencer. (2024a). *CE and waste.* https://corporate.marksandspencer.com/sustainability/plan-a-our-planet/waste-circular-economy

Marks and Spencer. (2024b). *Hand-me-down quality: M&S kidswear becomes the first major high street brand to join the Dotte Resale Collective.* https://corporate.marksandspencer.com/media/press-releases/hand-me-down-quality-ms-kidswear-becomes-first-major-high-street-brand-join

McCann-Erickson. (2007). *Can sustainability sell?* http://www.rumbosostenible.com/wp-content/uploads/MCCann_Can-sustainability-sell.pdf?257bb3

Next. (2023). *Corporate social responsibility report.* https://xcdn.next.co.uk/PDFS/next-cr-report.pdf

Primark. (2023). *Primark sustainability and ethics report 2021/22.* https://primark.a.bigcontent.io/v1/static/Primark-Sustainability-and-Ethics-Report-2021-22

Quality Clothing Group. (n.d.). *Social compliance, quality, the environment.* http://www.quantumclothing.com/sustainability/

Retail Economics. (2023). *UK top ten retailers: Clothing, footwear and fashion retailers.* https://www.retaileconomics.co.uk/top-10-retailers-clothing-and-footwear

Ropke, I. (2009). Theories of practice—New inspiration for ecological economic theories studies on consumption. *Ecological Economics, 68,* 10(10), 2490–2497. https://ideas.repec.org/a/eee/ecolec/v68y2009i10p2490-2497.html

Sutton, P. (2004). *A perspective on environmental sustainability.* https://www.donboscogozo.org/images/pdfs/energy/A-Perspective-on-Environmental-Sustainability.pdf

ukft. (2021). *UKFT's compendium of industry statistics and analysis 2020.* https://ukft.s3.eu-west-1.amazonaws.com/wp-content/uploads/2021/11/24095453/UKFTs-Compendium-of-Industry-Statistics-and-Analysis-2020-Executive-Summary.pdf.

United Nations Environment Programme. (2022). *The environmental costs of fast fashion.* https://www.unep.org/news-and-stories/story/environmental-costs-fast-fashion

van der Merwe, J., & Al Achkar, Z. (2022). *Data responsibility, corporate social responsibility, and corporate digital responsibility.* https://www.cambridge.org/core/journals/data-and-policy/article/data-responsibility-corporate-social-responsibility-and-corporate-digital-responsibility/B7749EC986BFF98EF32D3192E8D4F9D7

Walk Free Foundation. (2018). *The global slavery index 2018.* https://cdn.walkfree.org/content/uploads/2023/04/13181704/Global-Slavery-Index-2018.pdf

Williams, J. (2024). Greenwashing: Appearance, illusion and the future of green capitalism. *Geography Compass.* https://doi.org/10.1111/gec3.12736

ZoomInfo. (2023). *Top ten textiles and apparel manufacturing in UK by revenue.* https://www.zoominfo.com/companies-search/location-united-kingdom-industry-textiles-appare

Rachel Vieira (rvieira@glos.ac.uk) is Deputy Head of the School of Business, Computing and Social Sciences at the University of Gloucestershire, UK. Rachel holds a Ph.D. in Marketing (University of Gloucestershire) and an MA in Art History and Design (Birmingham City University). Rachel's research interests include the circular economy, fashion consumption, sustainability, and consumer behaviour. More specifically, her Ph.D. research explored the concept of alternative consumption with a focus on sharing and the circular economy. A key aspect was how fashion consumers construct identities through engagement in sharing. The use of a simulation method provided a rich site for data generation, highlighting challenges consumers faced around ownership and self-image. What remains to be seen is if alternative consumption is a niche practice or the start of a new wave of sustainable consumption.

Peter Jones (pjones@glos.ac.uk) is Emeritus Professor in the School of Business, Computing and Social Sciences at the University of Gloucestershire, UK. He previously served as Dean of the Business School at the University of Plymouth, and as Head of the Department of Retailing and Marketing at Manchester Metropolitan University. He has worked as a retail and academic consultant in a range of countries. He has active research interests in modern slavery in the service sector, in sustainable development in retailing, and more generally in corporate social responsibility. He has published his research work in a range of book chapters, academic journals and professional publications.

Chapter 3
France

Mehri Shabani and Erica Bicchi

Abstract The French textile and clothing industry (TLC) has been an important pillar of the national economy since the nineteenth century and has had a significant impact on global fashion. Despite its historical importance, the sector has faced challenges since the beginning of the twenty-first century, including a decline in production and employment due to automation and globalisation, complex supply chains and environmental impacts. In 2021, 2.8 billion TLC products were placed on the French market, highlighting the need to promote reuse and recycling to reduce waste, conserve resources and limit emissions. In France, the TLC sector is managed by an eco-organisation named *Re_fashion*, which ensures that companies meet their waste management obligations, including prevention, collection, sorting and recovery of end-of-life products. This study examines how French TLC companies integrate sustainability and circular economy (CE) practices through digital technologies. The analysis of the 10 SMEs with highest scores, based on the *Moralscore* French online platform, reveals key sustainability strategies, including local production, transparency and the use of more sustainable materials. France can thus be considered one of the European leaders in the adoption of CE, thanks to measures such as Extended Producer Responsibility (EPR), eco-design and consumer awareness initiatives. Lastly, digital technologies such as AI, IoT and 3D printing are supporting the transformation of the industry, optimising resources and reducing waste in this sector. By focusing on responsible production, sustainability and CE principles, the French TLC industry can reduce its environmental footprint and move towards a more sustainable future.

M. Shabani (✉) · E. Bicchi
CERADE, ESAIP La Salle, Saint-Barthélemy-d'Anjou, France
e-mail: mshabani@esaip.org

E. Bicchi
e-mail: ebicchi@esaip.org

M. Shabani
Group Analysis and Processes (GA&P), University of Angers, Angers, France

E. Bicchi
LPG (Laboratoire Planétologie et Géosciences), University of Angers, Angers, France

M. Wynn and T. Wiegand (eds.), *Sustainability, the Circular Economy and Digitalisation in the European Textile and Clothing Industry*, SDGs and Textiles,
https://doi.org/10.1007/978-981-97-9116-3_3

Keywords Textile and clothing industry · TLC · Eco-organisation · *Re_fashion* · *Moralscore* French online platform · Extended producer responsibility (EPR) · Environmental labelling

3.1 Introduction and Background

The textiles and clothing industry remains a vital sector of the French economy, with a rich history and global influence (Valet de Pique, 2019). French expertise has always been prominent in textiles. In the 1830s and 40s, the main exports from France to other countries were textiles made of cotton, silk, and wool of various grades (Blancheton, 2021). From the mid-nineteenth century, clothing emerged as a new area of French expertise made possible by Paris' global reach (Bécuwe & Blancheton, 2018). In the nineteenth century, France's passion for fashion was reaffirmed through the development of haute couture and the establishment of esteemed couturier houses, including Charles Frederick Worth's store on Rue de la Paix in Paris. The early twentieth century witnessed the rise of iconic fashion houses, notably Coco Chanel, whose innovative designs rejected uncomfortable garments like corsets and introduced popular loose and free-flowing styles in the 1920s (Bécuwe & Blancheton, 2018). This era saw eccentric male fashion trends characterised by large earrings, wide trousers, and musk-based fragrances. Silk textiles had a brief reign as the most popular French export following World War I, accounting for 12% of all exports in 1919 (Ministère de l'économie des finances et de la souveraineté industrielle et numérique, n.d.), But the economic added value of the textile sector fell to 23% between 1935 and 1938 (Bécuwe & Blancheton, 2018).

World War II severely impacted The French fashion business, with several designer houses collapsing. Much of France's fashion industry was taken over by Germany, and plans were to move haute couture to Berlin and Vienna. Due to a lack of materials and designers' adaptation to military requirements, practical clothes, like the "zazou suit", became quite popular. However, Christian Dior's post-World War II design revolution (1947), characterised by a waist cut, an A-line skirt, and a mid-calf length, revived the apparel industry after years of shortages (Still in fashion, n.d.).

The 1960s saw the emergence of other well-known designers including Pierre Balmain and Hubert de Givenchy. However, French youth culture started to reject conventional high fashion in those years, looking to London and embracing informal designs instead. French fashion is deeply ingrained in the country's culture, with individuals often describing themselves in terms of their clothing. Although facing increasing competition from global fashion centers like London, New York, Milan, and Tokyo, France's fashion industry remains influential, attracting designers from around the world who seek to establish their careers in this fashion capital (Valet de Pique, 2019).

Figure 3.1, which focuses on employment, production, imports, and exports, shows trends in the French textile sector from 1990 to 2022. Between 1990 and

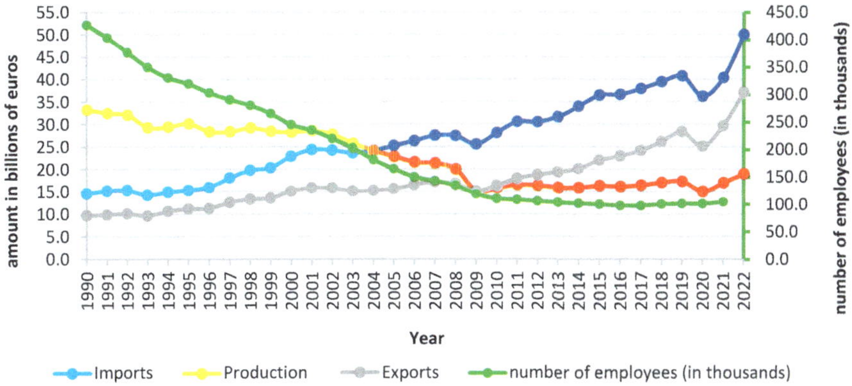

Fig. 3.1 TLC industry trends in France 1990–2022 (INSEE, 2023)

2005, the number of employees decreased significantly from around 400,000 to just over 100,000. This sharp reduction indicates that the industry underwent major structural change, most likely as a result of automation, outsourcing, or other factors that reduced the need for a big workforce.

The French textile industry's production continuously decreased between 1990 and 2009, going from about €35 billion in 1990 to about €20 billion in 2009. Production levels have remained mostly constant since then, with a few small variations through 2020 due to the COVID-19 pandemic. However, imports have shown a general upward tendency; they were steady at about €15 billion until 1996, when they began to rise sharply, reaching almost €40 billion in 2020, then followed by a decline due to COVID-19. This pattern emphasises France's increasing reliance on textile imports. Additionally, exports have increased gradually, rising from roughly €10 billion in 1990 to approximately €25 billion by 2022, which is indicative of France's significance in the world's textile supply chain. In 2023, the fashion industry's direct sales totaled €154 billion, of which €66.3 billion was for textiles and clothing, with imports and exports continuing to rise (Labeye, 2018).

In France, the textile and clothing sector is known as the TLC sector. The abbreviation is taken from the French term for the industry—"Textiles d'habillement, Linge de maison et Chaussures"—which means "the clothing, household linen, and footwear" industry. In 2021, the TLC industry comprised 3.3 billion items representing almost 12.2 kg of textiles and footwear per inhabitant per year (Re_fashion, 2022).

This sector is being managed and monitored in France by an eco-organisation named *Re_fashion* (formerly Eco-TLC, approved by interministerial decree on March 17, 2009) (Re_fashion, n.d.). This organisation ensures that companies comply with their obligations in terms of processing waste from their products. *Re_fashion*/Eco TLC is responsible for the prevention of waste and management (collection, sorting, and recovery) of the end-of-service life of products on behalf of the 5000 TLC companies.

This sector is facing several environmental, economic, and social challenges. In 2021, 2.8 billion products were put on the market in France. To prevent these items from becoming waste material, it is crucial to promote their reuse and recycling. This not only reduces waste but also preserves natural resources and limits greenhouse gas emissions and pollution. By reemploying and recycling, value is restored to items that would otherwise be discarded, supporting the growth of the circular economy (CE). However, the sector faces challenges related to the profitability and efficiency of recycling processes. Additionally, increasing collection and improved sorting can create jobs and foster local development and social inclusion; but ensuring worker qualifications and job stability within this sector remain significant issues to be addressed (Labeye, 2018).

The collection and trade of used TLC products have long been a well-established economic activity involving numerous participants from charitable organisations, businesses, and industries. This sector has a rich history of clothing trade and a tradition of donation and reuse (prior to recycling). There is a high level of synergy between the social and solidarity economy actors and those involved in textile collection, sorting, and recycling.

Based on the data in the year 2022, *Re_fashion* consists of 32 associates. The associates represent the whole textiles value chain and are organised into five "colleges": (a) general large retailers (e.g., Auchan, Carrefour, Monoprix); (b) fashion retailers (C&A, Galeries Lafayette, Etam); (c) direct sales and mail/online retailers (Damart, Groupe 3SI); (d) manufacturers and wholesalers (LVMH, Cotonflorlinge du lit); (e) apparel industry associations (Federation Nationale de l'Habillement, Union des Industries Textiles, and Union Sport & Cycle).

Re_fashion's aim (all the actors related to this organism are shown in Fig. 3.2) is to collect and recycle used textiles. In France, there are 44,830 voluntary drop-off points in public (or private) spaces or with associations. In 2021, 244,500 tonnes of used textiles were collected, and 190,550 tonnes were sorted by one of the 66 sorting centres under contract to the *Re_fashion* eco-organisation. After sorting, 58% of these textiles were destined for re-use, 33% for recycling and 9% for energy recovery as solid recovered fuel (SRF).

To establish a clear and understandable foundation for this research, it is essential to define the scale and boundaries of the French TLC sector under investigation. The TLC sector in France includes all clothing, household linens, and shoes that meet the following criteria:

- Placed on the French market: This encompasses products sold or given within metropolitan France as well as its overseas regions, including Guadeloupe, Martinique, Guyana, Reunion Island, Mayotte, Saint-Pierre-et-Miquelon, and Saint-Martin.
- Consumer Products: The sector includes products intended for consumer use.
- New Products: Only new items are included in the scope, excluding second-hand goods

The sector does not encompass:

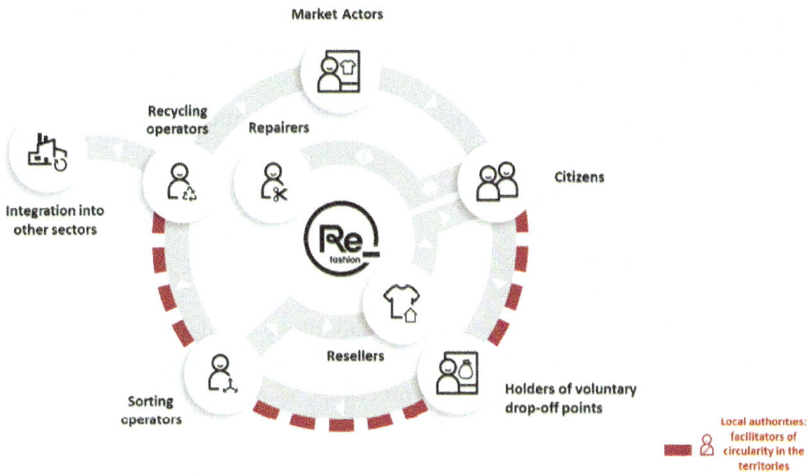

Fig. 3.2 Life cycle stages of textiles and shoes in France. *Source* Re_fashion (n.d.)

- Garments made of 100% leather or natural fur.
- Second-hand clothing, household textiles, and shoes Imported into France and originating from foreign markets.

The TLC sector in France is characterised by a wide range of company types, varying significantly in terms of size, commitment to environmental issues, production locations, and supply chain complexities. Key aspects of this diversity include:

- Companies vary widely in terms of size (from the very small to the very large), which can lead to differences in organisational resources.
- Companies have different levels of commitment to environmental issues, with the sector ranging from fast fashion to sustainable fashion, as well as companies with a niche positioning based on "Made in France".
- A large proportion of the companies marketing clothing in France have their production facilities outside Europe. This leads to higher impacts (e.g., impacts of more carbon-intensive energy mixes) with regulations that do not have the same levels of requirements in terms of emissions (air, water) as Europe, and which can also generate higher impacts.
- The supply chains of many players are complex, with at least three tiers in some cases with differences in terms and conditions and the nature of relationships; not all supplier relationships have the same level of durability. This creates potential difficulties in gathering the information needed to calculate environmental impacts (ADEME, 2022b).

Within this overall context, this chapter addresses three research questions (RQs):

RQ1 How are French TLC companies addressing sustainability in their corporate strategies and activities?
RQ2 What strategies and activities relating to the CE are being pursued in the French TLC sector?
RQ3 What role can digital technologies play in the transition to sustainability and the CE in the French TLC industry?

3.2 Research Method

For this chapter, a comprehensive research methodology was deployed to investigate how French textile and clothing companies address sustainability and the CE, and the role of digital technologies in these processes, as embodied in the three RQs noted above. The research design aims to provide a clear understanding of the scale and boundaries of the French TLC sector, focusing on small and medium-sized enterprises (SMEs) with staff headcounts of less than 250 and a turnover of ≤€50 million or a total balance sheet of ≤€43 million in the TLC sector. The methodology integrates both qualitative and quantitative approaches, leveraging existing data and systematic analyses to answer the three RQs posed in this study (see Table 3.1). These questions are designed to explore how companies are integrating sustainable practices into their corporate strategies, the specific CE activities they are pursuing, and the role of digital technologies in facilitating these transitions.

The study utilises a combination of literature review, secondary data analysis, case studies, and qualitative interviews undertaken by third parties. Each method is selected for its ability to provide relevant and reliable data. An existing database from

Table 3.1 Summary of research questions, data sources, and value categories

Research question	Data sources	Value categories
RQ1: How are French textile and clothing companies addressing sustainability in their corporate strategies and activities?	• Online platform of *Moralscore* (2024) • Scientific articles • Published reports • Specific case studies	Environment, working conditions, social contributions, fairness, governance, data treatment, innovation, technology enslavement, price, product quality
RQ2: What strategies and activities relating to the circular economy (CE) are being pursued in the French textile and clothing industry?		Environment, innovation, equity of relations
RQ3: What role are digital technologies playing in the transition to sustainability and the CE in the French textile and clothing industry?		Digital efficiency, usage, automation, and robotics

an online platform named *Moralscore* was used in combination with all published reports of case studies and keyword research on scientific articles. *Moralscore* (2024) is a platform that analyses companies through several criteria (listed in Table 3.1) based on the United Nations Sustainable Development Goals (SDGs). For each criterion, *Moralscore* considers a non-exhaustive list of information available on each company: NGO reports, media investigations, legal declarations, employee and consumer testimonials, terms of use, and charters, etc. For each RQ, the relevant data sources and value categories have been considered.

Table 3.2 presents the detailed information that has been searched for, by each category and for each criterion. A qualitative rating scale ranging from A to D is established for each category on the *Moralscore* online platform, which could define various meanings based on the associated category. The grading scale is designed to be factual, relying on concrete information rather than subjective judgments. As an example, for the category of legal and taxes, these scores have been assigned the following definitions. A: Pays all taxes in France, B: Pays all taxes in their country of origin, C: Pays a small portion of taxes in France, the rest being remitted to a country

Table 3.2 Detailed studied parameters for each value category

Value categories	Parameters	
Environment	• Water consumption and management • CO_2 emissions related to the company's overall activity • Manufacturing location • Raw materials used • Fast fashion and collection turnover • Advertisements and encouragement of over-consumption • Toxic waste • Animal welfare • Biodiversity	• Management of unsold products • Collection and upcycling of used products • Green server hosting • Means of transportation for manufactured products • Logistics and sourcing of raw materials • Sustainability and programmed obsolescence • Plastic packaging • Green energy supplier
Remuneration and working conditions	• Hourly income for legal working hours • Income visibility • Net average monthly income • Work conditions for workers • Surveillance of workers • Number of hours needed to earn minimum wage • Strikes: pay and working conditions	• Average number of hours worked per month • Time constraints • Multiple employment • Social security coverage • Job insecurity • Disabilities in the workplace
Social contributions	• Community involvement,	• Charitable activities
Equity in relationships with partners	• Responsibility towards its partners	• Responsibility for sourcing raw materials

(continued)

Table 3.2 (continued)

Value categories	Parameters	
Legal compliance and fiscal decency	• Discrimination and sexism • Corporate customers • Shared governance • Company employees • Job creation	• Gender parity • Taxation • Transparency • Ethical policies
Processing and use of personal data	• Data collection and anonymity • Ownership of collected data and access rights • Selling or granting access to customer data to external partners	• Spam • Service personalisation
Innovation	• New technologies,	• Business disruption
Impact of technologies on man and work	• Impact of automation, AI on work	
"Pragmatic" factors		
Price competitiveness	• Affordability of products	
Product quality	• Durability, customer satisfaction	

with favorable tax conditions or D: All income remitted to a low-tax country. For the category of remuneration and working hours the scoring system is as follows. A: Works the legal working hours, B: Works up to 1.2 times the legal working hours, C: Works up to 1.8 times legal working hours or D: Works more than 1.8 times the legal working time.

For RQ2, which examines the strategies and activities related to the CE, the analysis is limited to the parameters specifically related to innovation, equity, and the environment as defined below:

1. Environment: Contribute, without being legally obliged to do so, to sustainable development and the reduction of ecological impact.
2. Innovation, business disruption and new applications: Reinventing the company's core business and the way it delivers services; rethinking the offering; shaking up the status quo; introducing new technologies that make the product or service more relevant, easier or cheaper; introducing new uses; solving problems.
3. Equity of relations with partners: Treat partners, suppliers, and subcontractors with respect; offer them a fair share of income taking into account the risks, efforts and contributions of each person; do not abuse a possible dominant position.

After selecting the criteria based on the defined questions, the results of the analysis showed a list of companies based on their average score (ranging from 0 to 100). The first 10 companies categorised as SMEs were then extracted. This scoring system means that the same company might have different rankings and scores based on different chosen criteria.

To respond to RQ3, a systematic approach was pursued, assessing overall current digital technology deployment, followed by more detailed case studies. To have comparable data for all investigated companies and a deeper understanding of their approach to this domain, the criterion named "Drone automation (Dronification)" was selected from the platform. This criterion evaluated the impact of technology on people and work. Indeed, digital technology can have both positive and negative outcomes in the workplace: for example, the "dehumanisation" of work through the impact of algorithms and robotics, and the multi-faceted effects of digital technology on work and quality of services were all considered. The following parameters were evaluated:

- Digital Efficiency
- Digital Usage
- Automation and the use of robots to replace human workers

In summary, the research method includes a literature review, secondary data analysis, case studies, and interviews. Companies are scored on various sustainability and CE criteria, with a focus on SMEs, and the top-performing companies are further analysed, particularly regarding the impact of digital technologies on their operations.

3.3 Findings

3.3.1 Introduction

The idea of sustainable development, which is defined as "meeting the needs of the present without compromising the ability of future generations to meet their own needs" (World Commission on Environment & Development, 1987, p. 24) is associated with the term sustainability (United Nations General Assembly, 1987). Sustainability is seen as the central paradigm in a future in which environmental, societal, and economic considerations are balanced in the pursuit of an improved quality of life (Jeronen, 2013). Sustainable development is considered as the pathway to sustainability.

Key contextual factors that are shaping the approach of TLC businesses to sustainability include the following:

- French textile production has halved since 1976. Luxury and technical textiles persist, but the rest, notably ready-to-wear, are imported from Asia (50%) and Europe (33%). Today, only 3–5% of the clothing sold in France is produced in the country.
- The carbon footprint resulting from the consumption of imported textiles, which make up over 95% of the total used, represents 6 times the emissions of air transport, or nearly 30 million tonnes CO_2 equivalent.
- This carbon footprint continues to grow due to fast fashion and coal-fired production in Asia.

- Globalisation is leading to delocalised value chains, where it is difficult to identify and control environmental and social impacts.

For these reasons, TLC businesses are increasingly attempting to incorporate sustainability into their value chains and corporate strategies (Ministère de la Transition écologique et de la Cohésion des territoires, 2017). Fashion industry professionals are aware of the need to make profound changes to meet the challenge of ecological transition. In August 2019, 56 companies representing 250 fashion and textile brands committed to reducing the clothing sector's carbon footprint by signing the "Fashion Pact". This commitment is centred on three themes: the fight against CO_2 emissions, the preservation of natural environments, and the fight against ocean pollution caused by the discharge of plastic particles (INEC, 2018). For instance, businesses in the TLC sector cut their use of chemicals, water, and energy in the processing of cotton by 50% between 1990 and 2009 (Thiry, 2011). Though many businesses have either recently begun using sustainable practices or are currently in the process of incorporating sustainability into their corporate strategy, these advancements do not yet suffice to qualify the industry as sustainable (Peters & Simaens, 2020).

Corporate social responsibility (CSR) has taken a more important place in the life of French companies with the introduction of various laws and regulations. Indeed, the French government has imposed certain laws on companies, in particular, to monitor and follow-up on various actions and practices. A good example is the New Economic Regulations law (Légifrance, 2001), which requires companies to publish an annual CSR report, in which related activities concerning the environment and society are set out (Beji et al., 2021).

While sustainable development is a longer-term concept (based around the 17 UN SDGs), the CE has a more immediate and operational context. CE initiatives may involve several tools for action at different levels, from product design to consumer behavior, in a necessarily systemic approach to coordinating actions. It involves both local and global action. It aims to reduce and optimise resource consumption. It contributes to restoring a coherent place for the anthroposphere within the biosphere.

In France, these two concepts are interrelated, as the CE entails the use of a series of tools needed to move toward longer-term sustainability. Some of the SDGs are more relevant than others to the CE; for example, consuming less materials and energy (SDGs 12 and 13) is one key element of transitioning to a CE. Directly or indirectly, however, the CE also contributes to achieving other less obvious SDGs, encompassing ethical and social dimensions. In the following subsection, case studies of 10 French SMEs are discussed, based on their sustainability missions, with a focus on specific criteria to assess their approach to the CE.

3.3.2 RQ1. How Are French TLC Companies Addressing Sustainability in Their Corporate Strategies and Activities? (Case Studies of Approaches to Sustainability)

Based on the results provided by the *Moralscore* online platform (including all parameters), the top 10 SMEs of French origin are listed in Table 3.3.

The 10 companies listed in Table 3.3 meet all the criteria, but new companies are constantly being created or evolving in the sector. These brands are characterised by a diversity of commitments listed below:

- Enhancing the value of French or European industrial facilities and relocating certain activities to France—this is the commitment of Le Slip Français (2022a, 2022b) and 1083 (2020), for example.
- Fair trade, traceability, and total transparency throughout the value chain—Veja (2021) is a pioneer in this field.
- The design of a timeless, durable and functional offers to counter seasonal trends—as advocated by Hopaal (n.d.), for example.
- Enhancing the value of materials and highlighting unique fashion-related know-how—as Bleu de Chauffe does (2023).
- The circularity of the model to extend the lifespan of specific segments.

A more detailed evaluation is provided in Fig. 3.3. This evaluation provides a score (out of 100) for each company based on the 10 criteria set out in Table 3.2, indicating the companies' current situation in following the 17 SDGs. The aforementioned scores can be classified according to the qualitative scales (A-D), outlined in Sect. 3.2. The rating criteria used in this methodology follow the 17 SDGs.

The analysis shows that the type of product does not strongly correlate with the sustainability score, as both high and low scores are observed across categories like

Table 3.3 French SMEs in the TLC industry in the study

Company	Product	Worldwide sales	Number of employees
Veja	Shoes (all type)	€11.04 millions	35
Le Slip Français	Underwear, swimsuits	€ 20.7 millions	N.d.
Olly	Underwear, pyjamas	€45,600	N.d.
Bleu de Chauffe	Bags	€2.7 millions	25
Hopaal	Sport clothes	€300,000	6
1083	Jeans, clothes, shoes	€2 millions	65
Balzac Paris	Clothes, sneakers, handbags	€8.4 millions	15
Noyoco	Clothes, accessories	N.d.	<50
Sézane	Clothes, shoes, bags	N.d.	100
Atalaye	Swimsuits	N.d.	N.d

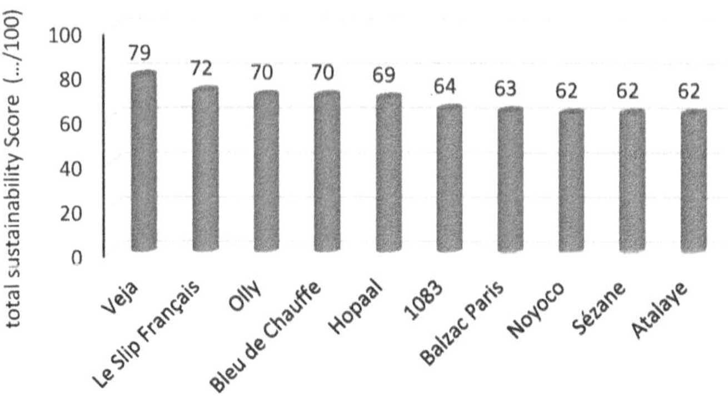

Fig. 3.3 Sustainability scores of 10 SMEs in France (Moralscore, 2024)

shoes, underwear, and outdoor clothing. SMEs, while constrained by resources, have the potential to leverage their smaller scale for innovative sustainability solutions. As regulatory and consumer demands for sustainability increase, companies with lower scores may need to improve their practices to stay competitive and meet these evolving expectations. In the following paragraphs, the five top-scoring companies are examined in more detail.

Veja is deeply committed to sustainable development goals through its extensive transparency practices, as detailed on its website. The company prioritises ethical and environmental considerations in its operations, focusing on reducing its carbon footprint and promoting fair labor practices. Veja ensures the traceability of its materials, including organic cotton and wild rubber, and works closely with its supply chain to uphold high standards of environmental stewardship and social responsibility. Their transparency extends to detailed reporting on the impact of their activities (Veja, 2021).

Le Slip Français (2022a, 2022b) is actively engaged in adopting the SDGs through its designation as a "mission-driven company". Their commitment is evident in their efforts to create a positive social and environmental impact. The company focuses on local production and sourcing, reducing its carbon footprint by maintaining a French manufacturing base and using eco-friendly materials. Le Slip Français (2022b) also emphasises social responsibility by ensuring fair labor practices and supporting local communities. Their mission-driven approach integrates sustainability into their core business model, aiming to balance economic performance with a strong commitment to environmental stewardship and social equity (Le Slip Français, 2022a).

Olly is dedicated to sustainable development goals by integrating ethical practices into its business model. The company focuses on producing high-quality, eco-friendly underwear and pajamas while maintaining transparency in its supply chain. Olly prioritises the use of organic and recycled materials, minimising environmental impact. Additionally, the brand emphasises fair labor practices and works closely with its suppliers to ensure ethical working conditions (Olly, n.d.).

Bleu de Chauffe emphasises the use of high-quality, sustainably sourced materials, including leather and canvas, to ensure longevity and reduce waste. The company's commitment to sustainability extends to its manufacturing process, which is designed to minimise environmental impact through responsible consumption and production practices. They also prioritise local production in France, supporting fair labor practices and reducing carbon emissions associated with transportation. By integrating these practices into their business model, the firm aims to promote responsible consumption and production, while supporting local economies and reducing environmental footprints (Bleu de Chauffe, 2023).

The **Hopaal** focuses on creating high-quality, durable clothing made from eco-friendly materials such as recycled polyester and organic cotton. Hopaal's approach emphasises minimalist design to avoid unnecessary waste and ensure that each product has a long lifespan. Their commitment extends to transparent supply chains, reducing their carbon footprint, and supporting ethical labor practices (Hopaal, n.d.).

Overall, these case examples indicate a strong commitment to the SDGs and sustainability in general across various companies. Each brand integrates sustainability into its core business model through transparent practices, ethical labor standards, and eco-friendly material sourcing.

3.3.3 RQ2. What Strategies and Activities Relating to the CE Are Being Pursued in the French TLC Sector?

Circular fashion and the circular fashion economy are essentially about reducing the amount of waste produced by the industry and encouraging reuse, recycling and the use of regenerative and sustainable materials. A circular textile system achieves this by innovating in textile design, adopting new technologies and renewable materials, increasing the reuse and recycling of old garments, and eliminating waste and pollution. Consuming less materials and energy is a key element of sustainable development. This is the main objective of the CE. Directly or indirectly, the CE also contributes to achieving other sustainable development objectives, including ethical and social dimensions.

Recently, the fashion industry has seen major changes with the development and implementation of certain CE practices. In parallel, academic researchers have proposed methods to develop a more circular and sustainable textile economy (Kozlowski et al., 2018), and changes across the life cycle of textiles have taken place. In addition, the textile industry faces fierce international competition due to globalisation. Firms are located in many parts of the country, covering the regions of Normandy to Picardy, the North, the Ardennes, Champagne, and Alsace. Law no 2020-105 (Légifrance, 2020) of February 10, 2020, on the CE and the fight against waste, gives a major boost to ecological transition. It marks an important milestone for the TLC sector, which is resolutely committed to sustainable development, and complements the initiatives already taken in this direction.

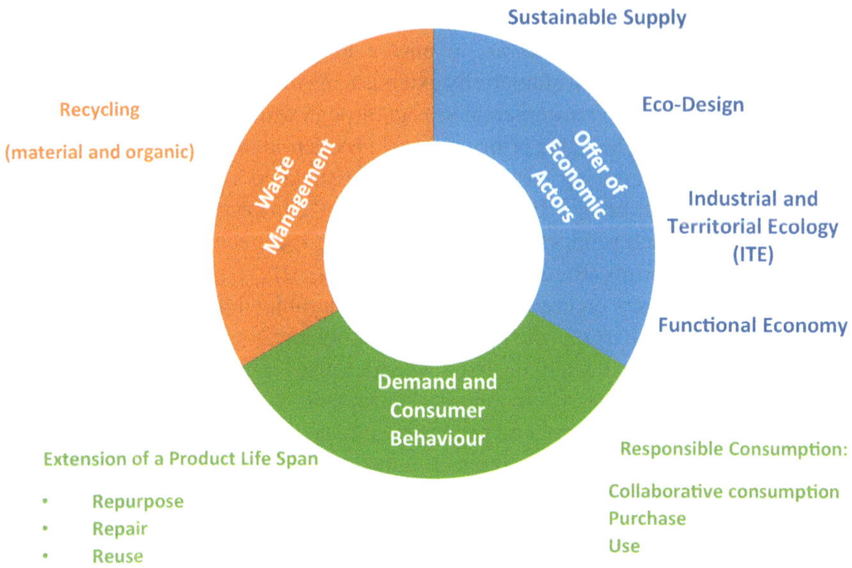

Fig. 3.4 The seven pillars of the transition to a CE in France (ADEME, 2020)

In France, the transition to a CE is officially recognised as one of the objectives of the energy and ecological transition and as one of the commitments of sustainable development. This includes three main action areas: waste management; the offer of economic actors; and consumer demand and behavior. To progress this transition, seven main objectives or "pillars" for change were identified (Fig. 3.4):

1. Sustainable sourcing: take into account the environmental and social impacts of the resources used, in particular those associated with their extraction and exploitation.
2. Eco-design: consider environmental impacts over the entire life cycle of a product and integrating them from its design.
3. Industrial and territorial ecology: create synergy and pooling between several economic players, encompassing the flows of materials, energy, water, infrastructure, goods, and services to optimise the use of resources.
4. The economy of functionality: favor use over possession, selling a service rather than a good.
5. Responsible consumption: consider environmental and social impacts at all stages of the product life cycle, and in purchasing choices, whether the buyer is public or private.
6. Extend the lifespan of products by resorting to repair, sale or purchase second-hand, or by donation, in the context of the reuse.
7. Improve waste prevention, management and recycling, including reinjecting and reusing waste materials into the economic cycle.

Businesses can play a key role in this transformation: optimising the life cycle of products put on the market, encouraging reuse, extending product life, improving collection (only 50% of packaging currently carries a sorting instruction), integrating a minimum rate of recycled material in products, preferring donation to destruction, no longer throwing away, and no longer wasting. Consumers also need to be better informed, so that they can become active players by making the right consumption choices and adopting the right responses. All these measures are in line with the objectives French ecological transition (French Government, 2021), and their implementation is still the subject of discussions with players in the TLC sector and will shortly be set out in regulations (decrees and orders). These necessary changes to control the environmental impact of products can, with support from all stakeholders, enable the fashion sector to gradually become more circular, and therefore more responsible. This has led to the emergence of new practices, such as (a) Extended Producer Responsibility (EPR), (b) Environmental labelling, (c) Use of the Triman Pictogram and Info-Tri logo, (d) Eco-design, (e) Extending the Duration of Use, and (f) Recycle and end-of-life practices. These are discussed in more detail below.

a. Extended Producer Responsibility

France has been a pioneer in introducing the principle of EPR, as evidenced in the 1975 Law on waste disposal and material recovery (Légifrance, 1975). France is one of the European countries that most extensively uses the EPR system, there being about twenty national EPR schemes (packaging, furniture, textiles, etc.), as well as schemes derived from European legislation, and voluntary schemes, where producer participation is not mandatory. EPR in the TLC sector is regulated by articles L. 541-10-27 and R. 543-214 to R. 541-219 of the environmental code (Légifrance, 2021) and Law No. 2020-105 of February 10, 2020, relating to the fight against waste and the circular economy (known as the "AGEC" law) (Légifrance, 2020).

According to the core EPR principles, manufacturers, distributors, and importers of textile products are responsible for managing the waste generated by these products. This requires companies selling these products in France to establish an individual waste management system or to contribute to the costs of collecting, sorting, and recycling used products through participation in an eco-organisation. Eco-organisations are private, non-profit companies authorised by public authorities to manage different categories of end-of-life products. They are funded by the producing companies, which thus transfer their responsibility under the principle of EPR. To finance the functioning of the EPR schemes, producers add an eco contribution to the sale price of their products. This contribution funds the eco-organisation responsible for the end-of-life management of the product and reflects its treatment cost. Sometimes, this contribution can be very low, which can be explained by the potential recovery value of materials from certain products (INEC, 2021).

To improve textile sorting performance for eco-organisations and individual systems as regards EPR in the TLC sector, specifications have been appended to the article of 23 November 2022, which came into force on 1 January 2023 (Légifrance, 2022a). This decree defines new objectives and organisational practices for the period 2023–2028, and has the following objectives:

1. To design more durable, more environmentally friendly textiles made from recycled textiles.
2. To develop the collection of used textiles, particularly in areas where it is lacking today.
3. To finance the sorting of collected used textiles.
4. To develop the reuse in France.
5. To promote textile repair by reducing its cost.
6. To contribute to the development of recycling of non-reusable textiles.

It also defines targets to be achieved in terms of the collection and recovery of used textiles:

1. Collection rate of used TLC: 60% by 2028.
2. Sorted used TLC to be reused: 120,000 tonnes in 2024.
3. Reuse rate less than 1,500 km from the collection location: 15% by 2027.
4. Recycling rate of used, non-reused TLC: 70% in 2024, 80% by 2027.
5. Recycling rate of used TLC incorporating at least 90% of non-reused synthetic plastic materials: 50% in 2025, 90% by 2027.
6. At most 0.5% of the TLC collected or supported is eliminated (storage, incineration without energy recovery).

Since it was set up, the TLC EPR sector has achieved an increase in the collection of used textiles of 40% since 2014 (Ministère de la Transition écologique et de la Cohésion des territoires, 2017). However, used textile collection remains insufficient: it stands at around 35%, with the rest of used textiles ending up in household waste for incineration or landfill. The sector is also dependent on exports for the reuse and recycling of used textiles. The TLC sector currently has a fund dedicated to financing repairs to encourage individuals to have their textiles repaired by approved repairers. It will support the rehabilitation of used TLC, the creation of new reuse/recycling structures and actions aimed at reducing the distance between the collection point and the reuse/recycling stage (via the reuse/recycling financing fund dedicated to SSE players and additional financial resources). Resources will be allocated each year for information and public awareness campaigns to encourage waste prevention, separate collection, repair, and reuse of TLC.

b. Environmental labelling

The Law of August 17, 2015 (Légifrance, 2015), regarding the energy transition for green growth (LTECV), established a requirement for transparency to provide consumers with clear and reliable information on the environmental impacts of the products and services offered to them. The CE Law (ADEME, 2020) goes further and introduces an environmental and social labelling system voluntarily. Primarily based on life cycle analysis, this labelling aims to raise consumer awareness of the impact of the products they purchase, but above all to encourage manufacturers to reduce the environmental and social impacts of the products they market. Thus, starting from January 1, 2022, after 18 months of testing, the implementation of information concerning the environmental characteristics of products became mandatory. These

Fig. 3.5 An example of environmental labelling (Ministère de la Transition écologique et de la Cohésion des territoires, 2019)

characteristics are established by prioritising an analysis of the entire life cycle of the textile product, from its design to its final disposal, and include the following criteria: durability; compostability; repairability; reusability; recyclability; presence of hazardous substances, precious metals, or rare earth elements; use of renewable resources; and incorporation of recycled materials (Ministère de la Transition écologique et de la Cohésion des territoires, 2019).

Furthermore, in order to more effectively disseminate information to consumers, the legislation requires the disclosure of bonuses and penalties paid by manufacturers to eco-organisations responsible for managing and processing the end-of-life of their products. This is based on the environmental performance criteria of the textiles marketed such as durability and/or recyclability of textiles in question. This mechanism, which applies eco-modulations (bonuses or penalties), aims to encourage brands to offer a more sustainable and eco-designed range of products. This system consists of an environmental rating ranging from A (the best rating) to E on the labels of textile products (Fig. 3.5).

Environmental labelling for textiles enables the calculation of the environmental impacts of each garment, considering all operations throughout its life cycle (e.g., cotton production, spinning, dyeing, finishing, transport, end of life). The method for calculating the "environmental cost" is an enhanced version of the European Product Environmental Footprint (PEF-wise approach). This method includes PEF indicators, weighting rules, and complementary indicators, and is based on eight key criteria: water consumption, physical durability of textiles, production conditions, use of pesticides and chemicals, microplastic emissions, valorisation of recycled materials, valorisation of refurbished textiles, and impact of fast fashion. In addition to these criteria, several product-specific parameters are included: type of product (pants, t-shirt, etc.), type of material (cotton, linen, polyester, microfibers, and microplastics), country of manufacture, toxicity of finishing processes, and end of life. The database used for environmental values (e.g., impact of cotton, 1 kWh, etc.) is the Empreinte database, supplemented by the Ecoinvent database (version 3.10) (Ecoinvent, 2023).

Some companies in the textile sector have already implemented environmental labelling on certain products. Notably, the children's clothing chain Okaïdi and Decathlon have worked in collaboration with ADEME and the Ministry of Ecological and Solidary Transition to establish a classification system. The AGEC law allows companies the freedom to inform consumers about the environmental or/and social impact of their products in various ways: through marking or labeling textiles in-store, dedicated displays, electronic information, or any other appropriate method.

Fig. 3.6 Example of Triman pictogram and Info-Tri logo (Your clothing and shoes have a future—donate them!) (Re_fashion, 2022a)

(c) The Triman pictogram and Info-Tri logo

To simplify sorting guidelines—ensuring that people can easily identify the correct container for a product—and to encourage more widespread participation in sorting (currently, only 50% of packaging includes sorting instructions), the L541-9-4 law proposed a modification to the Triman pictogram, the recycling symbol used in France and Europe (Légifrance, 2022b). The Triman pictogram, which was introduced in 2015, is straightforward and recognisable (featuring a figure with three arrows), but it has not garnered significant attention from consumers. The updated regulation now requires that this pictogram be accompanied by an explanatory sentence related to the Info-Tri logo. This sentence, which is recommended by ADEME, serves to inform consumers that the product is subject to specific sorting rules and, for products in the textile, clothing, and footwear (TLC) sector, indicates the appropriate collection point, such as public road bins, store collection boxes, or associations.

Such labels, combining the Triman pictogram and Info-Tri logo (Fig. 3.6), became mandatory on all TLC products marketed in France on February 1, 2023, with an additional 6 months allowed for products manufactured or imported before February 1, 2022.

(d) Eco-design: design and sourcing

Design choices can considerably reduce the environmental impact of a garment and improve its circularity. Firstly, the shape: in production, cutting the pieces of each garment generates between 20 and 30% of fabric waste. If designers rethink the cut of our garments, they can optimise the use of fabric, and even create zero-waste patterns, with no fabric scraps. 3D printers can also be a way of tailoring garments more closely to customer needs. Secondly, the choice of fabric, accessories (buttons, rivets, etc.) and patterns has consequences for the product's end-of-life (disassembly and recycling). Jeans can only be recycled up to 30% because of the many hard points (seams) and metal accessories. French jeans, as from the brand 1083, for example, offer children's jeans designed to be repaired, with legs that can be disassembled in the event of knee snags. In addition, favoring single-material fabrics facilitates recycling, as does substituting prints based on synthetic inks with woven or embroidered patterns.

Thirdly, sustainable sourcing also reduces environmental impact and facilitates product circularity. Conventional cotton is the second most widely used material after polyester. Its environmental impact can be reduced by using organic cotton, which uses fewer chemical fertilizers and pesticides. The outdoor clothes brand, Patagonia, began sourcing 100% of its organic cotton in 1996. Other less common natural fibres are much more environmentally virtuous: flax requires no irrigation and few inputs. This plant grows in northern France, Belgium, and the Netherlands. Hemp has similar properties. Among artificial fibers, lyocell is produced from eucalyptus cellulose and is preferable to viscose because the natural, non-toxic solvents used to transform wood into fibre are 99% recycled (Fondation Ellen Macarthur, 2017). As a result, pollutants are not discharged into wastewater. The use of synthetic fibers should be reserved for products for which they cannot be substituted, such as sports equipment and outdoor clothing. The use of recycled fibers is preferable, although this does not solve the problem of plastic microfibers and must therefore be accompanied by a strategy to recover microfibers from wastewater at all stages (production, use, waste). Eco-design is therefore an essential tool for limiting environmental damage during garment production and end-of-life management.

(e) Extending the Duration of Use

The production and transport of textile products are the phases with the highest environmental impact. One of the main action levers to reduce the sector's footprint is to decrease the purchase of new clothes and thus extend the usage duration of each garment to the maximum possible. The average lifespan of a garment worn in the EU in 2015 was reported to be 3.3 years (Gray, 2017).

The quality of the selected fabrics is a determining factor in the longevity of the garment. Brands must test the durability of their products based on objective and comparable criteria. For example, for a fabric to be strong, it is essential to choose long fibres. Some brands offer products with a warranty, thereby taking responsibility for the quality of the textiles they market. Some even provide a guarantee for refurbished clothing. Garments can also be designed to be multifunctional, worn on various occasions, and in different conditions, reducing the need to purchase multiple pieces.

Another way to extend the lifespan is through the growing market for second-hand clothing, encouraged by digital platforms. In France, 58% of collected textiles are resold online, and 6 out of 10 French people have already purchased second-hand items (Labeye, 2018).

(f) Recycle and end-of-life practices

Based on the annual report published by *Re_fashion* (2022b), out of 181 kilotonnes of TLC sorted, almost 60% was sorted for reuse, 31% was sorted for recycling, 8% was sorted for the production of Recovered Fossil Fuel (RDF) and only 0.5% was sorted for disposal in 2022 (Fig. 3.7). This figure also indicated the treated tonnes during the last five years as the result of EPR strategies in France. There was an increase in treatment from 189,510 tonnes in 2018 to a peak of 200,394 tonnes in 2019, followed by a significant drop to 156,499 tonnes in 2020, likely due to disruptions from the COVID-19 pandemic. Treatment measures recovered to 185,822 tonnes in 2021 but slightly decreased to 181,420 tonnes in 2022 (Re_fashion, 2022b).

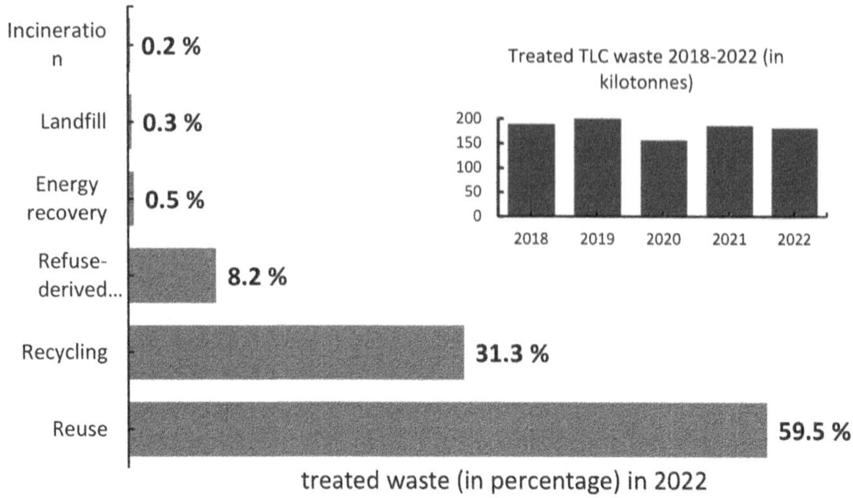

Fig. 3.7 Annual treated TLC wastes (2018–2022) and detailed treatment methods for the year 2022 (ADEME, 2022a; Re_Fashion, n.d.)

Committed to transforming the TLC Sector towards the CE, *Re_fashion* launched the Recycle platform in October 2021 (Re_fashion, 2021). A genuine "link weaver" between textile and footwear recycling professionals, the digital platform aims to industrialise the recovery of materials. It provides privileged access to players and solutions for recycling non-reusable textiles. *Re_fashion*'s Recycle offers professionals—material suppliers, processors and manufacturers using materials from recycling—the opportunity to reduce their carbon footprint by committing their business to preserving resources and biodiversity. Furthermore, it enables them to differentiate themselves from their competitors by employing innovative processes and materials, while ensuring the security of their supply.

Every year, 275 million pairs of shoes are placed on the market in France, representing 119,000 tonnes of leather, rubber, textiles and other plastics glued/sewn together. Shoes thus represent 19% of the tonnage of TLC put on the market, compared with just 6% of the tonnage of TLC entering sorting centres. Shoes are made of an average of 5 different materials, assembled with glues/seams that make them difficult to dismantle. Once dismantled and sorted, these materials have to be prepared (crushed, melted, purified) so that they can be integrated into new applications. There are thus many technical and costly steps for a result that is, unfortunately, less competitive than virgin material.

(g) Case studies

Following a narrowed evaluation of results provided by the *Moralscore* website, new scores were obtained for the same ten selected companies, based only on their performance as regards three circularity approaches: environment; innovation and business models; and partnerships (Fig. 3.8 and Table 3.4). Compared to Fig. 3.3

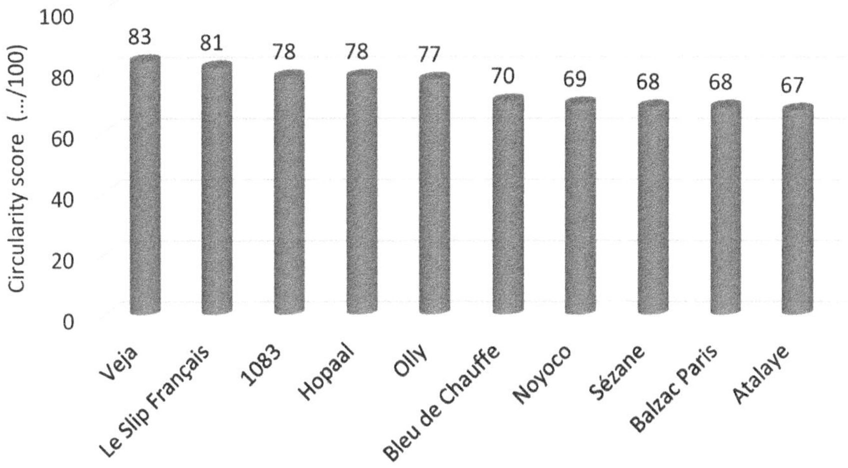

Fig. 3.8 Circularity scores of 10 selected SMEs in France. *Source* Based on data taken from *Moralscore* (2024)

(their sustainability score), some differences in scores are evident (minimum and maximum) in addition to detailed activities for all parameters.

The average scores out of 100 indicate that all companies are relatively strong in sustainability and innovation, with scores ranging from 67 to 83. Companies like Veja and Le Slip Français lead the rankings, scoring high across all criteria, particularly in environmental initiatives and innovation. Veja's use of agroecological cotton and the invention of B-Mesh, alongside Le Slip Français' commitment to carbon neutrality and local production, highlight their leadership in eco-friendly practices. Other companies like 1083 and Hopaal also demonstrate robust environmental practices, focusing on recycled materials and reducing transport emissions by manufacturing locally.

When comparing these companies, there is a general trend towards strong environmental responsibility, with all companies implementing CO_2 reduction measures and using sustainable materials. However, some differences emerge in their approaches. For instance, Veja's production in Brazil contrasts with the other companies' local or European production, potentially increasing its transport emissions. Innovation is another differentiator, where companies like Veja and Atalaye stand out for their pioneering products made from recycled materials. In contrast, others like Olly and Balzac Paris, while innovative initially, face challenges in maintaining their unique market position as competitors adopt similar practices.

Table 3.4 Company evaluation based on circularity approaches

Company	Detailed classification per criteria			Negative points
Veja (2021)	Environment	A−	• The company buys organic cotton grown according to agro-ecological principles • Production mainly in Brazil, with European clientele, impacting transport emissions • The company recycles (upcycles) "plastic bottles, recycled cotton from textile industry offcuts, recycled polyester" and have created a material called "B-Mesh": "a fabric made entirely from recycled polyester (polyethylene terephthalate or P.E.T.)"	Made in Brazil
	Innovation and new applications	A	• The company invented and created B-mesh: a fabric made entirely from recycled polyester. It takes 3 plastic bottles to create a pair of Veja shoes. B-Mesh is lightweight, breathable, and waterproof	
	Equity of relations with partners	A	• Purchases cotton at 3.04€/kg (2017), above market rate (1.73€/kg) • Pays ADEC-associated producers 50% upfront • Association receives 0.5€/kg fair trade premium for operational funding and certifications	
Le Slip Français (2022b)	Environment	A−	• The company is carbon neutral, minimising CO_2 emissions across all activities and subsidiaries • Uses recycled and responsibly sourced materials for most products • Products made in France, locally sold, significantly reducing transport-related CO_2 emissions	
	Innovation and new applications	A	• The company has a culture of constant innovation. It has mastered the art of digital communication by playing the "humor and derision" card • Le Slip Français is 100% made in France, which gives it an undeniable competitive edge over its competitors, even today	
	Equity of relations with partners	A−	• Company prioritises workers' conditions, with no reported strikes or grievances	

(continued)

Table 3.4 (continued)

Company	Detailed classification per criteria			Negative points
1083 (2020)	Environment	A	• Implemented eco-friendly practices since inception, reducing CO_2 emissions • Utilises recycled and organic materials for all his products • Manufactured in France, minimising transport emissions • Offers only two collections annually or via limited collections to avoid overproduction and waste	Unpublished results
	Innovation and new applications	A	• The company has a culture of constant innovation, enabling it to maintain a competitive edge. Its concept: clothing produced exclusively in France, manufactured responsibly	
	Equity of relations with partners	B	• Company website states implementation of supplier code of conduct • No controversial information suggests non-compliance with commitments to partners	
Hopaal (n.d.)	Environment	A	• Company initiates CO_2 reduction from inception (recycled products, local sourcing) • Uses recycled and organic materials for clothing production • 90% of clothes made in France, 10% in Portugal, reducing transport emissions • Offers two collections annually or "limited collections" to limit mass production, reducing environmental impact	Expensive
	Innovation and new applications	A	• The company sets itself apart by producing only clothing from 100% recycled products, with a strong commitment to the environment	
	Equity of relations with partners	B	• The company partners with certified suppliers with stringent social and environmental standards • Demonstrates concern for subcontractors' well-being and conditions	

(continued)

Table 3.4 (continued)

Company	Detailed classification per criteria			Negative points
Olly (n.d.)	Environment	A−	• Company implements effective CO_2 reduction measures since inception or for several years • The company's products are made in Hungary and, apart from the organic cotton which comes from India, the other materials come from Germany, Italy and France, considerably reducing the CO_2 emissions linked to very short-distance transport between the place of production and the place of sale • Uses recycled and responsibly sourced materials for all products	Expensive
	Innovation and new applications	B	• When the company was founded, it enjoyed success thanks to an innovative and unique concept: responsible lingerie made from GOTS (certified organic cotton) • Today, this competitive advantage is no longer as strong as it was in the early days, with many competitors offering a similar concept	
	Equity of relations with partners	A	• Company exclusively purchases from local partners for many years, fostering exemplary relationships • No controversies regarding suppliers have arisen, reflecting strong supplier relations	
Bleu de Chauffe (2023)	Environment	A−	• Exclusively uses sustainable materials and natural treatment since inception, yielding measurable impact • Thanks to the use of natural products, wastewater treatment is also carried out naturally (decantation), and the water discharged is as clean as initially • Responsibly sourced materials used for all products and plant-based treatments • Products made in France, minimising transport emissions	Expensive
	Innovation and new applications	B	• Since its inception, company has focused on respect for people and the environment • Describes itself as "local craftsmanship" on its website, offering vegetable tanning, and French artisanal production • Successful commercial enterprise, selling products from China to North America	

(continued)

Table 3.4 (continued)

Company	Detailed classification per criteria			Negative points
	Equity of relations with partners	A	• Company partners uphold core values of respect for humans and the environment • No controversies indicate non-compliance with partner commitments • Majority of partners located in France	
Noyoco (2022)	Environment	A−	• Defines itself as fully eco-responsible, focusing on low-impact materials and product durability • Selects high-quality recycled or natural materials, with a focus on minimising ecological impact • Uses biodegradable materials, avoiding petroleum plastics and microplastics; employs plant-based dyes • Clothing made in Romania, reducing transport-related CO_2 emissions. Offers only two collections annually or via limited collections to avoid overproduction and waste	Expensive
	Innovation and new applications	B	• The company is one of the first brands to be recognised as 100% eco-friendly. It is not a pioneer in this field, but it is clearly different from the major players in the sector	
	Equity of relations with partners	B	• No public charter, but evidence suggests a supplier code of conduct • Half of the workshops have exclusive contracts, fostering long-term partnerships and friendships, reflecting the company's focus on supplier relations	

(continued)

Table 3.4 (continued)

Company	Detailed classification per criteria			Negative points
Sézane (2022)	Environment	B+	• The company has reduced CO_2 emissions since its inception, using natural materials and vegetable tanning for some production. Operates clothing collection and recycling program • 40% eco-friendly materials, 50% recycled polyester, 50% organic cotton • 2/3 of the company's garments are made in Europe. Leather is produced in Italy, Portugal, and Spain. Suits, coats, and some dresses and blouses are produced in Eastern Europe (Bulgaria). The remaining third is produced in Asia (notably China and India), depending on the availability of raw materials, suppliers' production capacity, and know-how • Limited releases reduce environmental impact	Expensive
	Innovation and new applications	A	• From the start, the company was successful thanks to an innovative concept that no competitor had ever offered before. This is still today an undeniable competitive advantage. The concept: to offer limited-edition pieces online, without intermediaries, all backed up by outstanding customer service	
	Equity of relations with partners	B	• The company has a supplier code of conduct; no controversies regarding commitments • Values long-term, constructive relationships with suppliers, avoiding power imbalances • Implements rigorous standards, ensuring fair labor conditions, hygiene, and safety • Prohibits child labor, forced or clandestine work	
Balzac Paris (2022)	Environment	B+	• Implements CO_2 reduction measures since inception • Uses sustainable materials: organic cotton, Naia, Tencel, vegetable-tanned leather, and some recycled products • Many items made in France, reducing transport emissions • Offers limited releases to minimise environmental impact	Unpublished yearly review report

(continued)

Table 3.4 (continued)

Company	Detailed classification per criteria			Negative points
	Innovation and new applications	B	• When the company was founded, it enjoyed success thanks to an innovative and unique concept: clothing designed to be sustainable and in line with current styling trends. Many competitors offer a similar concept	
	Equity of relations with partners	B	• Company emphasizes trust-based relationships with suppliers, exclusively partnering with GOTS-certified ones • No controversies indicate non-compliance with partner commitments • Global organic textile standard	
Atalaye (2023)	Environment	A−	• Since inception, the company has effectively reduced CO_2 emissions, showing measurable positive impact • Products are made in Western Europe (Portugal and France), minimising transport-related CO_2 emissions • Uses recycled and responsibly sourced materials for all products	Expensive
	Innovation and new applications	A	• The company is a pioneer in the manufacture of swimwear from recycled materials sourced directly from the oceans (seabed waste)	
	Equity of relations with partners	C+	• No available information on a supplier code of conduct • No signs of strikes or controversies with suppliers • All suppliers and partners are located in Europe	

3.3.4 *RQ3. What Role Can Digital Technologies Play in the Transition to Sustainability and the CE in the French TLC Industry?*

Digital transformation allows companies to improve competitiveness, as well as response agility, flexibility, and adaptation to the current uncertain and changing environment. In addition, digitalisation allows companies to take advantage of new channels to offer a comprehensive and ubiquitous experience for their customers and the development of new business models. As regards sustainability, digital technologies help reduce overheads and improve efficiencies thereby enhance corporate sustainability. For example, artificial intelligence-based prediction and planning technologies facilitate improved resource utilisation, resulting in lower material and energy usage. By minimising the consumption of primary resources and consequently environmental impacts, digital systems enable a reduction in the quantity of waste and shrinkage. The waste produced by the physical production of samples and ongoing testing can also be decreased by developing digital products for testing, validation, and even pre-sale (DigiTVC, 2022).

In the French TLC industry, digitalisation has also improved product traceability as part of a digital sales strategy. This provides access to essential data that optimises the logistics service as a whole, facilitates checks on compliance with specifications, and ensures total transparency for customers and consumers, who can have confidence in the company. In similar vein, access to data on textile provenance, component origin and product "Trademark" guarantee supports the confidence of the company's partners and customers. In short, the digitalisation of the supply chain in the textile sector enables a wide range of benefits including reduced planning time; faster production; product customisation; the ability to get products to market much faster; improved customer service; anticipation of production line contingencies; increased visibility; and reduced environmental impact.

In the French TLC sector, SMEs in particular face various challenges in pursuing digitalisation. To help address these challenges, the European SUDOE project (incorporating the textile industries in France, Spain and Portugal) entitled "Digitalisation of the Textile Value Chain" (abbreviated to DigiTVC) (Interreg Sudoes, 2023) aims to guide the sector towards Industry 4.0 operations and strengthen the competitiveness of companies through digital transformation. Aimed at managers, consultants, and technology partners, the project aims to deliver recommendations, best practices, and methodologies for drawing up a digital development plan in textile companies, covering the current situation in the sector, the benefits and challenges of digitalisation, essential technologies and information systems, as well as providing success stories and expert recommendations. In addition, the project provides for the training of 10 experts per country (France-Spain-Portugal) to support companies in their digital transformation. The project is being carried out in France by UIT Sud (an association involving southern textile Industries) in the Occitanie and Nouvelle-Aquitaine regions.

In the updated results of this protocol in 2023 (Interreg Sudoes, 2023), the following technologies were identified as being key to digital transformation: the Internet of Things (IoT), 3D printing (or additive manufacturing), robotics (autonomous and collaborative), augmented reality (AR), simulation and virtualisation, big data and artificial intelligence (AI), cloud computing, vertical and horizontal integration of systems, cybersecurity, RFID (Radio Frequency Identification), and blockchain. The significance of these is not so much in how they are implemented individually, but rather in how they are combined to have the most impact and benefit in achieving their stated goals.

For example, 3D Computer-Aided Design (CAD), deployed in conjunction with IoT devices, AI and blockchain, can significantly impact the fashion and textile industry. In the clothing sector, enormous quantities of fabrics are often used and then discarded during the initial phases of creation. To address this overproduction of waste, companies must rethink their approach to product design and development. This is where such a combination of digital technologies can make a major impact.

3D CAD offers an exceptional opportunity to innovate in the design and development of textile products. It challenges traditional processes at every stage, from pattern-making to sampling, prototyping, and marketing. By optimising design time, reducing exchanges between departments, and minimising the number of sketches and prototypes needed, these technologies enable more efficient and sustainable design (IFTH, 2023). These advancements are especially important in response to new environmental regulations such as AGEC law (INEC, 2021). Companies are thereby compelled to review their practices to reduce their ecological impact and optimise the product lifecycle.

Textile sustainability has become a major concern for the fashion industry, and digital technology plays an essential role in this quest. Here are some key points on how digitalisation promotes textile sustainability:

- Design and Development: Digital tools enable the design of more sustainable textile products from the early stages of creation. They facilitate eco-design by integrating sustainability criteria from the outset.
- Supply Chain Transparency: Through digitalisation, brands can make their supply chains more transparent. This allows tracking of the origin of raw materials, monitoring production conditions, and ensuring responsible practices.
- Production Optimisation: Smart systems and software enable the optimisation of textile production. This reduces waste, costs, and environmental footprint.
- Traceability and Recycling: Digital passports for textile products are being developed. They will allow consumers to know the product's history, composition, and environmental impact. Additionally, the design phase now integrates recycling processes, thus promoting product circularity (Bidart, 2023).

Within the framework of the DigiTVC project, an analysis of the digital level of companies in the TLC sector was carried out. One of the elements evaluated that helps define the degree of digitalisation of companies and organisations in the sector is the level of implementation of enabling technologies and information systems. As mentioned, they are essential elements in digitalisation, so the degree of knowledge and implementation by organisations is a relevant indicator. Based on the results published in 2023, 66 companies in the TLC sector (26 French, 20 Portuguese and 20 Spanish) responded to the online diagnostic, which was structured around seven themes covering a 360° vision of the company's activities: purchasing/supplier relations; production; marketing/brand image; sales/customer relations; human resources; strategic development; and identification.

The impact of digital technology is generally considered (70%) to be very positive by companies, who see it as a factor for innovation and positive transformation. In fact, the majority (74%) of managers see a link between the development of their companies and the progress they have made in the digital field. The majority of companies are still under-equipped with digital tools for supplier and customer relationship management, as well as for production management and automation: 61% consider themselves to have a low to medium level of automation. The main constraints on the implementation of a digital transformation plan lie in the difficulty for managers, on the one hand, who may have a relatively short-term vision and, on the other, the need to establish a medium-term digital roadmap (strategic priorities for digital transformation).

Table 3.5 provides a summary overview of how and where the digital technologies can impact the TLC value chain, categorised by the life cycle stage.

Table 3.5 Potential application of digital technologies in the textile value chain. *Source* Present study, elaborated based on DiGiTVC report (DigiTVC, 2022)

Digital technology categories	Life cycle stages	Specific applications
Scan tools Technologies like RFID, barcodes, and AI-based scanning tools	Raw material handling, quality control, inventory management, production	Tracking inventory, verifying materials, detecting defects, and optimising production workflows
Digital fabric samples and 3D design	Design, prototyping, production planning	Creating virtual samples, simulating garment designs, and reducing the need for physical prototypes
Smart fabrics and IoT	Production, post-production, usage	Embedding sensors in fabrics, monitoring production processes, and tracking garment usage and performance
Blockchain and traceability	Supply chain management, production, post-production	Verifying the origin of materials, tracking production stages, and ensuring authenticity and compliance
Artificial intelligence (AI)	Design, production, quality control	Predicting trends, optimising production schedules, automating defect detection, personalising designs
Augmented reality (AR)/ virtual reality (VR)	Design, sales, marketing	Virtual try-ons, creating virtual showrooms, enhancing online shopping experiences, customer engagement
Cobotics and robotics	Cutting, sewing, packaging	Automating cutting and sewing processes, improving precision, reducing labor costs, enhancing productivity
Additive manufacturing (3D printing)	Prototyping, production	Customising designs, rapid prototyping, producing complex textile structures and accessories
Supply chain management (SCM) and product lifecycle management (PLM)	Supply chain management, production planning	Managing materials, optimising production processes, reducing waste, improving time-to-market

(continued)

Table 3.5 (continued)

Digital technology categories	Life cycle stages	Specific applications
E-commerce and social media integration	Sales, marketing, customer service	Personalising marketing campaigns, managing customer relationships, enhancing online sales channels
Phygital experiences	Sales, marketing, customer engagement	Interactive 3D catalogs, virtual product displays, enhancing in-store and online customer experiences
Quality control Advanced quality control with AI, machine vision, and NIR (near-infrared spectroscopy)	Production, post-production	Ensuring high standards of production quality, consistency checks, detecting and correcting defects in real-time

There are also negatives associated with digitalisation, most notably the possible replacement of humans by robotics and other technologies, the loss of worker autonomy and the negation of their know-how, and the reinforcement of task control and surveillance processes. *Moralscore* was used to evaluate the robotisation parameter on the ten selected SMEs. This found that for all SMEs there was nothing to suggest that factories are being robotised and that employees will be replaced by machines, or there was simply no public information available on the progress of robotisation within the company. The reason for this lack of information is likely due to the fact that digitalisation and, in particular, robotisation are practices that require a significant financial investment, which is not currently available to this size of company (Table 3.6).

By adopting digital technologies such as AI, IoT and 3D CAD, companies can optimise resource use, reduce waste and improve supply chain visibility. This leads to faster production, customised products and better customer service. However, challenges include limited adoption of digital tools and the potential negative impacts on workers. Despite these challenges, the majority of companies see digitalisation as

Table 3.6 Robotisation and automation level in selected SMEs (Moralscore, 2024)

Company	Product
Veja	• There's nothing to suggest that factories are being robotised and that employees will be replaced by machines
Le Slip Français	• No precise information on robotisation is available to indicate the company's position and progress in this area
Olly	• No public information is available on the progress of robotisation within the company

(continued)

Table 3.6 (continued)

Company	Product
Bleu de Chauffe	• No public information is available on the progress of robotisation within the company
Hopaal	• No public information is available on the progress of robotisation within the company
1083	• No public information is available on the progress of robotisation within the company
Balzac Paris	• No public information is available on the progress of robotisation within the company
Noyoco	• No information on the subject, except that the company selects its garment manufacturing partners for their expertise, and that it produces in small quantities, sometimes by hand
Sézane	• No public information is available on the progress of robotisation within the company
Atalaye	• There is no precise information on robotisation to indicate the company's position and progress on this subject

a positive force for innovation and growth, assuming financial resources are available to pursue such initiatives.

3.4 Conclusion

The French textiles and clothing industry has been a cornerstone of the nation's economy, with a rich history dating back to the nineteenth century when France was a global leader in textiles and haute couture. French fashion has influenced global trends, with iconic designers like Coco Chanel, Yves Saint-Laurent, and Christian Dior shaping modern fashion. However, the sector has faced challenges in the past, including a significant decline in production and employment since the 1990s due to automation and globalisation. Now, however, the industry faces new environmental, economic, and social challenges. In 2023, the French fashion industry's direct sales totaled €154 billion, of which €66.3 billion was for textiles and clothing, corresponding to approximately 2.6 billion items of clothing, household linen, and footwear put on the market each year (equivalent to around 624,000 tonnes, or about 9.5 kg per year per inhabitant).

The eco-organisation *Re_fashion,* responsible for waste management in the French TLC sector, plays a crucial role in promoting recycling and CE practices. Despite these efforts, the sector struggles with the profitability and efficiency of recycling processes. The industry's supply chains are complex, with many companies outsourcing production outside Europe, leading to higher environmental impacts. Recent research focuses on how French TLC companies address sustainability and

CE practices, particularly using digital technologies. This study highlights the importance of integrating sustainable practices into corporate strategies and the role of digital efficiency and automation in transitioning to a more sustainable and circular economy and addresses the three main RQs set out in Sect. 3.1.

Concerning RQ1, French TLC companies are addressing sustainability through diverse strategies integrated into their corporate activities. The case studies of 10 French TLC SMEs highlight the following key approaches:

1. Local Production and Relocation: Companies like Le Slip Français and 1083 emphasise the value of French or European industrial facilities, relocating certain activities back to France to reduce carbon footprints and support local economies.
2. Transparency and Fair Trade: Veja is a pioneer in ensuring transparency across its supply chain, prioritising ethical sourcing, fair labor practices, and environmental stewardship. This approach is seen as crucial for building consumer trust and meeting sustainability goals.
3. Timeless and Durable Design: Hopaal advocates for creating products that resist seasonal trends, focusing on durability and functionality to minimise waste.
4. Material and Craftsmanship Valuation: Bleu de Chauffe and others emphasise the use of high-quality, sustainably sourced materials and unique fashion-related craftsmanship to promote responsible consumption.

Each company integrates sustainability into its business model, addressing issues like ethical labor, eco-friendly materials, and minimising environmental impacts. Despite variations in product types, the commitment to sustainability is consistent across the industry, driven by local regulations, consumer expectations, and the broader goals of sustainable development such as the Law no. 2020-105 of February 10, 2020, on the fight against waste and the circular economy.

Concerning RQ2, the CE in the textile industry involves redesigning textiles, incorporating new technologies, and recycling old garments. France has made significant strides in transitioning to the CE, emphasising sustainable sourcing, eco-design, responsible consumption, waste management, and extending product lifespans. The French government has identified seven pillars for the CE, including sustainable sourcing, industrial ecology, and waste prevention. EPR is a key component of France's approach to the CE, which mandates that manufacturers manage the waste generated by their products. This is enforced through eco-organisations funded by an eco-contribution added to product prices. France has also introduced environmental labeling to inform consumers about the environmental impacts of their purchases and to encourage more sustainable production practices. Additionally, the Triman pictogram and Info-Tri logo have been introduced to simplify recycling and sorting for consumers.

Eco-design plays a critical role in reducing the environmental footprint of garments, with an emphasis on optimising fabric use, choosing sustainable materials, and designing for easier recycling. Extending the lifespan of garments through durable materials, multifunctional designs, and second-hand markets is also crucial. Recycling practices are being enhanced through platforms like *Re_fashion*, which aims to industrialise the recovery of materials from non-reusable textiles.

Overall, the transition towards a CE in the textile industry in France is marked by legislative support, innovative practices in design and production, and a strong focus on consumer awareness and responsibility. This shift is supported by the collaboration of various stakeholders, including businesses, consumers, and the government.

Concerning RQ3, the French TLC industry enhances competitiveness, flexibility, and sustainability by leveraging digital technologies such as AI, IoT, and 3D printing. These technologies optimise resource use, reduce waste, and improve transparency across the supply chain. The European SUDOE project supports SMEs in digital adoption, aiming for Industry 4.0 through training and best practices. Challenges include the to-date limited digital tools adoption, and the potential negative impacts on worker autonomy. However, the overall impact is seen as positive, driving innovation, faster production, and improved customer engagement.

Businesses play a key role in this transformation of textile sector: optimising the life cycle of products put on the market, encouraging reuse, extending product life, improving collection (only 50% of packaging carries a sorting instruction), integrating a minimum rate of recycled material in products, preferring donation to destruction, no longer throwing away, no longer wasting. Consumers also need to be better informed, so that they can become active players by making the right consumption choices and adopting the right reflexes. All these measures are in line with the ecological goals, and their implementation is still the subject of discussions with players in the TLC sector and will shortly be set out in regulations (decrees and orders). These necessary changes to control the environmental impact of products can enable the fashion sector to gradually become more circular, and therefore even more responsible.

Reducing the ecological footprint of the products means first and foremost optimising the use of resources and production processes at every stage of the life cycle. In the textile industry, this mainly entails due consideration of issues linked to climate change, water consumption and pollution (eutrophication, use of toxic chemicals, etc.). To do this, the following actions are of particular importance: choose raw materials with less impact, giving preference to recycled, bio sourced and/or high environmental performance materials; adopt more responsible production methods and technologies, particularly for dyeing and finishing processes; and design textiles that are more durable, resistant to washing and deformation, to extend their lifespan and avoid the need for frequent renewal; promote the reuse and recycling of products at the end of their life cycle. If these guidelines can be actively pursued by all stakeholders, the TLC industry in France can speed its adoption of sustainability principles and its continued transition to the circular economy.

References

1083. (2020). *Le jean fabriqué en France, 100% coton bio | 1083*. https://www.1083.fr/jeans.html

ADEME. (2020). *Économie circulaire: définition, enjeux et concepts*. https://economie-circulaire.ademe.fr/economie-circulaire

ADEME. (2022a). *Filière des Textiles d'habillement, linge de maison et chaussures - Données 2022*. La Librairie ADEME. https://librairie.ademe.fr/7224-filiere-des-textiles-d-habillement-linge-de-maison-et-chaussures-donnees-2022.html

ADEME. (2022b). Rapport au Parlement sur le bilan de l'expérimentation affichage environnemental textiles d'habillement et chaussures de la loi Anti-Gaspillage pour une Economie Circulaire, article 15. In *2022*.

Atalaye. (2023). *Accueil - Bain de Biarritz*. https://atalaye.com/fr/?v=11aedd0e4327

Balzac Paris. (2022). *Eco-responsible clothing*. https://balzac-paris.com/en

Bécuwe, S., & Blancheton, B. (2018). French textile specialisation in long run perspective (1836–1938): Trade policy as industrial policy. *Business History*.

Beji, R., Yousfi, O., Loukil, N., & Omri, A. (2021). Board diversity and corporate social responsibility: Empirical evidence from France. *Journal of Business Ethics, 173*(1), 133–155. https://doi.org/10.1007/S10551-020-04522-4

Bidart. (2023). *Revue Annuelle des Nouveaux Usages | Numérique & durabilité textile | Association Nationale Recherche Technologie*. ANRT. https://www.anrt.asso.fr/fr/revue-annuelle-des-nouveaux-usages-numerique-durabilite-textile-36820

Blancheton, B. (2021). French fashion and textile during COVID-19. Once again, flying to high quality and innovation to survive. *Fashion and Textiles, 8*(1), 43. https://doi.org/10.1186/s40691-021-00278-1

Bleu de Chauffe. (2023). *Le made in France*. https://www.bleu-de-chauffe.com/fr/module/tb_presentation/madeinfrance

DigiTVC. (2022). *Protocol—DigiTVC*. https://digitvc.eu/protocol/

Ecoinvent. (2023). *Ecoinvent v3.10*. https://ecoinvent.org/ecoinvent-v3-10/

Fondation Ellen Macarthur. (2017). *A new textiles economy: Redesigning fashion's future*.

French Government. (2021). *France Relance: Transition écologique*. https://www.ecologie.gouv.fr/actualites/plan-france-relance-transition-ecologique-objectif-strategique

Gray, S. (2017). *Mapping clothing impacts in Europe: The environmental cost* (pp. 1–41). WRAP.

Hopaal. (n.d.). *Pour Hopaal, la performance ne s'embarrasse pas du superflu*. Retrieved August 25, 2024, from https://hopaal.com/blogs/news/pour-hopaal-la-performance-ne-s-embarrasse-pas-du-superflu

IFTH. (2023). *Mode/habillement: l'IFTH met le numérique au service de la transition écologique - IFTH - Institut Français du Textile et de l'Habillement*. https://www.ifth.org/2023/11/30/mode-habillement-lifth-met-le-numerique-au-service-de-la-transition-ecologique/

INEC. (2018). *L'économie circulaire dans l'industrie textile*. https://www.insee.fr/fr/statistiques/1281038

INEC. (2021). *La responsabilité élargie du producteur et la loi AGEC*.

INSEE. (2023). *Les comptes de la Nation en 2023 Insee Résultats*. https://www.insee.fr/fr/statistiques/8068749

Interreg Sudoes. (2023). *Digitalization protocol of textile value chain*.

Jeronen, E. (2013). Sustainability and sustainable development. *Encyclopedia of Corporate Social Responsibility*, 2370–2378. https://doi.org/10.1007/978-3-642-28036-8_662

Kozlowski, A., Searcy, C., & Bardecki, M. (2018). The reDesign canvas: Fashion design as a tool for sustainability. *Journal of Cleaner Production, 183*, 194–207. https://doi.org/10.1016/J.JCLEPRO.2018.02.014

Labeye, B. (2018). L'industrie textile en France: une production mondialisée, sauf pour les produits de luxe et les textiles techniques. *Insee Première*. https://www.insee.fr/fr/statistiques/3632345#consulter

Le Slip Français. (2022a). *Company with mission - Le slip français*. https://www.leslipfrancais.fr/en/pages/entreprise-a-mission

Le Slip Français. (2022b). *Fabrication française - Le slip français*. https://www.leslipfrancais.fr/pages/fabrication-francaise

Légifrance. (1975). *Loi n° 75-633 du 15 juillet 1975 relative à l'élimination des déchets et à la récupération des matériaux*. https://www.legifrance.gouv.fr/loda/id/JORFTEXT000000888298

Légifrance. (2001). *Loi n° 2001-420 du 15 mai 2001 relative aux nouvelles régulations économiques*. https://www.legifrance.gouv.fr/loda/id/JORFTEXT000000223114

Légifrance. (2015). LOI n° 2015-992 du 17 août 2015 relative à la transition énergétique pour la croissance verte (1). *Journal Officiel "Lois et Décrets."* https://www.legifrance.gouv.fr/loda/id/JORFTEXT000031044385

Légifrance. (2020). *LOI n° 2020-105 du 10 février 2020 relative à la lutte contre le gaspillage et à l'économie circulaire (1)*. https://www.legifrance.gouv.fr/jorf/id/JORFTEXT000041553759/

Légifrance. (2021). *Article L541-9-5 - Code de l'environnement*. https://www.legifrance.gouv.fr/codes/article_lc/LEGIARTI000043979280

Légifrance. (2022a). *Arrêté du 23 novembre 2022 portant cahiers des charges des éco-organismes et des systèmes individuels de la filière à responsabilité élargie du producteur des textiles, chaussures et linge de maison (TLC)*. https://www.legifrance.gouv.fr/jorf/id/JORFTEXT000046600083

Légifrance. (2022b). *Article L541-9-4 - Code de l'environnement*. https://www.legifrance.gouv.fr/codes/article_lc/LEGIARTI000044244398

Ministère de la Transition écologique et de la Cohésion des territoires. (2017). *Produits textiles (TLC)*. https://www.ecologie.gouv.fr/politiques-publiques/produits-textiles-tlc

Ministère de la Transition écologique et de la Cohésion des territoires. (2019). *L'affichage environnemental des produits*. https://www.ecologie.gouv.fr/politiques-publiques/laffichage-environnemental-produits-services-hors-alimentaire

Ministère de l'économie des finances et de la souveraineté industrielle et numérique. (n.d.). *Tableaux du commerce extérieur de la France | economie.gouv.fr*. Retrieved August 29, 2024, from https://www.economie.gouv.fr/saef/tableaux-commerce-exterieur-france

Moralscore. (2024). *Découvrez les marques qui collent à votre éthique*. https://moralscore.org/domains/habillement/

Noyoco. (2022). *Noyoco—Sustainable meets cool*. https://noyoco.com/

Olly. (n.d.). *Olly Lingerie: Lingerie Ethique en Coton Bio & Dentelle Recyclée - VOLAGO*. Retrieved August 25, 2024, from https://volago.fr/marque-ethique/olly/

Peters, J., & Simaens, A. (2020). Integrating sustainability into corporate strategy: A case study of the textile and clothing industry. *Sustainability*, *12*(15), 6125. https://doi.org/10.3390/SU12156125

Re_fashion. (n.d.). *Refashion Pro pour une filière 100% circulaire*. Retrieved August 25, 2024, from https://refashion.fr/pro/fr

Re_fashion. (2021). *Recycle platform-Re_fashion*. https://recycle.refashion.fr/

Re_fashion. (2022a). *La signalétique info-tri pour la Filière Textile est validée!* https://refashion.fr/pro/en/sorting-info-textile-sector-validated

Re_fashion. (2022b). *Rapport d'activité 2022*. https://refashion.fr/rapport-activite/2022/index.html

Sézane. (2022). *Derrière l'étiquette*. https://www.sezane.com/fr/derriere-l-etiquette/

Still in fashion. (n.d.). *Christian Dior revived Haute Couture after the Second World War*. Retrieved August 29, 2024, from https://stillinfashion.com/christian-dior-revived-haute-couture-after-the-second-world-war/

Thiry, M. C. (2011). Staying alive: Making textiles sustainable. *AATCC Review*. www.unwater.org/statistics_res.html

United Nations General Assembly. (1987). *Report of the world commission on environment and development: Our common future towards sustainable development 2. Part II. Common challenges population and human resources 4*. 1.

Valet de Pique. (2019). *How did France become the fashion capital of the world?* https://www.valetdepique.fr/en/blogs/made-in-france/comment-la-france-est-devenue-la-capitale-mondiale-de-la-mode
Veja. (2021). *Transparency | VEJA*. https://project.veja-store.com/fr/single/transparency
World Commission on Environment and Development. (1987). *Our common future*. https://digitallibrary.un.org/record/139811

Mehri Shabani (mshabani@esaip.org) is an assistant professor at higher education institute of ESAIP La Salle, France, specialising in life cycle assessment (LCA) and circular economy and an associate researcher in the Faculty of Science, University of Angers, France. Mehri is in charge of a specialisation entitled as Environmental and Energy Transtion (T2E) for engineering students. Her research interests include sustainability, life cycle assessment (LCA), and bioremediation (by the use of microbial fuel cell) of industrial wastewater and polluted water. Currently, she is involved in various regional and international research projects on the previously mentioned topics. Her teaching topics include sustainable development, environmental analysis including LCA and circular economy.

Erica Bicchi (ebicchi@esaip.org) is an assistant professor at higher education institute of ESAIP La Salle, France, specialising in risk prevention and environmental management, and an associate researcher at the university of Angers, Laboratoire de Planétologie et Géosciences, LPG UMR 6112, France. Erica is in charge of a specialisation entitled "Environment and Circular Economy". Her main research topics focus on the use of marine micro-organisms (foraminifera) as tools, in the fossil record, to reconstruct paleoenvironments and date them, and in present-day record, to use them as bioindicators of ecological quality of the marine environment. More recently, Erica has specialised in circular economy issues related to environmental pollution (i.e., marine plastic pollution and municipal solid waste management).

Chapter 4
Türkiye

Semra Ocak and Sevil Acar

Abstract This chapter examines the Turkish textile and clothing (T&C) industry's perspective on sustainability and its journey toward a Circular Economy (CE). Currently, the industry makes a significant contribution to the manufacturing sector as a whole in the country, as well as to the general economy in terms of employment and trade. The findings indicate the high resource dependence and considerable environmental impacts of the sector. Using company-level data, the chapter concludes that CE actions are predominantly integrated into company sustainability policies. Company reports show that there is a limited number of companies that follow a roadmap that implements the best available technology in the value chain for a fully circular transition. However, notable efforts can be observed in corporate documents, especially those incorporating thorough disclosures on sustainability and corporate social responsibility. The results highlight the need for improved collaboration among the stakeholders and for a strengthened ambition in the sector toward a sustainable CE business model. Limitations evident in the chapter highlight the need for further research and national and industry-specific analyses that investigate both the possibilities and deficiencies within the industry concerning sustainability, circularity, and digital transformation.

Keywords Circular economy · Digitalization · Sustainability · Turkish T&C industry

S. Ocak · S. Acar (✉)
Bogazici University, Istanbul, Türkiye
e-mail: sevil.acar@bogazici.edu.tr

S. Ocak
e-mail: semra.ocak@std.bogazici.edu.tr

M. Wynn and T. Wiegand (eds.), *Sustainability, the Circular Economy and Digitalisation in the European Textile and Clothing Industry*, SDGs and Textiles,
https://doi.org/10.1007/978-981-97-9116-3_4

4.1 Introduction and Background

The textile and clothing (T&C) industry is of significance in Türkiye in terms of its contribution to macroeconomic indicators such as gross domestic product, share in manufacturing industry, industrial production, exports, net foreign exchange input to the economy, employment, and investment. Despite a decline in its prominence in recent years, it remains a significant industry in terms of generating added value (Alüftekin et al., 2009).

Textile manufacturing, with its origins tracing back to the Seljuk era (1050–1300) and the Ottoman Empire (1285–1923), was widespread and advanced, particularly during the sixteenth and seventeenth centuries. Carpet, rug, and felt production in the Turkish culture is one of the oldest traditional handicrafts. It is recorded that carpets, rugs, and silk textiles produced in Anatolia were sent to various parts of the world at the beginning of the thirteenth century. The number of silk looms in Bursa province reached a thousand in the fourteenth century. The development of the textile and clothing industry occurred during the Ottoman Empire, with production in small enterprises in Denizli and Tokat provinces for weaving and in Bursa province for silk products. In 1915, 18 of the 22 public industrial enterprises, 10 of the 28 joint-stock companies, 45 of the 214 private sector companies, and 73 of the total 264 industrial workplaces were operating in this industry (Öngüt, 2007; Uludağ İhracatçı Birlikleri, 2018; Uyanık & Çelikel, 2019).

Following the foundation of the Turkish Republic in 1923, Sümerbank was established as a state entity in 1933, under which all textile and clothing enterprises and workshops were gathered. Sümerbank, although a state body, played a key role in the private sector through its investments and staff training, subsequently facilitating the transfer of accumulated knowledge to the private sector. After the 1950s, the private sector developed, and the production of synthetic fibers began after the 1960s. The period between 1960 and 1980 witnessed a rapid acceleration in technological advancements within this sector. This progress was facilitated by the import substitution policy and government incentives implemented during the planned period within the economy. After the implementation of the free-market economy and privatisation policies during the 1980s, significant increases in textile and ready-made clothing exports occurred. In the early 1980s, Türkiye primarily exported textile products such as yarn, fiber, and fabric. Starting from 1984, there was a shift toward exporting more ready-made clothing products (Öngüt, 2007; Uludağ İhracatçı Birlikleri, 2018). By the 1990s, the textile and ready-made clothing sector accounted for 40% of total exports. Subsequently, the industry developed further, and apart from silk and cotton, textiles produced from wool, hemp, and linen have become industries in various cities (Esi, 2017; Sayar, 2020).

The textile industry in Türkiye has emerged as a prominent manufacturing sector, particularly with the implementation of neoliberal economic strategies during the 1980s. Turkish governments have established free trade zones and provided incentives, such as export subsidy credits and exemptions of taxes and duties on imported machinery, to attract foreign investment. Consequently, Türkiye has become a manufacturing area for European companies due to its advantageous geographical location and cheap labor compared to Europe. In the 1990s, the Turkish textile and clothing sectors emerged as the largest industries in terms of exports, accounting for 38.5% of Turkish exports (Divrik & Baykal, 2022).

As part of the new market-oriented economic policies that started in the 1980s, investments of public institutions operating in the T&C sectors were restricted. The privatisation process, which was initiated in the 1990s, had a notable impact on employment rates in public enterprises in the ready-made clothing sector, which was 3.9% in 1988 and decreased to 1.5% in 2000 (Alüftekin et al., 2009). With the implementation of the Customs Union Agreement with the European Union (EU), which was enacted in 1996, corporations embarked on substantial capital ventures and augmented their manufacturing capabilities. After the signing of the Agreement, the opportunity to export to the EU market without restrictions was made possible. Between 1990 and 1995, US$6 billion was allocated exclusively to purchase textile machinery (Bostan et al., 2010; The Ministry of Industry and Technology, 2021a).

Currently, the textile, ready-made clothing, and leather sectors contribute significantly to the manufacturing industry in terms of employment and foreign trade. These sectors account for 15% of the production value in the manufacturing industry and contribute 16% to the overall added value, that is, approximately 6% of the country's GDP. Türkiye was the 5th largest exporter in the country-based ranking with a 3.4% share in world textile exports in 2020 and the 6th largest exporter with a 3.6% share in world ready-made clothing exports (The Ministry of Industry and Technology, 2021a).

The number of companies in the textile sector, which consists mainly of small and medium-sized enterprises (SMEs), is around 17,800 according to 2020 data (Table 4.1) (The Ministry of Industry and Technology, 2021a). This number increased to 19,256 in 2021, including textile, home textile, and carpet companies. The number of companies operating in both the textile and clothing industries is reported to be around 60,000, where more than 1.1 million people are employed, corresponding to 7.5% of total employment and 20.3% of the manufacturing labor force (The Ministry of Industry and Technology, 2021a; The Ministry of Trade Easy Export Platform, 2024).

The T&C industry of Türkiye exported approximately 65% of its production in 2020, which contributed to nearly 15% of Türkiye's total exports, and accounted for roughly one-seventh, corresponding to US$25 billion, of the total export earnings of Türkiye, which amounted to approximately US$170 billion (The Ministry of Industry and Technology, 2021a).

Ready-made clothing and apparel, with a share of 9% in general exports, ranks 4th after the automotive industry with a share of 13%, chemicals and chemical products with a share of 11.2%, and steel with a share of 9.9%. According to the Turkish Exporters Assembly data, Türkiye's exports of ready-made clothing and apparel reached a total value of US$21 billion in 2021, corresponding to an increase of 20% compared to the year 2020. When comparing export figures between 2011 and 2021, the increase was recorded as 36% (Fig. 4.1) (İHKİB, 2021, 2022). However,

Table 4.1 Production, exports, and employment values of clothing and textiles in Türkiye in 2020

	Production value (Billion TL)	Exports (Billion US$)	Employment	Number of enterprises
Textile	234.28	12.34	460,399	17,751
Ready-to-wear clothing	153.60	14.99	608,587	35,934
Leather/Leather products	20.91	1.37	66,785	6,557
Total	408.78	28.70	1,135,771	60,242

Source The Ministry of Industry and Technology (2021a)

Türkiye's ready-made clothing and apparel exports decreased by 9.2% from 2022 to 2023, dropping to US$19.3 billion, of which US$11.6 billion was exported to EU countries (İHKİB, 2023). In addition, a general decline in numbers was observed despite minor rises in import statistics (Fig. 4.1) (İHKİB, 2022).

In 2019, Türkiye's exports of carpets, ready-made clothing and apparel, textiles, and raw materials constituted 16% of the total country exports with US$28 billion. In 2020, exports in the same group decreased by US$1 billion to 27 billion dollars, which corresponds to 18% of the total country exports. While Türkiye's ready-made clothing exports amounted to approximately US$18.3 billion in 2021, they reached US$19.5 billion in 2022, with an increase of 6.43%. Exports to Germany, which is Türkiye's largest export market of ready-made clothing and apparel, amounted to US$3.08 billion in 2021, increasing by 11.33% in 2022 and reaching US$3.43 billion. As Fig. 4.2 indicates, Türkiye exported 69.5% of its ready-made clothing to Germany (17.6%), Spain (13%), the UK (10.1%), Netherlands (8.4%), France (5.2%), USA (4.2%), Italy (3.6%), Iraq (2.5%), Denmark (2.5%), and Israel (2.4%) in 2022. In

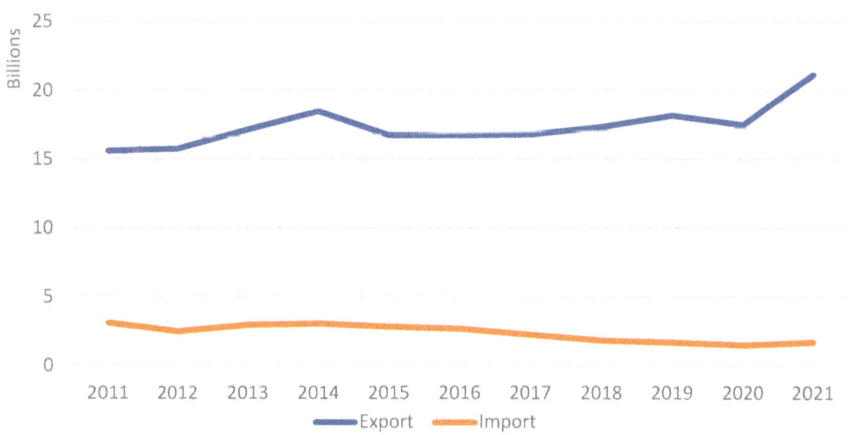

Fig. 4.1 Ready-made clothing and apparel foreign trade of Türkiye between 2011–2021 (in billion US$). *Source* İHKİB (2022)

Fig. 4.2 Top 10 countries with the highest value of ready-made clothing and apparel imports from Türkiye (in thousand US$). *Source* The Ministry of Trade (2024)

comparison with 2021, there were small increases and decreases in exports to several countries, including Spain, the UK, the Netherlands, France, Iraq, the USA, Italy, Poland, and Saudi Arabia (The Ministry of Industry and Technology, 2021b; The Ministry of Trade, 2024).

Between 2022 and 2023, the top 10 countries to which Türkiye exported textiles and raw materials were Italy, Germany, the United States, Spain, the United Kingdom, Morocco, Iran, and Egypt. While the total exports of textiles and materials to EU member countries in 2022 was US$5.6 billion, this figure decreased by 17.6% in 2023. Total textile and raw materials exports were US$12.9 billion in 2022 and US$11.6 billion in 2023 corresponding to a 11% decline (İTKİB, 2024). While Türkiye's ready-made clothing and apparel imports amounted to US$1.04 billion in the first six months of 2022, they increased by 35.8%, reaching US$1.41 billion in the same period of 2023 (İHKİB, 2024).

The textile industry is primarily based on agriculture, which requires the use of natural fibers such as cotton and wool, leading to an upward trend in cotton production. While 750 thousand tons of fiber cotton was produced in 2015, the production volume reached 950 thousand tons in 2019 (The Ministry of Trade Easy Export Platform). Cotton holds a significant position in Türkiye reflecting the significance of textile exports which were worth in excess of US$18 billion in 2020. The International Cotton Advisory Committee (ICAC) states that Türkiye's position in the global cotton market for the 2019–20 season was ranked 11th in terms of plantation area, 5th in terms of yield, 6th in terms of total production, 4th in terms of consumption, and 5th in terms of imports (Can et al., 2021).

There has been a significant growth in the cultivated area for cotton in the last decade in the "South Eastern Anatolian Project (GAP)" region, which currently contributes 62.5% of Türkiye's cotton production. In the 2020–2021 season, the surface area dedicated to cotton production declined by 24.8% compared to the previous season, covering an area of 3,592 hectares. Additionally, uncertainties derived from the pandemic, weather conditions, a decrease in cotton prices, an increase in production costs, and the shift in cotton production areas accelerated this decline (The Ministry of Trade, 2021a). In addition, the tragic earthquake which occurred in 2023 caused an increase in cotton yarn prices, since more than 50% of manufacturers were located in the region affected by the earthquake (Erdoğan, 2023).

Prominent purchasing groups within the T&C industry are primarily engaged in high value-added activities along the chain, namely design and marketing. Moreover, this industry exerts influence over purchasing decisions, product selection, and supply processes. Countries engaged in production, such as Türkiye, the Far Eastern countries, and Southeast Asian countries, are mostly involved in the labor-intensive cutting and sewing stages (The Ministry of Trade Easy Export Platform, 2024). In Türkiye, the majority of enterprises operating within the sector are classified as SMEs. According to the Social Security Institution data, the T&C and leather industries in Türkiye employ a considerable workforce, with approximately 1.1 million individuals officially registered. The number of workers working in the ready-made clothing industry is reported as 602,800; however, the rate of unionisation stands only at 6.8%. On the other hand, approximately 29.1% of paid employees in the textile industry were not registered to any social security institution in 2019 (Hisarlı & Demir, 2022).

More than half of the workforce in the T&C industry are women. In addition, a growing percentage of workers are engaged without a formal status, which is mostly common in home-based and small workshops, but also in factories at times. The challenges faced by women engaged in home-based work encompass various aspects, including low wages with no job security, the absence of collective bargaining agreements, unpredictable and lengthy working hours, and detrimental effects on women's health, such as lower back pain, anxiety, and stress. In addition, child and immigrant labor are widely used in the sector to reduce labor costs (Hisarlı & Demir, 2022).

The T&C industry poses physical, chemical, and psychosocial risks that threaten the health of employees. Chemical substances (solvents, optical brighteners, anti-wrinkle agents, flame retardants, heavy metals, pesticides, and anti-microbial agents) used in the textile industry cause many health risks. As a result of the use of these chemicals, they can cause diseases such as pneumonia, asthma, lead poisoning, eczema, lung cancer, and leukemia in employees due to skin-contact or inhalation. Besides, since the textile industry is a labor-intensive sector, musculoskeletal system diseases are widely experienced. Another occupational health and safety risk is noise, which may cause problems such as decreased productivity and concentration, fatigue, hearing problems, headaches, arrhythmias, muscle contractions, and narrowing of blood vessels (İş Güvenliği, n.d.; Rençber & Ceylan, 2022). Work overload, occupational stress, inadequate salary, no participation in decision-making, and lack of control over work could be listed as psychosocial risks derived from workplaces (Ramos-Galarza & Acosta-Rodas, 2019). In addition, mobbing, job insecurity, gender discrimination, and abuse are observed.

The global production of garments has experienced an almost two-fold increase over the past 15 years. A result of this is the prevailing trend of "fast fashion," coupled with the availability of inexpensive clothing and a rise in per capita purchases. This trend implies a "buy-and-throw-away philosophy" which leads to overconsumption, thereby giving rise to excessive use of raw materials and natural resources (Papamichael et al., 2023). Currently, the manufacturing, distribution, and use of apparel are based on a linear model. This situation results in externalities generated by the textile industry, which is one of the most environmentally detrimental and resource-dependent sectors (Ellen MacArthur Foundation, 2017).

Pollution from cotton cultivation and textile garment production are other issues contributing to the environmental burden. Conventional cotton production requires significant amounts of water. In addition, the use of fertilizers and pesticides causes eutrophication and toxicity. Pesticides in cotton production account for 25% of global use, even though the crop covers only 2% of agricultural land (Chen et al., 2021a, 2021b; James & Lings, 2018). On the other hand, organic cotton is produced without the use of synthetic pesticides and fertilizers. In organic production, farmers use compost, animal manure, or green manure, which improves soil structure and water infiltration. Türkiye ranks 4th in organic cotton production with a 10% share following India (50%), China (12%), and Kyrgyzstan (12%). The other countries among the top seven countries are Tanzania (5%), Tajikistan (4%), and the U.S. (3%) (Textile Exchange, 2022).

Other phases of textile production, specifically chemical processes, also create a negative environmental impact. Dyeing is one of the most polluting phases in which dyes, wetting agents, and softeners are used (Chen et al., 2021a, 2021b). In addition, numerous chemicals are used during the final stages of production, such as water and dirt resistance in outdoor garment, wrinkle prevention or stiffening agents, as well as biocides to diminish odors caused by bacteria or the growth of mold during transportation and storage (Manshoven et al., 2019). These chemicals cause water pollution. In Türkiye, the Ergene River Basin is one such case, which suffers from enduring pollution over the years derived from untreated textile wastewaters and a significant reduction of groundwater level (Alkaya & Demirer, 2014).

The T&C industry is a significantly resource-intensive and waste-generating industry, responsible for 3–10% of global CO_2e emissions (McKinsey&Company, 2022). 1.2 billion tons of CO_2 are produced annually by the textile industry globally. The apparel business uses about 60% of all textiles, and the fashion sector is estimated to be responsible for 5% of world emissions (The Ministry of Industry and Technology, 2021b). After the iron and steel and cement industries, the textile sector is reportedly the 3rd most energy-intensive industry in Türkiye (Alkaya & Demirer, 2014). In a study conducted in the Marmara region, the total carbon footprint resulting from production for a textile company producing fabric and carpets was calculated as 31.2 kg-CO_2e per kg-fabric (Coşkun & Doğan, 2021).

The sector predominantly depends on finite resources, amounting to 98 million tons each year. These resources encompass oil, fertilizers used for cotton cultivation, as well as chemicals used for the production, dyeing, and finishing of fibers and textiles. Textile manufacturing consumes approximately 93 billion m^3 of water annualy, thereby exacerbating issues in regions that suffer from water scarcity. In addition, this linear system fails to benefit from economic opportunities, deteriorates natural resources, and generates adverse societal consequences (Ellen MacArthur Foundation, 2017). These negative externalities originate from the current linear "take-make-dispose" model incorporating minimum rates of reuse, repair, fiber-to-fiber recycling of textiles, and low emphasis on product's durability and recyclability (European Commission, 2022).

It is estimated that approximately 600 thousand tons of 1.8 million tons of products are wasted annually in Türkiye. The process of collecting used clothing and reintroducing it to the market faces certain challenges. However, the recent introduction of clothing collection boxes has resulted in the collection and sorting of approximately 40,000 tons of clothing per year, out of which 10,000 tons can be recycled. If all the discarded clothing is gathered and recycled, Türkiye could generate an additional value of US$2.5 billion (Türkiye Tekstil Sanayii İşverenleri Sendikası, 2022).

Given the resource-dependent and pollutive characteristics of the T&C industry, it has been identified as a prominent sector for the implementation of sustainable practices in accordance with the policies outlined in the European Green Deal (European Commission, 2020).[1] The European Green Deal (EGD) was initiated in 2019 with an aim to achieve climate-neutrality by 2050, resource-efficiency, and a competitive economy in the European continent. Following the announcement of the European Green Deal, new strategies and action plans were announced within the scope of a rapid transition to the new order in many different areas with the participation of different stakeholders. These strategies aim to promote the adoption of circular business models, and empower consumers toward sustainability. In March 2020, the European Commission announced the Circular Economy Action Plan (CEAP) with the aim of reducing pressure on natural resources and promoting sustainable growth and employment opportunities. As outlined in the CEAP, the Commission introduced new regulations to enhance the environmentally friendly, circular, and energy efficient physical goods available in the EU market. These regulations are relevant for the entire life cycle of these goods, encompassing aspects such as design, daily use, repurposing, and end-of-life management (European Circular Economy Stakeholder Platform, 2022; European Commission, 2020; Krämer, 2020).

Given the complexity of the textile value chain as well as its negative externalities on nature and the environment, the European Commission announced the EU Strategy for Sustainable and Circular Textiles under the umbrella of the EGD. The strategy focuses on innovation, collaboration, and competitiveness between stakeholders to achieve more sustainable and circular production. It covers "empowering business and private consumers to choose sustainable textiles and easy access to reuse and

[1] Other sectors that fall under this policy are plastics, packaging, battery and tools, construction materials, information technologies, and the food industry (European Commission, 2020).

repair services; introducing circular materials and production processes, achieving high levels of separate collection of textile waste, boosting the sorting, re-use and recycling of textiles" (European Commission, 2020, p. 13).

On the other hand, EGD is not only an environmental strategy, but also a road map aiming at achieving new forms of international trade and division of labor. Following the commencement of the EGD, business and government officials in Türkiye have developed strategies as they have strong economic relationships with the EU (Aşıcı & Acar, 2022). Within the scope of the EGD, an action plan has been announced to ensure Türkiye's adaptation to the EGD by supporting the green transformation of Türkiye's economy and industry, maintaining competitiveness in trade with third countries, especially with the EU, and improving integration into the transforming global value chains (The Ministry of Trade, 2022). Türkiye's Green Deal Action Plan encompasses the actions planned to be carried out under 9 headings, including 32 goals and 81 actions. The main themes are listed as follows: "Carbon Border Adjustment, Clean, Affordable and Secure Energy Supply, Fight against Climate Change, Green and Circular Economy, Green Finance, Sustainable Agriculture, Sustainable Smart Mobility, Diplomacy, and Informative and Educational Activities" (Aşıcı & Acar, 2022, p. 49).

Given the strategic importance of information and knowledge, the European Commission is introducing data governance policies to be able to support sustainability transformation (Piétron et al., 2023). The CE is based on a significant amount of data such as product life cycle stages, wastes, and by-products, as well as the use of raw materials, reuse techniques of waste materials, and logistic requirements. The emergence of digitalisation introduces novel approaches to accessing data that CE systems need (Antikainen et al., 2018; Rajput & Singh, 2019). Digital technologies have the potential to facilitate the CE within the T&C industry to achieve the objectives of both the EGD and the CEAP.

This chapter examines how the Turkish T&C industry views sustainability and assesses the measures it is taking to move toward the CE. In this context, the chapter examines how the selected companies in the Turkish T&C industry have approached sustainability and the CE to date. More specifically, the chapter addresses the following research questions (RQs):

RQ1: How is the Turkish T&C industry inter-related with the other sectors of the economy in terms of its input-output connections, employment generation, resource use, and greenhouse gas (GHG) emissions?

RQ2: How are Turkish T&C companies addressing sustainability in their corporate strategies and activities?

RQ3: What strategies and activities relating to the CE are being pursued in the Turkish T&C industry?

RQ4: What role are digital technologies playing in the transition to sustainability and the CE in the Turkish T&C industry?

This chapter has five main sections. Following this introduction, Sect. 4.2 introduces the research method. This is followed by Sect. 4.3, in which the findings are set out and analysed, and the RQs are addressed. Section 4.4 then discusses these findings and related themes. Finally, Sect. 4.5 provides a conclusion to the chapter.

4.2 Research Method

The research adopts a mixed methods approach, which makes use of mainly secondary data to analyse input–output relationships pertaining to the T&C sector in Türkiye, a literature review, and an overview of case studies from the sector. As such, this research is a combination of both qualitative and quantitative approaches.

Existing literature examining the contributions of the T&C sector to the overall economy and economic sectors generally focuses on estimating economic impacts through input–output (IO) analysis or computable general equilibrium (CGE) models. Due to its comprehensiveness and flexibility, IO analysis is a good tool for investigating an economic structure like T&C that is highly interconnected with input–output relationships with other sectors. Using the IO methodology, the direct and indirect effects of demand changes in a particular sector on sectoral and economy-wide production (supply), added value, employment, and GHG emissions can be analyzed (Kim & Kim, 2015).

In the following section, the IO relationships of the T&C sector are investigated in order to identify and compare the sustainability-related impacts of the sector. An IO table shows the interdependence between economic sectors and industries. Each sector uses intermediate goods for the use in other sectors, and each sector provides intermediate goods for the use in other sectors. This analysis technique was developed by Leontief (1991), a Nobel Prize-winning economist, and has been used by many economists. It is one of the most suitable methods for researching inter-sectoral relations. It is also useful for development planning and estimating the impact of economic decisions, shocks, and policies.

The data for the IO analysis conducted in this study are based on TURKSTAT National Income Accounts and IO tables. The 24-sector version of the 2012 IO table published by TURKSTAT, aggregated (at fixed prices, million TL) and updated for the year 2018 by Acar et al. (2022) is used in the analysis (see Appendix for the details of sectoral aggregation). The T&C industry that is investigated here is an aggregation of the NACE Rev.2 codes C13—Manufacture of textiles, C14—Manufacture of wearing apparel, and C15—Manufacture of leather and related products (Eurostat, 2008).

Using matrix algebra in IO analysis, production in an economic structure with n sectors can be expressed as follows:

$$X = AX + Y \tag{4.1}$$

where X shows gross output, Y is total demand, which represents an $(n \times 1)$ vector, and A is an $(n \times n)$ matrix that stands for the inter-sectoral relationships. Equation (4.1) can be simplified as follows:

$$X = (I - A)^{-1}Y \tag{4.2}$$

Total demand consists of household consumption C, fixed capital investment I, government expenditure G, and exports EX, such that $Y = C + I + G + EX$. The impact of a final demand contraction on sectoral output can be calculated as follows:

$$\Delta X = (I - A)^{-1}\Delta Y \tag{4.3}$$

The sum of any column j of the matrix ΔX gives the effect of the change in final demand of sector j on the level of output in the economy as a whole.

The direct and indirect requirements matrix Z can be found in Eq. (4.4). With Eq. (4.4), it is possible to separate the effects of changes in the total demand of any sector on output (X), value added (VA), employment (EMP), and greenhouse gases (GHG) on a sectoral basis. In general notation, K is expressed as an $(n \times n)$ matrix of diagonal coefficients as follows:

$$Z = K(I - A)^{-1}Y \tag{4.4}$$

$K_{VA} = \frac{VA_j}{X_j}$ is the ratio of the added value created by sector j to the gross output of that sector. Similarly, $K_{EMP} = \frac{EMP_j}{X_j}$ represents the employment coefficients of sector j and $K_{GHG} = \frac{GHG_j}{X_j}$ shows the GHG emission coefficients of sector j.

Following the IO analysis, 25 T&C companies from Türkiye (Table 4.2) were researched as they have sustainability reports or information uploaded on their web pages. The main criteria when choosing the companies on the list were the diversity of textile products and location and the existence of efforts toward sustainable production. The companies examined in the article are listed in Table 4.2. Company names are not used in the text due to ethical considerations and potential confidentiality issues. Companies were randomly numbered and evaluated in terms of their activities regarding sustainability and the CE. Activities, targets, and statements of the companies based on the information obtained from their websites and reports provide insights for this study and allow an evaluation of their sustainability approaches and CE activities.

4.3 Findings

This section now directly addresses the four research questions using the methods and techniques outlined above. Translations were made from Turkish to English by the first author and double-checked by the co-author.

Table 4.2 Türkiye-based T&C companies covered in this study

Company name	Products	Location (in Türkiye)	Company name	Products	Location (in Türkiye)
Ereks Garment	Denim garment	Tekirdag, Istanbul	İpeker	Fabric	Bursa, Istanbul
Sarp Jeans	Denim garment	Izmir	Abiteks	Knitwear, denim, leather, accessories, high fashion, woven	Istanbul
Maritaş Denim	Fabric, yarn processing, woven	Kahramanmaras	MEM Textile	Yarn, fabric	Kahramanmaras, Istanbul
İpekyol Group	Womenswear, shoes, accessory	Edirne, Istanbul	Atlas Denim	Denim fabric	Adana, Istanbul
Akın Textile	Fabric, garment	Kırklareli, Istanbul	Kipaş Textiles	Yarn, fabric	Kahramanmaras
Gamateks Textile	Fabric, outerwear, sleepwear, sportswear, kids wear, clothing	Batman, Denizli, Afyon, Istanbul	Koton	Clothing	Istanbul
Yünsa	Woven fabric	Istanbul, Tekirdag	Mavi	Denim garment	Istanbul
Çalık Denim	Yarn, fabric	Malatya, Istanbul	Almer Textile	Yarn	Kayseri
Melike Textile	Yarn	Gaziantep, Bursa	Isko Denim	Denim fabric	Bursa
Orta Anadolu	Denim fabric	Kayseri, Istanbul	İpekiş Mensucat	Fabric	Bursa
Altınyıldız	Yarn, woven	Izmir	Bossa Denim	Yarn, denim fabric	Adana
Cross Textiles	Denim garment, medical textile	Istanbul, Tekirdag, Tokat	LC Waikiki	Clothing, shoes, accessory, home textile	Istanbul
System Denim	Denim garment	Kırklareli, Istanbul			

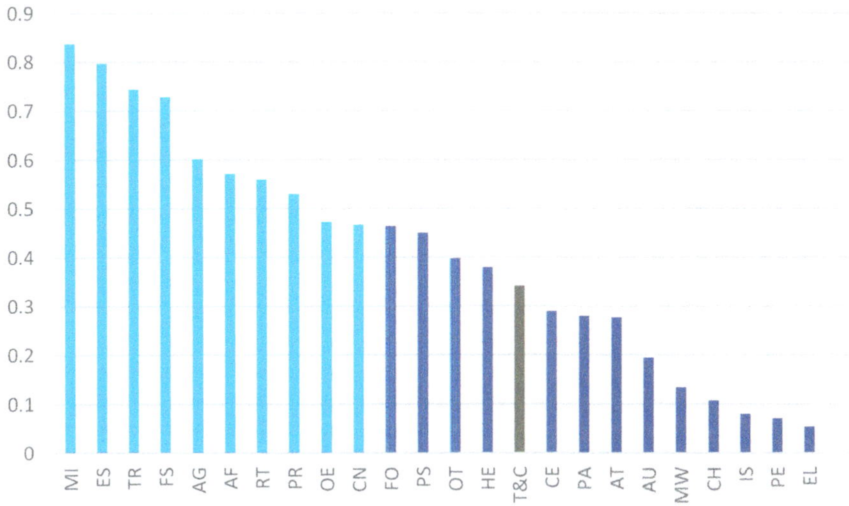

Fig. 4.3 Sectoral value-added coefficients (million TL/million TL)

4.3.1 RQ1. How Is the Turkish T&C Industry Inter-related with the Other Sectors of the Economy in Terms of Its Input-Output Connections, Employment Generation, Resource Use, and GHG Emissions?

Figures 4.3, 4.4, and 4.5 show the added value, employment, and greenhouse gas emission coefficients of the sectors, respectively. (All the figures were derived and illustrated by the authors using TURKSTAT data.) As shown in Fig. 4.3, mining (MI) is the sector with the highest value-added coefficient, creating an added value of 0.08 units per unit production, whereas T&C is the 16th sector creating 0.3 units of added value per unit production among all sectors.[2]

Figure 4.4 indicates that agriculture (AG) is the sector with the highest employment coefficient, creating employment for 13 people per million Turkish Liras (TL) of agricultural production. T&C is the eighth among all sectors, creating employment for 4 people per million TL of production.

According to the environmental results of the IO analysis, cement (CE) and electricity (EL) are the sectors with the highest emission coefficients, emitting 0.0006 and 0.0003 tons of GHG emissions per 1 TL of production, respectively. They are

[2] The abbreviations of the sectors in the figures are as in below: AG: Agriculture, MI: Mining, FO: Food Processing, T&C: Textiles, Clothing, OE: Other Economy, PA: Paper Products, PE: Petroleum Products, CH: Chemicals, CE: Cement, IS: Iron and Steel, MW: Machinery, White Goods, AU: Automotive, EL: Electricity, CN: Construction, RT: Retail trade, TR: Transportation, AT: Air Transport, PS: Postal and Courier Services, AF: Accommodation and Food Services, PR: Professional Services, FS: Financial and Real Estate Services, TS: Tourism, ES: Education Services, HE: Health Services.

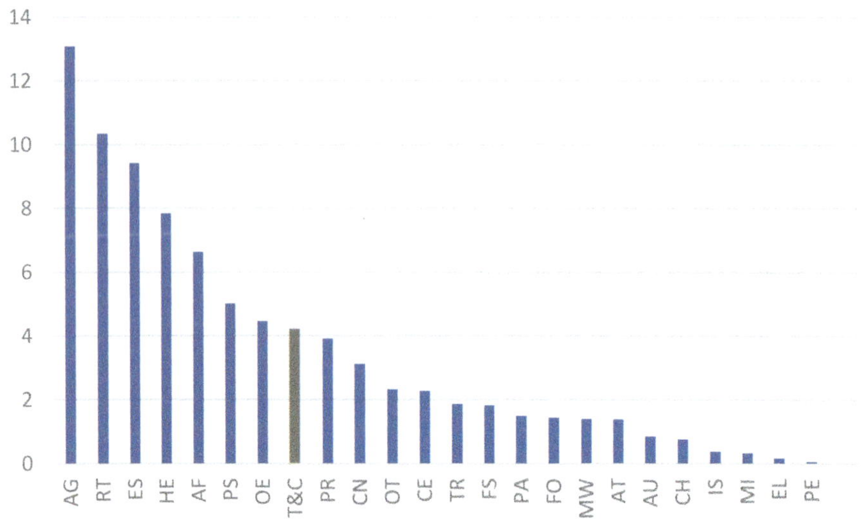

Fig. 4.4 Sectoral employment coefficients (number of employees/million TL)

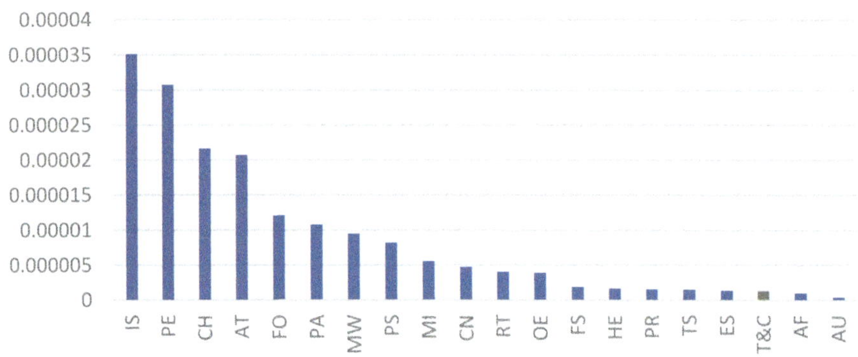

Fig. 4.5 Sectoral GHG coefficients (million tons/million TL)

followed by agriculture (AG) and transport (TR) in the third and fourth order. These four sectors are excluded from Fig. 4.5, as the other sectors emit far less GHG than these four. T&C is one of those sectors whose emission coefficient is relatively low (1.3827E-06) when compared to sectors like iron and steel (IS) and petroleum products (PE).

Manufacturing of textiles requires water, chemicals, and energy. One way to detect the resource intensity of the T&C sector in Türkiye could be to examine the intermediate input use of the sector from other sectors (Fig. 4.6). Accordingly, in 2018, once the intermediate input provision of the T&C sector to itself is excluded from the figure, T&C used intermediate inputs mostly from the chemicals (CH), other economy (OE), agriculture (AG), financial and real estate services sectors (FS), and

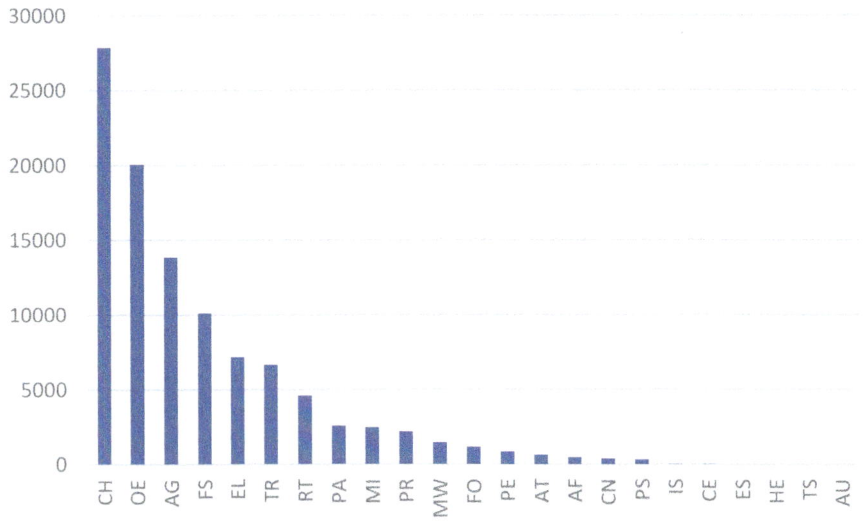

Fig. 4.6 Sectoral intermediate input use of the T&C sector (million TL at basic 2018 prices)

electricity (EL). It is noteworthy that chemicals and agriculture are highly resource-intensive sectors, and the electricity sector in Türkiye is highly dominated by fossil fuels. Figure 4.6 evidences the fact that, although the T&C sector is not highly carbon-intensive compared to other sectors, it uses considerable amounts of raw materials and natural resources in its production processes.

These findings imply that although the sector might not be primarily affected by domestic or cross-border carbon pricing policies, such as the CBAM, or carbon trade markets, such as the EU ETS, circular economy standards of the EU might have considerable consequences for the Turkish T&C exporters as the EU is a major export market for Türkiye.

4.3.2 *RQ2. How Are Turkish T&C Companies Addressing Sustainability in Their Corporate Strategies and Activities?*

Companies in Türkiye were selected from various sectors including yarn, fabric, garment, clothes manufacturers, and retailers, some of which export their products (Table 4.2). Of the 25 companies, 12 companies have issued sustainability or corporate social responsibility reports, while others provide information on sustainability and the CE on their websites. When selecting companies, factors such as their size and location, operations at different stages of textile production, as well as different perspectives on the importance of sustainability and the circular economy, were taken into account in order to ensure a comprehensive coverage of heterogeneous examples.

Companies #5, 6, 7, 8, 14, 15, 21, and 22 are among Türkiye's top 500 industrial enterprises in all sectors, according to the report prepared by the Istanbul Chamber of Industry in 2022 (Istanbul Chamber of Industry, 2022). Among the companies not included in the top 500 industrial enterprises, companies #1, 3, 11, 23, and 24 are categorised as small and medium-sized enterprises because they had less than 250 employees. The remaining 12 companies are deemed to be large enterprises in terms of their number of employees. It is also observed that SMEs appear to have initiatives on sustainable textile production and CE, as stated on their websites and sustainability reports.

Issues regarding sustainability are divided into subsections such as carbon or GHG levels, water, energy, and chemical use to account for environmental indicators, while occupational health and safety, gender equality, and fair working conditions are identified as social indicators in their open-source reports and websites. The main themes extracted from the reports/sources are summarised in Table 4.3.

Sustainability Vision

Sustainability is addressed as a core element in most companies' policies. Upon reviewing the content available on their websites, it is evident that these companies incorporate sustainability strategies through explicit statements and objectives. Company#22, one of the companies known for adhering to sustainable production practices, articulates its vision as creating a working environment which is safe and healthy. Another fabric manufacturer, Company#14 distinguishes itself by emphasising sustainable solutions and innovations operated by a sustainability department. It is observed that companies with specialised teams or departments tend to prioritise the dissemination of their sustainable initiatives as open-source, as well as actively promoting such endeavors. Additionally, numerous garment and clothing companies, including Company#9 and Company#1, highlight the concept of "sustainable fashion" (Abiteks; Ereks, 2020) within their publications. Company#11 portrays itself as a "green factory" (İpeker), while Company#7 undertook comprehensive sustainability risk analyses utilising a matrix to identify risks and propose corresponding corrective measures (Cross Textiles, 2022, p. 34).

Adoption of the UN SDGs

Based on the data gathered from their websites and reports, 52% of the investigated companies set out their goals in alignment with the UN SDGs (Table 4.3). It is noteworthy that Company#14, Company#6, Company#8, Company#15, Company#16, and Company#25 have integrated SDG6, SDG8, SDG9, and SDG13, respectively, into their strategic framework, while Company#7 demonstrated a comprehensive coverage of all the SDGs in its sustainability report. SDG12 (Ensure sustainable consumption and production patterns) and SDG17 (Strengthen the means of implementation and revitalise the Global Partnership for Sustainable Development) emerge as the most frequently addressed goals by companies seeking compliance with their policies.

Table 4.3 Companies' sustainability approaches and activities addressed in their reports/websites

Actions of companies	#1	#2	#3	#4	#5	#6	#7	#8	#9	#10	#11	#12	#13	#14
Sustainability vision, activities	X	X	X	X	X	X	X	X	X	X	X	X	X	X
Targeting UN SDGs				X		X	X	X	X		X		X	X
Supply chain management	X			X	X	X	X	X	X		X			X
Environmental indicators														
Low carbon footprint	X		X	X		X	X	X	X	X	X		X	X
Low water footprint	X	X	X	X		X	X	X	X	X	X			X
Low GHG emissions	X		X	X		X	X	X	X	X	X			X
Energy efficiency	X	X	X	X	X	X	X	X	X	X	X	X	X	X
Eco-labeled, sustainable material use	X		X	X		X	X	X	X		X	X	X	X
Less chemical use	X		X	X	X	X	X	X	X	X	X		X	X
Social indicators														
Occupational Health and Safety (OHS) measures		X		X	X	X	X	X	X		X			X
Gender equality				X	X	X	X	X			X		X	X
Fair working conditions	X	X		X	X	X	X	X			X		X	X
No child labor	X			X	X	X	X	X			X			X
Animal rights							X				X			
Certificates (BCI, GOTS, GRS, RCS, OEKOTEX, etc.)		X	X		X	X	X	X	X	X	X	X	X	X
Corporate social responsibility—projects with NGOs	X		X	X		X	X	X	X	X	X	X	X	X

Actions of companies	#15	#16	#17	#18	#19	#20	#21	#22	#23	#24	#25
Sustainability vision, activities	X	X	X	X	X	X	X	X	X	X	X
Targeting UN SDGs	X	X	X		X						X
Supply chain management	X	X		X	X			X	X		
Environmental indicators											
Low carbon footprint	X	X	X	X	X			X	X	X	X
Low water footprint	X	X	X	X	X		X	X	X	X	X
Low GHG emissions	X	X		X	X			X	X	X	X
Energy efficiency	X	X	X	X	X	X	X	X	X	X	X
Eco-labeled, sustainable material use	X	X	X	X	X		X	X	X	X	X
Less chemical use	X	X	X	X	X			X	X	X	X
Social indicators											
Occupational Health and Safety (OHS) measures	X	X	X		X	X		X	X		X

(continued)

Table 4.3 (continued)

Actions of companies	#15	#16	#17	#18	#19	#20	#21	#22	#23	#24	#25
Gender equality	X	X	X		X			X	X		X
Fair working conditions	X	X	X		X			X	X		X
No child labor		X			X			X			
Animal rights					X			X			
Certificates (BCI, GOTS, GRS, RCS, OEKOTEX, etc.)	X	X	X	X	X	X	X	X	X	X	X
Corporate social responsibility—projects with NGOs	X	X	X		X			X	X	X	X

Source Authors' illustration of the compiled data

Environmental Indicators

The textile industry is one of the main resource-intensive and polluting industries. Therefore, there is a growing emphasis on implementing strategies aimed at mitigating the environmental impacts of textile production, which is prominently featured in the sustainability policies of many companies. In order to identify environmental indicators and actions, not only concrete data on production processes, but also companies' declarations regarding sustainable production methods are taken into account.

Generally speaking, companies mostly emphasise their efforts to reduce water, energy, and chemical consumption, alongside their waste management strategies as part of their commitment to the CE. A number of companies, including Company#1, Company#7, Company#11, Company#18, Company#19, and Company#22 conduct Life Cycle Assessment (LCA) studies to evaluate the environmental impacts of their production processes. Reports predominantly focus on aspects such as chemical use, water and energy consumption, and measures aimed at minimising the use of natural and conventional resources. Renewable energy sources, rainwater harvesting, and wastewater use are also specified incorporating environmentally friendly production methods. For instance, Company#9 collects rainwater from rooftops and reuses it within its systems (Ereks), while Company#16 has introduced a "respect for water" collection (Koton, 2022, p. 9). Company#25 achieved the target for reducing total water draft by 21% and water consumption per meter of fabric by 11%. Similarly, Company#15 reduced water consumption during the yarn dyeing phase by 95% in the denim finishing process.

Chemical use is one of the most harmful activities within the textile industry. Only four companies fail to incorporate measures concerning chemical utilisation in their reports or on their websites. Company#2, Company#5, Company#6, Company#7, Company#9, Company#10, Company#11, Company#14, Company#16, Company#19, Company#22, and Company#25 affirm their compliance with the REACH (Registration, Evaluation, Authorization, and Restriction of Chemicals) and the ZDHC (Zero Discharge of Hazardous Chemicals Certification and Testing Programs) accords. In addition, non-hazardous materials and dyestuff are used to reduce chemical use. While Company#8 uses electrospray technology for the dyeing process, many denim manufacturers have laser machines for bleaching and engraving to optimise the utilisation of chemicals and water. Eco-labeled dyestuff and eco-corrosion printing are offered as alternatives to conventional ones by Company#11. Company#4 strives to save chemicals, as well as energy, obtaining spontaneously colored cotton through colored cotton planting.

Water consumption plays a pivotal role in the textile industry. Water use and wastewater generation create significant effects on the environment. Therefore, sustainability issues focus as much on water use as on chemical use. The vast majority of enterprises emphasise the need to reduce water use, with some incorporating wastewater treatment processes to attain this objective. Company#4 shows its motives through the utilisation of sustainable piece dyeing techniques and claims that four million liters of water are saved in denim washing. On the other hand, energy efficiency and efforts for reducing energy consumption are mentioned in all the reports and websites, frequently highlighting investments made in renewable energy sources and/or the use of them. Company#6 adopts dry irrigation methods to reduce water consumption, while cultivating genetically modified organism (GMO)-free colored cotton.

Policies aimed at reducing the carbon footprint are mentioned in 20 reports, while GHG emissions are discussed in 18 reports. Carbon footprint and GHG emissions calculations are employed by Company#1, Company#7, Company#14, Company#16, Company#22, and Company#25. An alternative approach to decreasing GHG emissions involves the utilisation of sustainable materials in the logistic supply chain. Company#4 has initiated the use of plastic boxes for packaging made of recycled materials as an alternative to cardboard boxes, resulting in the reduction of carbon emissions as well as the benefit of approximately 2,000 hectares of forest per year.

The use of sustainable materials and products is also one of the main measures to reduce environmental impacts. Company#1 affirms its commitment to providing sustainable products produced using sustainable practices through the utilisation of the Green Map Tool, which entails a preliminary analysis conducted in partnership with manufacturers and suppliers (Abiteks). In line with the increasing attention to sustainable and organic products among customers, companies promote sustainable products as well as those approved by certification processes. Various companies prioritise the use of organic cotton certificated by standards such as Organic Content Standard (OCS) and the Global Organic Textile Standard (GOTS). Recycled and upcycled materials are also put into practice as part of circular and sustainable production methods.

Social Indicators

The social rights of workers are predominantly discussed in the corporate social responsibility (CSR) reports of companies. Occupational health and safety (OHS) measures in the textile industry are specifically taken by companies due to the high probability of accidents and hazards during the production processes. In addition, the Law of Occupational Health and Safety (OHS)[3] was enacted in 2013 in Türkiye, which mandated all industry sectors to take OHS measures. However, only 17 of the companies studied were identified as explicitly mentioning OHS measures in their workplaces, as reflected in the reports and websites. The right of customers to purchase healthy clothes is also encompassed within social rights, such that, for instance, in Company#25 Ready-to-Manufacture (RTM) and Green-to-Wear (GTW) audits are carried out twice a year. Gender equality, fair working conditions, and the prohibition of child labor are discussed in 15, 17, and 11 of the company reports, respectively, mostly aligning with the UN SDG objectives. While companies generally exhibit a positive stance on animal rights with various projects in support, only four companies dedicated separate parts of their reports to animal rights. For example, Company#11 introduced the Vegan Textile Process to ensure that there is no product containing animal matter in the whole process from obtaining raw materials to the final product.

Social Responsibility Activities

Companies encourage socially responsible activities by taking initiatives focusing on social, environmental, and economic issues. Collaboration between companies and NGOs raises awareness on various crucial topics such as poverty, gender equality, animal rights, education rights, forestation, and solidarity with people in need. For instance, Company#23 launched a special afforestation project carried out in the arid regions in Türkiye, while Company#19 demonstrated support with the affected people and animals during the wildfires in the southern region of Türkiye in 2021. Similarly, Company#7 launched a project for food and animal shelter donation. Company#23 also embarked upon a project to integrate those released from prisons into society. Company#16 focuses on social gender equality within its social responsibility campaigns. Moreover, campaigns such as providing dry food for street animals as well as the collection of unused clothes in good condition and sending them to village schools were organised.

4.3.3 RQ3. What Strategies and Activities Relating to the CE Are Being Pursued in the Turkish T&C Industry?

While several companies in Türkiye have developed sustainability strategies, the CE is still only included in strategic reports to a limited extent. Since the CE is seen in company supply chain management as a novel approach, knowledge and implementations are limited. Nevertheless, various companies engage in CE policies by referring to fundamental pillars of the CE such as reduce, reuse, repair, refurbish, remanufacture, repurpose, recycle, and recover.

[3] Occupational Health and Safety Law: https://www.resmigazete.gov.tr/eskiler/2012/06/20120630-1.htm.

In particular, activities such as recycling, regenerating, and reusing material are highlighted as essential components of CE activities. Company#14, for example, has initiated a project utilising 50% recycled or reused materials, including pre-consumer cotton and recycled polyester. Postconsumer plastic bottles are used to convert them into polyester in their recycled polyester facility. Similarly, Company#17 includes products made of recycled cotton and polyester in its collections. Company#25 innovatively uses fish scales, a significant waste product from the fishing industry, to produce textile fibers, preventing the harmful effects of burying fish scales in the ground and reducing the use of natural resources for raw materials. Company#19 produces buttons from hazelnut shells, which are by-products of hazelnut production, and designs back labels using a mixture of recycled polyester and olive seeds. Company#16 prefers cardboard labels and recycled plastic materials with Forest Stewardship Council (FSC) certification and actively uses these materials in the production of product labels. Company#1 and Company#4 implement the Green Office practice of reducing consumption, using resources efficiently, minimising waste generation, collecting waste separately at the source, and recycling waste.

Recycling and reuse of materials are the main CE practices mentioned in the reports; however, durability, extended product life, and traceability do not receive sufficient attention as part of CE schemes. While only 40% of the companies emphasise traceability, collaboration to implement circular strategies is mentioned by 48% of the companies. The Green Map Tool, innovated by Company#1, stands as a notable endeavor that addresses traceability through a customised end-to-end supply chain framework for both suppliers and logistics partners. All CE activities are performed in line with the reduction of waste and emissions targets. The majority of companies spell out GHG reduction and carbon neutrality targets in line with the international agreements.

The collection of pre- and post-consumer textile wastes is another issue to be addressed in the near future. The EU will impose a separate collection of textile waste as of 2025. It will also be prohibited to burn these wastes or throw them into household garbage. This regulatory measure is expected to significantly reduce the environmental impact associated with textile waste and address the prevailing trend of fast fashion, aligning with the CEAP (EPRS, 2024; Türkiye Tekstil Sanayii İşverenleri Sendikası, 2022). The targets also entail actions to ensure the consumption of long-lived, recyclable, and non-hazardous materials (EPRS, 2024). Current practices aim to minimise waste generation and promote the reuse of textile wastes, aligning with the goals of the EGD and the CEAP. Company#22 focuses on post-consumer textile wastes by developing partnerships to recycle old jeans to make recycled fabric. Company#4 also sends all waste clothes and fabrics collected in stores to the recycling facilities. It also prohibits these products from being burned or thrown into household garbage. It is anticipated that this practice will contribute significantly to textile recycling activities. On the other hand, only seven companies mention the durability and extended life cycle of a product. However, recycled materials are considered durable and functional in Company#14's report.

4.3.4 RQ4. What Role Are Digital Technologies Playing in the Transition to Sustainability and the CE in the Turkish T&C Industry?

A number of examples of how digital technologies are being deployed in support of sustainability or the CE are referred to in these reports and on company websites. Table 4.4 summarises the CE activities of companies regarding waste reduction and collection of pre- and post-consumer wastes, durability and traceability of products, collaboration with stakeholders to leverage efforts to build a circular system, and digitalisation. Company#8, for instance, points out in its sustainability report that digital transformation and Industry 4.0 facilitate communication with suppliers and consumers, prioritising data security amid digitalisation. Company#22 developed a digital decision-making tool, the Circle Fashion Tool, to assess the end-of-life options and circular alternatives to achieve more circular fashion. Company#19 shows a willingness to improve the communication channels via digitalisation to offer better services to the customers as of 2020. Company#16's strategic goals include achieving full traceability in its supply network by 2030 and leveraging technology to digitalise the network.

Through the integration of digital technologies, such as Blockchain, the Internet of Things (IoT), Big Data, and Cloud Computing (as part of Information and Communication Technology (ICT) systems as a whole), the industry can enhance transparency and traceability across the supply chain, thereby simplifying the process of monitoring and validating the origin, as well as providing considerable support to supply chain managers. Internet of Things (IoT) technologies have the capability to employ algorithms and platforms to assess circular data as a key enabler for decision-making processes to apply circular business models (Hora et al., 2023; Voulgaridis et al., 2022). The wider fashion industry has also developed digital transformation approaches. Several fashion brands launched such initiatives through Artificial Intelligence (AI) algorithms in the context of private-labels, 3D body scanning, customer experience monitoring, and virtual stores to capture changing customers' preferences and improve the value chain (Akhtar et al., 2022).

Digital technologies have the potential to deliver the benefits of circular strategies at three levels: Processes including "robotization, additive manufacturing, digital design, sensor technologies, and machine learning" (Berg et al., 2020, p. 6); products based on technologies such as Internet of Things (IoT), blockchain, and digital twins to facilitate the tracking and tracing of products and components and optimise the value chain; and lastly, platforms to make connections between consumers and producers through apps and websites (Berg et al., 2020). To make this transition, a Strategic Research and Innovation Agenda for the circular economy has recently been published as part of the H2020 project CICERONE, which puts emphasis on digitalisation at the forefront of numerous key innovation fields such as waste management, industrial symbiosis, and product traceability (Berg et al., 2020; CICERONE, 2020). To align with the EU Strategy for Sustainable and Circular Textiles, particular measures will be on the agenda such as eco-design requirements for textiles, more

Table 4.4 Circular economy and digitalisation actions of companies in the study

Company No.	Reduction of emissions and wastes	Collecting-using pre- and/or post-consumer waste	Reusing products/ materials	Recycled material use—regenerate	Extend product use phase/ durability	Digitalisation in CE implementations	Traceability	Collaboration with stakeholders
#1	X		X	X	X	X	X	
#2	X							
#3								
#4	X	X	X	X	X			
#5	X		X	X	X			
#6	X	X	X	X				X
#7	X					X	X	X
#8	X	X	X	X	X	X	X	X
#9	X	X	X	X				X
#10	X						X	
#11	X		X	X	X			X
#12			X	X	X			
#13	X	X	X	X				
#14	X	X	X	X	X		X	X
#15	X		X	X		X	X	X
#16	X	X	X	X		X		X
#17	X		X	X			X	
#18	X		X	X				X
#19	X	X	X	X	X	X	X	X

(continued)

Table 4.4 (continued)

Company No.	Reduction of emissions and wastes	Collecting-using pre- and/or post-consumer waste	Reusing products/ materials	Recycled material use—regenerate	Extend product use phase/ durability	Digitalisation in CE implementations	Traceability	Collaboration with stakeholders
#20		X	X	X				
#21								
#22	X	X	X	X		X		X
#23	X	X	X	X			X	X
#24	X							
#25	X	X	X	X			X	

Source Authors' illustration of the compiled data

transparent information, and a Digital Product Passport (European Circular Economy Stakeholder Platform, 2022). The Digital Product Passport (DPP) has been introduced, aimed at providing reliable and standardised data to stakeholders on "track and trace" of the product origin, components, alternative ways of repair and disassemble processes, as well as its handling at the end of the product's service life (Adisorn et al., 2021).

At the governmental level, the "Digital Transformation and Industrial Move" program of the 2023 Industry and Technology Strategy has been put forward by the Ministry of Industry and Technology in Türkiye. The program includes introducing green production approaches to manufacturing industry, supporting new investments based on technology-intensive modernisation and cleaner production of infrastructure and enterprises in Organized Industrial Zones (OIZs), and establishing the Industrial Registry Information System to develop the "Economically Valuable Waste Monitoring System" within the scope of the CE (The Ministry of Trade, 2021b, p. 8).

Digital technologies in textile production have been implemented not only for traceability throughout the production chain, but also in many areas such as creating virtual product catalogs, virtual fashion shows, real-time monitoring of waste collection systems, and remote monitoring of machine performances (Altun Kurtoğlu, 2022). Training and educational programs have been carried out by digital tools. Company#19 improves digital resources for these activities by creating a Digital Training Library and Manager Development Programs. During the pandemic, meetings were held using digital platforms. Company#7 set targets encompassing digitalisation-based transformation in the whole supply chain and operations. Collections are sent to customers as digital links instead of actual products through shipments since 2022, thereby creating a digital collection.

4.4 Discussion

The results presented above show that sustainability and CE are on the agenda of the companies studied at various levels. Firstly, sustainability actions have been implemented as science-driven targets to reduce the environmental impacts of manufacturing processes. These sustainability strategies are carried out by either specialised departments or as part of companies' corporate social responsibility policies. The growing awareness among suppliers and customers, along with the international agenda on sustainability, has been influencing companies in their attitudes toward sustainable and clean production techniques. There is also a notable corporate inclination toward adopting environmentally conscious practices in production and distribution processes.

Secondly, CE actions are predominantly integrated into sustainability policies. Given that many Turkish companies exporting to Europe prefer sourcing raw materials from CE-compliant suppliers, some prominent textile firms in Türkiye have embraced CE practices within the realm of sustainability efforts (Balbay et al.,

2021). The initiatives aimed at fostering CE primarily revolve around recycling and reusing materials to reintroduce them into production cycles. However, various other aspects of CE such as refurbishment, reducing consumption, product durability, and resource sharing are not widely adopted in the T&C industry. The reutilisation of second-hand textile products stands out as a pivotal method for prolonging product lifespans, with projections indicating that second-hand sales will surpass fast fashion by 2029. Despite this trend, the second-hand market in Türkiye remains relatively small compared to those in Europe and the United States. In addition, such practices where customers borrow a certain number of products by paying a monthly fee also offer important opportunities for entrepreneurs in Türkiye (Altun Kurtoğlu, 2022).

To mitigate the negative environmental effects of textile production, numerous approaches have emerged as noteworthy for achieving sustainable production. Recycling and reuse, in particular, come to the fore in terms of promoting both sustainable production and consumption. Additionally, the reduction of overproduction and consumption, the extension of the product lifespan, and the enhancement of the design of products for circularity are essential components of sustainable practices (McKinsey&Company, 2022). In that sense, the CE offers a comprehensive framework to tackle the challenges associated with textile production, incorporating issues such as natural resource utilisation, energy consumption, water usage, and the generation of pollutants and waste.

The CE in the T&C industry aims to reduce the excessive use of resources through the implementation of a circular business model encompassing not only recycling but also product upcycling and reusing materials including pre-consumer and post-consumer wastes. The CE concept further promotes ecologically responsible practices in production and consumption.

The CE business model in the textile industry primarily requires reducing material and energy intensity, minimising the dispersion of toxic substances, improving recycling capabilities, maximising the utilisation of renewable resources, prolonging the durability of products, as well as the reproduction and redistribution of new markets (Chen et al., 2021a, 2021b; Koszewska, 2018; Maity et al., 2020; Oliveira Neto et al., 2024). Although it is challenging to quantify the benefits of circular transition in the textile industry, it is estimated that the overall benefit to the world economy could be approximately EUR 160 billion (US$ 192 billion) in 2030 if the fashion industry embraces circular practices. Additionally, textile recycling and repair practices may create green jobs (Ellen MacArthur Foundation, 2017; Schumacher & Forster, 2022).

Thirdly, collaboration among stakeholders needs to be improved with the aim of fostering more traceable and transparent processes, while simultaneously empowering all involved partners across the value chains. Several reports highlight endeavors by companies under the statement of "creating a positive impact for/with stakeholders," "embracing diversity and differences" to create a "responsible value chain" (ISKO, 2022, p. 47; Koton, 2022, p. 55; Mavi, 2024, p. 104; Orta Anadolu, 2013, p. 21). The main focus of such partnerships may also generate more innovative products with less harmful impacts on the environment, sustainable raw material procurement, and R&D activities. The concept of industrial symbiosis stands out as a prime example of successful collaboration within this framework.

Fourthly, although most of the companies declare their commitments to international and national legislation and environmental protection standards, the mechanisms for auditing and enforcing compliance to prevent pollution and excessive resource consumption remain inadequate. In-depth analysis of their sustainability reports and websites reveals that some companies fail to achieve their environmental conservation objectives. On the other hand, the presence of international agreements and the tendency of consumers to demand more organic/healthy/environmentally friendly products force manufacturers to comply with the legislation. In that sense, most of the companies transform their supply chains through alignment with the EGD and the CEAP.

Fifthly, companies that are ambitious about implementing CE activities improve their R&D and innovation facilities to gain an advantage. Digitalisation is one of the technological advancements to make connections to emerging markets and new customer behaviors. Additionally, it enables companies to secure advantageous positions in the global market (Divrik & Baykal, 2022). Digital transformation also allows companies to ensure their resilience to overcome potential uncertainties and crises in the market (Yıkılmaz & Kör, 2023). However, it is observed that only limited reports mentioned digitalisation as part of CE activities. Lack of qualified personnel and experts, inadequate budget, and lack of separate departments dedicated to digitalisation are impediments to digital transformation (Yıkılmaz & Kör, 2023). Companies that have sustainability departments and specialised policies in that field have various digital practices including Internet sales, QR codes, using software for designing products, etc.

Examination of the strategies and reports of companies has illustrated that sustainability, as a broader concept, has attracted more attention compared to CE and digitalisation. On the other hand, as sustainable business models prevail, companies incorporate more circular practices to reduce resource consumption and waste, emphasising renewable energy use, adopting zero waste policy in their production chain. As discussed above, companies in the textile sectors, most of which are SMEs, implement ambitious policies to promote sustainable and circular production approaches. The sector has advantages such as the high level of awareness among customers and companies, thereby engendering a willingness to progress circular transition, advancing harmonisation processes to the international environmental agenda which is supported/imposed by national agencies, and the presence of best practices. On the other hand, it is evident that digitalisation activity fails to fully maintain digital transformation. Auditing mechanisms performed by national authorities, and lack of transparency in sharing the results, also prevent identification of harmful production policies. In that sense, reporting and publishing internal auditing results may contribute to transparency for both suppliers and customers. Most of the companies examined in this study publish their policies, practices, and targets to show their sustainability actions and strategies. These strengths and deficiencies are summarised in Fig. 4.7.

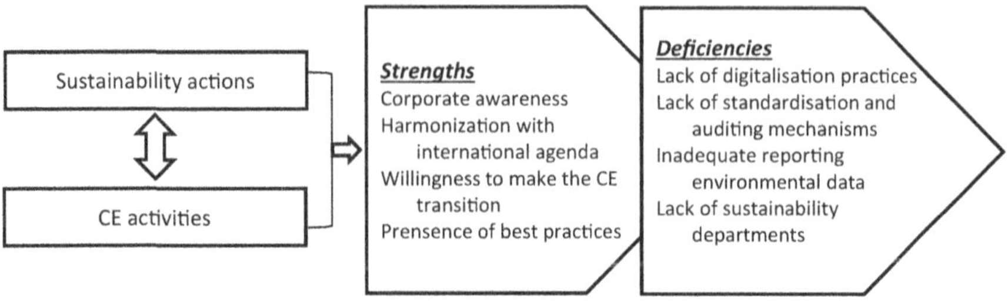

Fig. 4.7 The strengths and deficiencies of sustainability and CE actions in the Turkish T&C industry

4.5 Conclusions

This chapter has examined how companies in the Turkish T&C industry are approaching the issues of sustainability and the CE and evaluated the role that digital technologies are playing in the transition to a sustainable and circular T&C industry. The results suggest that both circularity practices and digital transformation are limited in the Turkish T&C industry. Despite the presence of a robust textile sector and a well-established industrial basis, Türkiye heavily relies on foreign technology. This situation limits the applicability of Industry 4.0 in the sector (İlhan, 2019). Nevertheless, there are companies that adopt CE objectives and digital technologies at production and consumer levels. The Turkish Textile Industry Employers' Union has determined 10 priority Industry 4.0 applications for the textile industry as follows: Tracking Machines in Real Time, Making Real-Time Efficiency Calculations, Blue Collar Tracking and Business Optimisation, Digitising the Maintenance and Repair of Machines, Accelerating Shipment Times and Providing Transparency in Processes, Detecting Errors Early with Digital Twin, Digitising Suppliers, Producing 3D Digital Prototypes, Making Personalised Production, and Using Artificial Intelligence Support in Product Designs (Türkiye Tekstil Sanayii İşverenleri Sendikası, 2019).

The research reported here, based on publicly available data and surveys, indicates that input circularity in the T&C industry in Türkiye is around 16% and output circularity is 23%. Best practices could be engendered through modifications in tax regulations concerning the handling of waste, surplus materials, scrap, and unsaleable items. On the other hand, financial incentives for both promoting CE implementation and digitalisation could significantly contribute to digital transformation in the sector (SKD, 2022). Similarly, examination of company reports and websites demonstrated that there is an effort toward embracing circular practices in the supply chains, adhering to industry standards in accordance with international agreements, exporting goods, and meeting the demands of both customers and suppliers. In this context, processes from raw material procurement to design, from presenting the product to the end consumer to waste management, should be reconsidered with an understanding of circularity, and digital tools should be put into effect to facilitate this process. Particularly, in design and supply chain processes, energy efficiency, repairability, upcycling, durability, and recyclability of products are identified as potential strategies for resource and cost savings (Balbay et al., 2021).

Currently, the use of digital technologies as enablers for competitive advantage in global and local markets is limited. However, significant initiatives are evident in the corporate reports, particularly in those that include comprehensive sustainability and corporate social responsibility disclosures. Initiatives such as the introduction of QR codes, virtual exhibitions, and digital tools for ensuring traceability may be viewed as the initial stages of the digital transformation process. It is estimated that mandatory applications such as Digital Product Passport (DPP) could accelerate this transformation. Making appropriate investments and establishing specialised departments devoted to sustainability and circularity could notably facilitate digitalisation across the value chain in favor of circularity.

This study aimed to reveal the current situation in the Turkish T&C industry regarding sustainability vision, circularity implementation, and digitalisation. However, the study has its limitations since it is solely based on publicly available reports and websites and does not encompass all Turkish T&C companies. Future research could include a more in-depth analysis of the potential and shortcomings within the sector regarding sustainability, circularity, and digitalisation. Furthermore, national and sector-specific studies could focus on digitalisation to encourage a transition that further enables sustainability and the implementation of CE principles. Best practice exemplar cases could be developed to highlight the benefits of circular, sustainable, and environmentally friendly production.

Appendix: Details of Sectoral Aggregation in the IO Model

Model sectors	NACE Rev. 2 product code
AG: agriculture	A01–A03
MI: mining	B05, B06–B09
FO: food processing	C10–C12
TE: textiles, clothing	C13–C15
OE: other economy	C16, C30–C33, E36–E39, G45, G46, N80–N82, O84
PA: paper products	C17, C18
PE: petroleum products	C19
CH: chemicals	C20–C22
CE: cement	C23
IS: iron and steel	C24
MW: machinery, white goods	C25–C28
AU: automotive	C29
EL: electricity	D35

(continued)

(continued)

Model sectors	NACE Rev. 2 product code
CN: construction	F41–F43
RT: retail trade	G47
TR: transportation	H49, H50
AT: air transport	H51, H52
PS: postal and courier services	H53
AF: accommodation and food	I55, I56
PR: professional services	J58–J63, M71–M75, N77–N78, S94–S96, T
FS: financial and real estate services	K64–K66, L68, M69–M70
OT: other tourism	N79, R90–R93
ES: education services	P85
HE: health services	Q86–Q88

References

Abiteks. (n.d.). *Abiteks sustainability*. Retrieved July 17, 2024, from https://www.abiteks.com.tr/sustainable-fashion-by-abiteks

Acar, S., Aşıcı, A. A., & Yeldan, A. E. (2022). Potential effects of the EU's carbon border adjustment mechanism on the Turkish economy. *Environment, Development and Sustainability, 24*, 8162–8194. https://doi.org/10.1007/s10668-021-01779-1

Adisorn, T., Tholen, L., & Götz, T. (2021). Towards a digital product passport fit for contributing to a circular economy. *Energies, 14*(8). https://doi.org/10.3390/en14082289

Akhtar, W. H., Watanabe, C., Tou, Y., & Neittaanmäki, P. (2022). A new perspective on the textile and apparel industry in the digital transformation era. *Textiles, 2*(4), 633–656. https://doi.org/10.3390/textiles2040037

Alkaya, E., & Demirer, G. N. (2014). Sustainable textile production: A case study from a woven fabric manufacturing mill in Turkey. *Journal of Cleaner Production, 65*, 595–603. https://doi.org/10.1016/j.jclepro.2013.07.008

Altun Kurtoğlu, S. (2022). Türk Tekstil ve Hazır Giyim İşletmelerinde Döngüsel Ekonomiye Geçiş-Kisa Bir Değerlendirme. *Mühendislik Bilimleri ve Tasarım Dergisi, 10*(3), 1107–1116. https://doi.org/10.21923/jesd.988416

Alüftekin, N., Taş, A., Çakar, G., & Bayraktar, F. (2009). Küresel Krizden Çıkışta Kümelenme Modeli: Tekstil ve Hazır Giyim Sektörü Örneği. [In English: Clustering model in coming out of the global crisis: Textile and ready-made clothing industry example]. In *ZKÜ Sosyal Bilimler Dergisi, Cilt* (Vol. 5, Issue 10). https://dergipark.org.tr/tr/pub/ijmeb/issue/54831/750683

Antikainen, M., Uusitalo, T., & Kivikytö-Reponen, P. (2018). Digitalisation as an enabler of circular economy. *Procedia CIRP, 73*, 45–49. https://doi.org/10.1016/j.procir.2018.04.027

Aşıcı, A. A., & Acar, S. (2022). Channels of cooperation between the EU and Turkey on green transformation. *Ankara Avrupa Çalışmaları Dergisi, 21*(1), 43–67.

Balbay, Ş., Sarıhan, A., & Avşar, E. (2021). Dünyada ve Türkiye'de "Döngüsel Ekonomi / Endüstriyel Sürdürülebilirlik" Yaklaşımı. *European Journal of Science and Technology*. https://doi.org/10.31590/ejosat.971172

Berg, H., Le Blévennec, K., Kristoffersen, E., Strée, B., Witomski, A., Stein, N., Bastein, T., Ramesohl, S., & Vrancken, K. (2020). *Digital circular economy as a cornerstone of a sustainable*

European industry transformation. https://www.era-min.eu/sites/default/files/publications/201023_ecera_white_paper_on_digital_circular_economy.pdf

Bostan, A., Ateş, İ., & Ürüt, S. (2010). Türkiye Tekstil ve Hazır Giyim Sektörünün Rekabet Gücü: Avrupa Birliği Ülkeleri ile Bir Karşılaştırma [In English: Competitiveness of Turkish textile and ready-made clothing sector: A comparison with European Union countries]. *Journal of Management and Economics Research, 8*(13), 43–58. https://dergipark.org.tr/tr/pub/yead/issue/21818/234537

Can, F., Mart, C., Ödemis, B., & Akişcan, Y. (2021). Cotton growing around the Mediterranean. In *Pest management in cotton: A global perspective* (pp. 245–263). CABI International. https://doi.org/10.1079/9781800620216.0013

Chen, F., Ji, X., Chu, J., Xu, P., & Wang, L. (2021a). A review: Life cycle assessment of cotton textiles. *Industria Textila, 72*(1), 19–29.

Chen, X., Memon, H. A., Wang, Y., Marriam, I., & Tebyetekerwa, M. (2021). Circular economy and sustainability of the clothing and textile industry. *Materials Circular Economy, 3*(1). https://doi.org/10.1007/s42824-021-00026-2

CICERONE. (2020). *Strategic research and innovation agenda on circular economy*. http://cicerone-h2020.eu/wp-content/uploads/2020/09/Circular-Economy-SRIA-2020.pdf

Coşkun, S., & Doğan, N. (2021). Tekstil Endüstrisinde Karbon Ayak İzinin Belirlenmesi. *Süleyman Demirel Üniversitesi Fen Bilimleri Enstitüsü Dergisi, 25*(1), 28–35. https://doi.org/10.19113/sdufenbed.670336

Cross Textiles. (2022). *2022 annual report on sustainability*. https://crosstextiles.com/2022-gri-sustainability-report/

de Oliveira Neto, G. C., Santos, R. A. R., Pinto, L. F. R., Flausino, F. R., de Oliveira, D. E. P., & Seri, M. N. (2024). Innovative circular practices integrating business model for textile industry. *Journal of Engineered Fibers and Fabrics, 19*. SAGE Publications Ltd. https://doi.org/10.1177/15589250241226481

Divrik, B., & Baykal, E. (2022). A qualitative study about internationalization of Turkish textile & clothing industries. *Vlakna a Textil, 29*(3), 8–21. https://doi.org/10.15240/tul/008/2022-3-002

Ellen MacArthur Foundation. (2017). *A new textiles economy: Redesigning fashion's future*. https://ellenmacarthurfoundation.org/a-new-textiles-economy

EPRS. (2024). *Digital product passport in the textile sector*. http://www.europarl.europa.eu/thinktank

Erdoğan, Ç. (2023). *Cotton and products annual, country: Turkey*.

Ereks. (n.d.). *Responsible production*. Retrieved July 22, 2024, from https://www.ereksgarment.com/pages/responsible-production

Ereks. (2020). *Sustainability agenda*. https://www.ereksgarment.com/pages/our-vision

Esi, B. (2017). Türk Tekstil Endüstrisi ve Gelişimi [In English: Turkish textile industry and development]. *Journal of Awareness (JoA), 2*(Special), 643–663.

European Circular Economy Stakeholder Platform. (2022). *Green deal: New proposals to make sustainable products the norm and boost Europe's resource independence*. https://circulareconomy.europa.eu/platform/en/news-and-events/all-news/green-deal-new-eu-proposals-make-sustainable-products-norm-and-boost-europes-resource-independence

European Commission. (2020). *Circular economy action plan for a cleaner and more competitive Europe*. https://op.europa.eu/en/publication-detail/-/publication/45cc30f6-cd57-11ea-adf7-01aa75ed71a1/language-en/format-PDF/source-170854112

European Commission. (2022). *EU strategy for sustainable and circular textiles: COM(2022) 141 final*. https://environment.ec.europa.eu/strategy/textiles-strategy_en

Eurostat. (2008). NACE Rev. 2 statistical classification of economic activities in the European community. https://ec.europa.eu/eurostat/documents/3859598/5902521/KS-RA-07-015-EN.PDF

Hisarlı, D., & Demir, B. (2022). *Türkiye's garment industry profile and the living wage*. https://www.temizgiysi.org/wp-content/uploads/2017/03/CP_EN.pdf

Hora, S. T., Bungau, C., Negru, P. A., & Radu, A.-F. (2023). Implementing circular economy elements in the textile industry: A bibliometric analysis. *Sustainability, 15*(20), 15130. https://doi.org/10.3390/su152015130

İHKİB. (2021). *2021 Ocak-Aralık Hazırgiyim ve Konfeksiyon Sektörü İhracat Performans Değerlendirmesi* [In English: 2021 January–December ready-made clothing and apparel sector export performance evaluation].

İHKİB. (2022). *Türkiye Hazırgiyim ve Tekstil Dış Ticareti 2011–2021 (Yıllık)* [In English: Türkiye ready clothing and apparel foreign trade 2011–2021 annual]. www.ihkib.org.tr

İHKİB. (2023). *2023 Ocak-Aralık Hazırgiyim ve Konfeksiyon Sektörü İhracat Performans Değerlendirmesi* [In English: 2023 January–December ready-made clothing and apparel sector export performance evaluation]. https://www.ihkib.org.tr/fp-icerik/ia/d/2024/02/05/2023-4-hazirgiyim-ve-konfeksiyon-sektoru-ocak-aralik-donemsel-bilgi-notu-202402051538470363-BA94D.pdf

İHKİB. (2024). *Türkiye'nin En Fazla İthalat Yaptığı Ülkeler Genel, Hazırgiyim ve Tekstil 2022–2023 (6 Aylık)* [In English: Türkiye's top import countries general, ready-made clothing and textile 2022–2023 (6 months)]. https://www.ihkib.org.tr/tr/bilgi-bankasi/istatistikler/istatistikler/i-69

İlhan, İ. (2019). Concept of Industry 4.0 in textile manufacturing processes. *Pamukkale University Journal of Engineering Sciences, 25*(7), 810–823. https://doi.org/10.5505/pajes.2018.69851

İpeker. (n.d.). *Green factory! Sustainable production for future generations.* Retrieved July 17, 2024, from https://www.ipeker.com/

İş Güvenliği. (n.d.). *Tekstil İşkolunda İş Sağlığı Güvenliği.* Retrieved April 18, 2024, from https://www.isguvenligi.net/iskollari-ve-is-guvenligi/tekstil-iskolunda-is-sagligi-guvenligi/

ISKO. (2022). *Sustainability impact report.* https://isko.cdn.prismic.io/isko/d4b20a48-2eb1-485f-a6d1-ee44d2628f7a_AAFF-ISKO+Sustainability+Impact+Report+2022+V7+LR.pdf

Istanbul Chamber of Industry. (2022). *Türkiye'nin 500 Büyük Sanayi Kuruluşu 2021* [In English: Türkiye's 500 largest industrial enterprises 2021]. https://www.iso.org.tr/sites/1/upload/files/iso500-2021_opt-11231.pdf

İTKİB. (2024). *Türkiye Tekstil ve Hammaddeleri İhracatında İlk 50 Ülke* [In English: Top 50 countries in Türkiye's textile and raw materials exports].

James, K., & Lings, J. (2018). *Life cycle management and circular economy challenges for the textile sector: Session wrap up.* Springer.

Kim, H., & Kim, B.-G. (2015). Economic impacts of the hotel industry: An input-output analysis. *Tourism Review, 70*(2), 132–149.

Koszewska, M. (2018). Circular economy—Challenges for the textile and clothing industry. *Autex Research Journal, 18*(4), 337–347. https://doi.org/10.1515/aut-2018-0023

Koton. (2022). *Sustainability report 2022.* https://koton.mncdn.com/static/2022SustainabilityReportEn.pdf

Krämer, L. (2020). Planning for climate and the environment: The EU green deal. *Journal for European Environmental and Planning Law, 17*(3), 267–306. https://doi.org/10.1163/18760104-01703003

Leontief, W. (1991). The economy as a circular flow. *Structural Change and Economic Dynamics, 2*(1), 181–212.

Maity, S., Singha, K., Pandit, P., & Ray, A. (2020). Circular economy in fashion and textile from waste. In *Recycling from waste in fashion and textiles* (pp. 235–251). Wiley. https://doi.org/10.1002/9781119620532.ch11

Manshoven, S., Christis, M., Vercalsteren, A., Arnold, M., Nicolau, M., & Lafond, E. (2019). *Textiles and the environment in a circular economy.* https://www.eionet.europa.eu/etcs/etc-wmge/products/etc-wmge-reports/textiles-and-the-environment-in-a-circular-economy

Mavi. (2024). *2023 annual report.* https://www.mavicompany.com/i/assets/documents/2024/Q4/MAVIANNUALREPORT2023-final.pdf

McKinsey&Company, A. F. & L. G. (2022). *Scaling textile recycling in Europe-turning waste into value.* https://www.mckinsey.com/industries/retail/our-insights/scaling-textile-recycling-in-europe-turning-waste-into-value

Öngüt, Ç. E. (2007). *Türk Tekstil ve Hazır Giyim Sanayiinin Değişen Dünya ve Rekabet Şartlarına Uyumu* [In English: Adaptation of the Turkish textile and ready-made clothing industry to the changing world and competition conditions]. Devlet Planlama Teşkilatı İktisadi Sektörler ve Koordinasyon Genel Müdürlüğü.

Orta Anadolu. (2013). *Corporate social responsibility report.*

Papamichael, I., Chatziparaskeva, G., Voukkali, I., Navarro Pedreno, J., Jeguirim, M., & Zorpas, A. A. (2023). The perception of circular economy in the framework of fashion industry. *Waste Management and Research, 41*(2), 251–263. https://doi.org/10.1177/0734242X221126435

Piétron, D., Staab, P., & Hofmann, F. (2023). Digital circular ecosystems: A data governance approach. *GAIA—Ecological Perspectives for Science and Society, 32,* 40–46. https://doi.org/10.14512/gaia.32.S1.7

Rajput, S., & Singh, S. P. (2019). Industry 4.0—Challenges to implement circular economy. *Benchmarking, 28*(5), 1717–1739. https://doi.org/10.1108/BIJ-12-2018-0430

Ramos-Galarza, C., & Acosta-Rodas, P. (2019). Stress and productivity in workers of textile companies. *Journal of Fashion Marketing and Management: An International Journal, 23*(1), 17–29.

Rençber, S. Y., & Ceylan, A. (2022). Bir Tekstil Fabrikasında Çalışan İşçilerin Karşılaştıkları Riskler ve İş Kazası Geçirme Durumlarının Değerlendirilmesi (In English: Evaluation of risks encountered by workers working in a textile factory and occupational accident situations]. *Gümüşhane Üniversitesi Sağlık Bilimleri Dergisi, 11*(2), 748–759.

Sayar, R. (2020). Türk Tekstil Sektörünün Tarihi, Ekonomideki Yeri, Yapısı ve Ekosistemi: Yeni Girişimcilik Örnekleri [In English: History of the Turkish textile sector, its place in the economy, structure and ecosystem: New entrepreneurship examples]. In A. Kara & Y. Esmer (Eds.), *Girişimcilik Perspektifinde Türk Tekstil ve Giyim Sektörü.* https://www.researchgate.net/publication/342360627

Schumacher, K. A., & Forster, A. L. (2022). Textiles in a circular economy: An assessment of the current landscape, challenges, and opportunities in the United States. *Frontiers in Sustainability, 3,* 1038323.

SKD. (2022). *Preliminary research on circularity potential of five sectors in Turkey.* https://www.skdturkiye.org/images/icerik/Preliminary_Research_Report_on_Turkeys_Circular_Economy_Potential_in_5.pdf

Textile Exchange. (2022). *Organic cotton market report 2021.* https://textileexchange.org/knowledge-center/reports/organic-cotton-market-report/

The Ministry of Industry and Technology. (2021a). *Tekstil, Hazırgiyim ve Deri Ürünleri Sektörleri Raporu* [In English: Textile, ready-made clothing and leather products sectors report].

The Ministry of Industry and Technology. (2021b). *Tekstil Sektörü Analiz Raporu ve Kılavuzu TR32 Bölgesi (Denizli, Aydın, Muğla)* [In Turkish: textile sector analysis report and guide TR32 region (Denizli, Aydın, Muğla)]. www.sanayi.gov.tr

The Ministry of Trade. (2021a). *Türkiye Country Report, 79th Plenary Meeting of the ICAC,* 06–09 December.

The Ministry of Trade. (2021b). *Yeşil Mutabakat Eylem Planı* [In English: Green deal action plan].

The Ministry of Trade. (2022). *Yeşil Mutakabat Çalışma Grubu Yıllık Faaliyet Raporu* [In English: Green deal working group annual report].

The Ministry of Trade. (2024). *Ticaret Bakanlığı İhracat Genel Müdürlüğü Hazır Giyim Sektör Raporu* [In English: The Ministry of Trade general directorate of export ready-to-wear industry report]. https://ticaret.gov.tr/data/5b87000813b8761450e18d7b/Haz%C4%B1r%20Giyim%20Sekt%C3%B6r%20Raporu%202022.pdf

The Ministry of Trade Easy Export Platform. (n.d.). *Tekstil.* Retrieved April 17, 2024, from https://www.kolayihracat.gov.tr/sektorler/tekstil-hali-ve-ev-tekstili-dahil

Türkiye Tekstil Sanayii İşverenleri Sendikası. (2019). *Türkiye Tekstil Sektörünün Endüstri 4.0 Durum Değerlendirmesi ve Yol Haritası* [In English: Industry 4.0 situation assessment and roadmap of the Turkish textile sector].

Türkiye Tekstil Sanayii İşverenleri Sendikası. (2022, December 1). *Tekstilde "geri dönüşüm" trend değil, bir zorunluluk!* [In English: "Recycling" in textile is not a trend, it is a necessity!].
Uludağ İhracatçı Birlikleri. (2018). *Uludağ İhracatçı Birlikleri Türkiye Tekstil Sektörü ve Bursa* [In English: Turkish textile sector and Bursa]. Ekim. https://uib.org.tr/tr/kbfile/turkiye-tekstil-sektoru-ve-bursa
Uyanık, S., & Çelikel, D. C. (2019). *Türk Tekstil Endüstrisi Genel Durumu.* [In English: General situation of Turkish textile industry]. Teknik Bilimleri Dergisi.
Voulgaridis, K., Lagkas, T., Angelopoulos, C. M., & Nikoletseas, S. E. (2022). IoT and digital circular economy: Principles, applications, and challenges. In *Computer networks* (Vol. 219). Elsevier B.V. https://doi.org/10.1016/j.comnet.2022.109456
Yıkılmaz, İ., & Kör, B. (2023). Digital transformation In SMEs: A focused review of the research literature. *Journal of the Human and Social Science Researches, 2023*(2), 661. https://doi.org/10.15869/itobiad.1267517

Semra Ocak (semra.ocak@std.bogazici.edu.tr) is a Ph.D. student at the Institute of Environmental Sciences at Bogazici University. She holds a B.Sc. in environmental engineering and an M.Sc. from the Institute of Environmental Sciences. Her Master's thesis focused on environmental conservation failure in watershed management. Her research interests include circular economy, life cycle assessment, cleaner production, environmental assessment, environmental governance, watershed management and environmental planning. She is currently working on the circular economy and life cycle assessment in the textile sector. She is currently investigating the potential for a circular economy transition in denim production.

Sevil Acar (sevil.acar@bogazici.edu.tr) works as Professor of Economics at the Faculty of Managerial Sciences at Bogazici University. She holds a BA degree from Bogazici University (2000–2005), MA degree from Istanbul Technical University (2005–2007) and a Ph.D. degree in Economics from Marmara University (2007–2011). Her research focuses on environmental and resource economics, particularly sustainable development indicators, climate change, and the resource curse. Awarded a Swedish Institute scholarship, she conducted research at the Centre for Environmental and Resource Economics (CERE), Umeå University, Sweden during her Ph.D. studies. In collaboration with the Global Subsidies Initiative (GSI-IISD), she investigated fossil fuel subsidies and renewable energy in Türkiye. She is Lead Author for the Intergovernmental Science-Policy Platform on Biodiversity and Ecosystem Services (IPBES) Regional Assessment for Europe and Central Asia and the Transformative Change Assessment. Her articles have appeared in various journals including Structural Change and Economic Dynamics, Ecological Economics, and Energy Policy. She also authored and edited three books (The Curse of Natural Resources: A Developmental Analysis in a Comparative Context; Macroeconomics of Climate Change in a Dualistic Economy; Handbook of Green Economics).

Chapter 5
Italy

Guilherme Hörner Bussolo, Gianmarco Bressanelli, Filippo Visintin, and Nicola Saccani

Abstract Italy holds a significant position in the European Textile and Clothing (T&C) industry, renowned for its economic influence, sustainability initiatives, and technological innovations. The sector's prominence is underscored by Italy's substantial share of the EU T&C market, employing a significant portion of the workforce and driving economic activity. In particular, the Italian T&C industry is anchored in local industrial clusters, rich in skilled human capital and artisanal traditions. Prato, a key textile industrial district, exemplifies Italy's integration of tradition and modernity, showcasing a dynamic blend of small enterprises and innovative practices. The district's history of adaptation to market changes, coupled with recent shifts towards circular economy (CE) principles, highlights the resilience and forward-thinking approach of the Italian T&C industry. This chapter examines the significant contribution of Italy's T&C industry, emphasising its impact on the fashion sector, its commitment to sustainability, and the evolving landscape of CE and digitalisation practices. A grey literature analysis, coupled with multiple case studies from the Prato district, illustrates successful implementations of sustainable practices, digital innovations, and collaborative efforts to promote a CE. By leveraging digital tools and sustainable strategies, the Italian textile industry sets a benchmark for global standards, balancing environmental responsibility with economic viability.

Keywords Circularity · Textile industry · Sustainability · Italy

G. H. Bussolo (✉) · F. Visintin
University of Florence, Florence, Italy
e-mail: guilherme.hornerbussolo@unifi.it

F. Visintin
e-mail: visintin.filippo@unifi.it

G. Bressanelli · N. Saccani
University of Brescia, Brescia, Italy
e-mail: gianmarco.bressanelli@unibs.it

N. Saccani
e-mail: nicola.saccani@unibs.it

M. Wynn and T. Wiegand (eds.), *Sustainability, the Circular Economy and Digitalisation in the European Textile and Clothing Industry*, SDGs and Textiles,
https://doi.org/10.1007/978-981-97-9116-3_5

5.1 Introduction

Italy is renowned globally and stands as a cornerstone of the European Textile & Clothing (T&C) industry, not only due to its significant economic contributions but also through its influence on fashion, sustainability, and technological advancements (Coppola et al., 2023). In 2023, the EU T&C industry reported a turnover of €169 billion, with Italy playing a significant role in these figures (EURATEX, 2024). Italy is responsible for producing over 40% of the EU's apparel (European Commission & Centre for Industrial Studies (CSIL), 2021). This substantial share underscores Italy's importance in the European T&C industry, particularly in the high-quality and luxury segments, which are pivotal to the EU's competitive positioning globally. The Italian T&C industry's unique charm is often attributed to a blend of non-measurable and intangible factors such as creativity, talent, a tradition of good taste, and a dedication to beauty. These elements collectively define the "Made in Italy" label, synonymous with elegance and impeccable style (Colombi, 2015). "This sector is commonly found within local industrial clusters that benefit from a well-trained workforce and a rich heritage of craftsmanship" (Belfanti, 2015; Motta et al., 2024). Additionally, the close relationship with artisanal traditions further enhances the uniqueness of these sectors.

Currently, the textile industry in Italy is navigating a critical period of transformation, aiming to balance modernisation and technological advancements with its enduring emphasis on creativity and craftsmanship (Dominidiato et al., 2023; McKinsey & Company & The Business of Fashion, 2024). Issues related to labour practices, as seen in the case of Chinese migrants working in Italian Industrial Districts (IDs), emphasise the intersection of agency, regulations, and working conditions (Ceccagno, 2017). In terms of sustainability, innovation, and efficiency, Italy has been at the forefront internationally. As the industry progressed, challenges such as environmental sustainability and labour practices emerged, leading to an urgent need for a shift towards circular economy (CE) principles (Sandvik & Stubbs, 2019). This transition is vital for mitigating the environmental footprint of textile production and consumption. By aligning with circular practices, the T&C industry can enhance resource efficiency, reduce environmental impacts, and foster resilience in the face of interconnected challenges (Naidoo et al., 2021). Industrial clusters, such as Prato, have been crucial in fostering innovation and sustainability within the Italian T&C industry. These clusters have been pioneers in introducing CE practices and sustainable strategies, setting benchmarks for other regions (Mazzoni, 2021). However, the transition to a CE requires significant changes in production and consumption models, emphasising the importance of sustainable practices within the industry (Bressanelli et al., 2022). In addition, textile IDs such as Prato face various challenges in the areas of sustainability, circularity, and digitalisation (Bressanelli et al., 2022; Saccani et al., 2023).

This chapter aims to provide a detailed overview of the challenges and opportunities for sustainability, circularity, and digitalisation in the Italian textile industry, using the district of Prato as an illustrative example to demonstrate the practical implementation of CE principles and the challenges that emerge therein. More specifically, the chapter addresses the following research questions (RQs):

RQ1 How are Italian T&C companies addressing sustainability in their corporate strategies and activities?
RQ2 What strategies and activities relating to the CE are being pursued in the Italian T&C industry?
RQ3 What role are digital technologies playing in the transition to sustainability and the CE in the Italian T&C industry?

The role of Italy's T&C industry within the broader European context is discussed, highlighting its economic impact, commitment to sustainability, and the evolving landscape of circularity and digital innovation. The research utilises a grey literature review to gather data from non-conventional sources such as industry reports, government documents, and non-peer-reviewed journals. This approach enables the analysis of a wide array of current practices and recent innovations that may not yet be documented in academic literature. This research also utilises a case study method to collect data from three companies located in the district of Prato, referred to as Company A, Company B, and Company C. The chapter is organised as follows. Section 5.2 details the research method. Section 5.3 presents the findings, which include an overview of the Italian T&C industry landscape (Sect. 5.3.1), an analysis of the Prato Textile Industrial District (ID) (Sect. 5.3.2), and the case studies (Sect. 5.3.3). Section 5.4 directly addresses the three RQs relating to sustainability, the CE, and digitalisation challenges and opportunities for the Italian T&C industry. Section 5.5 then discusses some emergent issues worthy of note. Finally, Sect. 5.6 provides a conclusion to the chapter, pointing out limitations but also summarising the contribution of this research initiative.

5.2 Research Method

The primary objective of this study is to ascertain a comprehensive understanding of the current transition towards sustainability and circularity within the Italian T&C industry. Additionally, this research sought to delineate the contributory role of digital technologies in facilitating this transition, underscoring their potential impact on industry practices. Thus, the research methodology combines the use of grey literature with multiple case studies, focusing on three companies situated in the Italian industrial district of Prato.

The incorporation of grey literature is fundamental to this methodology, due to its ability to provide the most up-to-date and comprehensive information about the Italian T&C industry as a whole, which often precedes that found in traditional academic publications. This type of literature is invaluable in enriching the research foundation with a wide array of perspectives, particularly those emanating from government reports, industry-specific white papers, and technical documents. This comprehensive approach addresses and fills the gaps that are often left by peer-reviewed sources alone, thereby enhancing the overall quality and relevance of the research outcomes. Moreover, the accessibility of grey literature in diverse formats facilitates a more flexible and thorough exploration of the subject matter. This flexibility is essential in adapting the research process to emerging trends and current developments, which are characteristic of the rapidly evolving T&C industry. Such an approach enables the researcher to maintain a dynamic perspective that aligns closely with the actual conditions within the industry. Table 5.1 offers a comprehensive catalogue of the various reports utilised throughout the research process, sourced from an extensive range of databases and reputable institutions. It serves as an essential reference point, detailing the scope of literature and data incorporated into the study.

In addition to grey literature analysis, multiple case studies have been used to capture a comprehensive and nuanced understanding of the diverse operational environments and strategic approaches of different organisations in the value chain. Three companies situated in the Italian ID of Prato, referred to herein as Company A,

Table 5.1 Grey literature resources utilised in this research

Report	Source	Year
Circular solutions for nature	Sitra	2024
Data on the EU textile ecosystem and its competitiveness—final report	European Commission Centre for Industrial Studies	2021
Euratex—facts and key figures	Euratex	2024
How brands can embrace the sustainable fashion opportunity	Bain & Company, Inc. WWF—Italy	2022
Just fashion transition	The European House—Ambrosetti & Confindustria Venezia	2022
L'industria tessile-abbigliamento in Italia - il quadro generale	Confindustria Moda/Sistema Moda Italia	2023
Piano Nazionale di Ripresa e Resilienza #nextgenerationitalia	Italian Government	2021
Prato textile-fashion district	Confindustria Toscana Nord	2024
Rapporto sull'economia circolare in Italia	Network Gruppo di lavoro del Network e della Fondazione per lo sviluppo sostenibile	2023
STEP2030 Sustainable Textiles European Partnership	Euratex	2023
The Italian textile industry a key economy sector	ITMA services	2022
The export of Lucca, Pistoia and Prato in 2023	Confindustria Toscana Nord	2024
The state of fashion	McKinsey & Company & The Business of Fashion	2024

Table 5.2 Companies involved in the multiple case study analysis

Company	Main activity	Employees (number)	Turnover (m€)
Company A	Textile yarns and fabrics supplier	50–100	Around 97
Company B	Textile yarns and fabrics supplier	<30	Around 11
Company C	Apparel and fashion brand	<30	<4

Company B, and Company C, have been involved in the study. The selection included companies recognised for their excellence both within and outside the ID, exhibiting varying degrees of integration, and being of different sizes and positions within the supply chain. To gather comprehensive insights, interviews were conducted with the CEOs or top managers of these companies, and the information obtained was triangulated with secondary sources, such as national reports and sustainability reports. The primary objective of these interviews was to understand the specific CE initiatives each company is pursuing and their impact on the ID and on overall business performance.

By analysing these varied case studies, the research can identify unique and common factors influencing business practices, innovation, and sustainability efforts across different organisational contexts. This multiplicity of case studies ensures a richer, more robust analysis, leading to more generalised and transferable insights that can inform broader industry practices and policies. Table 5.2 summarises the profiles of the companies involved. Company A and Company B provide distinct perspectives due to their varying scales and financial performances, with Company A employing between 50 and 100 people and generating a turnover of approximately €97 million, while Company B employs fewer than 30 people and has a turnover of around €11 million. The inclusion of Company C, an apparel and fashion brand employing fewer than 30 people with a turnover of less than €4 million in 2022, adds another layer of diversity, highlighting the differences in challenges and strategies faced by smaller enterprises compared to different industry sectors.

5.3 Findings

This section explores key aspects of the Italian T&C industry, beginning with an overview of the industry's landscape to provide context and understanding of its current state. Italy, with its rich history of artisanal craftsmanship and industrial innovation, has managed to maintain its leadership in the T&C sector, particularly through the resilience and adaptability of its industrial districts. The textile industrial district of Prato is then examined, and noted as a significant hub for its innovative practices and contributions to the T&C sector, Europe's largest and a global leader in recycled wool production, which exemplifies this leadership. Despite challenges like global competition and economic shifts, Prato has maintained its relevance by integrating CE principles and sustainability practices. This is followed by detailed case studies of three prominent companies, referred to as A, B, and C, highlighting their unique approaches, challenges, and successes within the industry. Company A leads in sustainable fabric innovation, Company B excels in transforming textile waste into valuable products, and Company C, a sustainable fashion start-up, is pioneering eco-design and customer engagement. These companies illustrate Prato's dynamic approach to sustainability, setting a benchmark in the global textile industry. These sections collectively offer a comprehensive analysis of the Italian T&C sector and its evolving dynamics.

5.3.1 *The Italian Textile and Clothing Industry Landscape*

Italy holds a significant position within the European T&C market, underscoring its strategic relevance in this sector. In 2022, the Italian T&C industry generated 37.4% of the total turnover within the EU27 countries, solidifying its role as a leader in the region. Its relevance in the European and global T&C market is underscored by its substantial contribution to EU turnover, robust export performance, positive trade balance, and strong positioning as a key supplier to major international markets. Italy has upheld its position as one of the leading textile-producing countries worldwide, alongside China, India, Germany, and Turkey (Coppola et al., 2023).

Over the past five years, the Italian T&C market has experienced significant fluctuations, reflecting broader economic trends and specific sectoral dynamics. The industry saw a robust recovery and growth, particularly in the years following the initial impact of the COVID-19 pandemic. In 2022, the turnover increased by 17.8% year-on-year, reaching €62.5 billion (Confindustria Moda & Sistema Moda Italia, 2023). Exports have been a significant driver of growth. In 2022, Italian T&C exports rose by 18.9%, reaching €38.5 billion (Confindustria Moda & Sistema Moda Italia, 2023). Imports also surged, with a notable 32.4% increase, totalling €28.7 billion (Confindustria Moda & Sistema Moda Italia, 2023). The export market showed robust performance across various regions, with the United States showing the most substantial increase among the top 20 destinations; EU markets, including France and Germany, also exhibited significant growth (Confindustria Moda & Sistema Moda Italia, 2023). The Textile segment within the industry grew by 19.3% in 2022, while the Clothing segment saw a 17.1% increase (Confindustria Moda & Sistema Moda Italia, 2023). In particular, the Clothing segment is the most significant within the Italian T&C industry (Moda & Sistema Moda Italia, 2023). It generates €25.7 billion in revenue, accounting for 41.1% of the total industry revenue. Additionally, this segment comprises 61.0% of the companies and 54.1% of the employees, highlighting its dominant role within the sector. Other segments also make substantial contributions to the industry's revenue and employment. For instance, the Wool Textiles segment generates €5.9 billion in revenue, which represents 9.5% of the total industry revenue. It employs 28,546 workers, accounting for 7.7% of the industry's workforce. Similarly, the chemical and other textiles segments contribute significantly. Employment within the Italian T&C industry is widely spread across various segments. The textile industry employs 115,018 individuals, which is 30.9% of the total workforce in the industry. In contrast, the clothing industry employs 257,576 individuals, making up 69.1% of the workforce. This indicates a balanced distribution of the workforce, with a significant concentration in the clothing segment (Confindustria Moda & Sistema Moda Italia, 2023). Looking ahead, the industry is expected to continue its positive trajectory, although growth rates may be moderate. For instance, the first half of 2023 saw a 7.1% increase in sectoral revenue, with a more modest 1.3% growth in the second quarter. This cautious optimism reflects the sector's resilience and adaptive strategies amidst global economic uncertainties. The industry's ability to navigate post-pandemic challenges positions it for continued, albeit more moderate, growth in the near future.

Italy's T&C landscape is marked by an impressive array of approximately 42,000 companies, a testament to its expansive and vibrant industry. Tuscany and Lombardy are the leading regions in terms of the number of T&C companies, with Tuscany hosting 8,121 companies and Lombardy 7,353 companies (Confindustria Moda & Sistema Moda Italia, 2023). These two regions account for 19.6% and 17.8% of the national total, respectively. Most regions saw a decline in the number of companies compared to the previous year, with Basilicata experiencing the highest decrease at −11.8%, followed by Molise at −13.9%. Campania was unique in maintaining the same number of companies as the previous year, indicating stability in this region. The Italian T&C industry is predominantly concentrated in regions known for their production, as seen in Fig. 5.1, such as the northern regions, which are renowned for their rich textile heritage and continue to be pivotal in defining the standards and innovations of the industry. In recent years, textile districts have also begun to emerge in Southern Italy, particularly in Campania (ITMA Services, 2022). Although these newer districts in the southern regions have not yet reached the prominence of their northern and central counterparts, they signify a growing diversification and expansion of Italy's textile sector across the country.

The T&C sector's share of regional manufacturing companies varies significantly. For example, in Tuscany, T&C companies make up 18.8% of the regional manufacturing companies, highlighting the sector's importance in that region. In contrast, in Lombardy, they represent only 8.5% of regional manufacturing companies, indicating a more diversified industrial base (Confindustria Moda & Sistema Moda Italia, 2023). The sector is characterised by a vast majority of small and micro-sized enterprises (SMEs), which form the backbone of this traditional yet innovative industry. According to Ministero dello Sviluppo Economico, 2011 SMEs are characterised as follows: (a) less than 250 employees and (b) an annual turnover not exceeding €50 million or total sales in the accounts not exceeding €43 million. Despite their small size, these companies thrive largely due to their strategic location within IDs (ITMA Services, 2022). The IDs are geographic areas that contain clusters of small SMEs specialising in various stages of the production process. These districts are characterised by a dense network of economic and social connections, allowing businesses to collaborate and support each other within the local ecosystem (Becattini, 2017). The structure of districts not only fosters high levels of productivity but also yields economic results that frequently surpass industry averages through

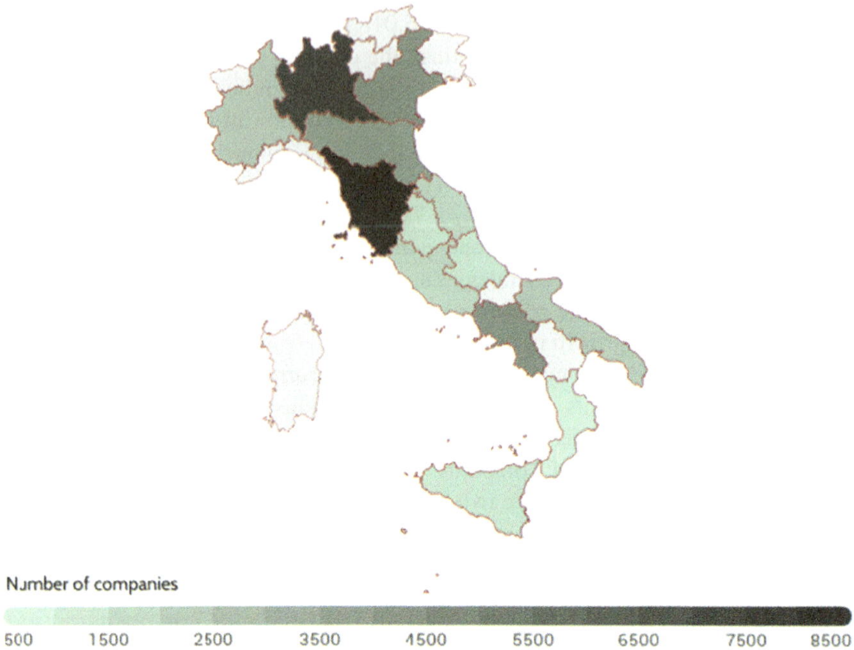

Fig. 5.1 Distribution of T&C companies by region. *Source* Chapter authors, based on Confindustria Moda & Sistema Moda Italia (2023)

the efficient organisation of resources, workforce specialisation, and collaborative innovation among local businesses.

The unique context of Italy played a relevant role in this evolution: other countries were unable to meet the demand for specialised goods, and Italy had a robust tradition of artisanal production, which ensured the availability of a skilled workforce. In response, smaller factories began to proliferate, producing a wider array of goods and frequently organising themselves into clusters (Mazzoni, 2021). The collective strength of these enterprises is enhanced by their shared historical background and a unified production focus, which promotes synergy and cooperative growth. Within these districts, businesses often specialise in specific stages of textile production such as spinning, warping, weaving, dyeing, or finishing (Lazerson & Lorenzoni, 1999). This specialisation allows each enterprise to enhance its craft, contributing to the overall efficiency and quality of the district's output. According to an assessment conducted by The European House—Ambrosetti & Confindustria Venezia (2022), the readiness for transition to sustainable practices is directly proportional to the size of companies. Large companies are more active in reporting, performance monitoring, and certification, while smaller companies lag behind. An analysis of responses from 167 companies to a voluntary sustainability report highlights this disparity. The assessment, which covered various aspects of sustainability management, revealed that 70% of the companies had turnovers under €30 million.

There is a direct correlation between the size of the companies and positive sustainability outcomes. Larger companies are more likely to have sustainability management tools in place, with significant differences observed in areas such as reporting and emissions measurement. However, even among larger companies, less than 50% conduct suppliers' human rights assessments. Among supply chain companies, 49% measure their performance, but only 18% publish a sustainability report. This trend is more pronounced in larger companies. For instance, only 7% of companies with turnovers under €50 million publish a sustainability report, and this scenario drops to 0% for companies under €5 million.

A significant barrier to sustainability transition is the lack of standardised tools. According to 85% of companies, a common standard could greatly enhance sustainability efforts, particularly among those already monitoring their performance. This percentage increases to 88% among companies actively engaged in sustainability efforts. Despite these challenges, 75% of companies believe they have sufficient financial resources for sustainable transition, and 86% feel they have adequate in-house skills to manage the change. The Italian T&C industry is facing increasing pressure to transition to sustainable and circular models. This transition is influenced by a range of factors, including the role of supply chain actors such as yarn mills and manufacturers (Kreci & Enegbuma, 2023). Key themes in this transition include the need for new design processes that reduce waste and pollution, the role of knitwear and woven producers, and the importance of governance, material sourcing, and partnerships in sustainable fashion practices. The Italian T&C industry, with its interconnected small and medium-sized enterprises, is particularly well-positioned for this transition (Montagna et al., 2022).

Italian T&C companies are proactively embracing sustainability in their corporate strategies and day-to-day operations. They are integrating sustainability into their core business and effectively monitoring and measuring their economic, social, and environmental impacts (Peters & Simaens, 2020). Companies are focusing on sustainable supply chain management practices, with internal orientation being a key factor in driving these initiatives. This shift involves adopting a variety of approaches that enhance sustainability across the production and supply chain. Environmental sustainability initiatives are also being widely embraced. These include corporate social responsibility (CSR), green supply chain management, and eco-design, all of which contribute to the industry's sustainability efforts (Islam et al., 2022). Companies are exploring innovative solutions like blockchain technology and the Internet of Things (IoT) to facilitate the adoption of a CE within the industry (Alves et al., 2021; Chae & Hinestroza, 2020). To drive the CE in the T&C sector, companies are rethinking production processes, maximising product use and reuse, promoting redistribution to new markets, and extending the lifespan of products (Chen et al., 2021). Moreover, Italian T&C companies are exploring innovative solutions like blockchain technology to address sustainability concerns, particularly in waste management (Colucci & Vecchi, 2021; Dursun et al., 2022).

On the other hand, sustainable regulation acts as a driver for transformation in the fashion industry. The Extended Producer Responsibility (EPR) regulation for textiles aims to improve environmental quality, legality, and eco-design in production processes. For example, the Italian government has decided to invest €150 million from the Recovery Fund to promote the CE through flagship projects on textile waste collection. In the Prato district, a local multi-utility company invested €30 million (partly coming from the Recovery Fund) in a project that envisages the construction of an automatic sorting plant for post and pre-consumer textile waste, with a capacity of approximately 33,000 tonnes per year. In this plant, artificial intelligence and highly advanced infrared technology will be used to automatically sort various textile materials by colour and composition. Collaboration across the value chain is essential for transforming the apparel system (Comune di Prato, 2023). Sustainable regulation can then empower designers to drive change in the fashion industry.

Lastly, the Italian T&C industry offers insightful examples of how companies are incorporating CE principles to promote sustainability. A study by Colucci and Vecchi (2021) identified a set of practices aligned with three core principles of the circular economy (CE): reuse, recycling, and product-life extension. Additionally, the study introduced a fourth category, termed resource preservation, which arises from implementing these CE practices and involves the more efficient use of resources. Resource preservation is a direct outcome of these CE practices, and it can be integrated into a company's broader sustainability strategy to reduce environmental impact, highlighting practices beneficial for the environment.

Candiani Denim exemplifies this trend through its closed-loop systems, recycling initiatives, water recovery efforts, and emphasis on extending product life cycles. By reprocessing over 20,000 tonnes of denim waste and reducing water usage by 50% per pair of jeans, Candiani significantly minimises its environmental footprint (Candiani Denim, 2024). WRÅD, another notable player, combines product innovation with consumer education. Utilising 80% recycled materials in its product line, WRÅD also engages in educational programmes reaching over 10,000 participants and operates take-back programmes that have collected over 5,000 garments. These efforts highlight WRÅD's commitment to reducing resource consumption and promoting sustainable fashion practices (Wrad, 2024).

Gucci also started in 2023 a Circular Hub, supported by Kering, which significantly advances the integration of CE principles within the luxury fashion sector. By redefining the entire value chain from raw materials to production optimisation and logistics, the hub fosters a comprehensive transformation towards sustainable practices. It aims to maximise the use of recycled materials and enhance product durability, repairability, and recyclability, thus reducing waste and environmental impacts. This initiative aligns with Italy's National Recovery and Resilience Plan and the EU's 2030 emissions reduction targets, highlighting its commitment to environmental sustainability. The involvement of over 700 direct suppliers and 3,500 sub-suppliers indicates a broad, systemic shift towards circularity, potentially influencing other industry players through its open innovation platform (Gucci & Kering Group, 2023). Moreover, the Circular Hub's focus on research and development, logistics efficiency, and industrial partnerships underscores the critical role of digitalisation in achieving CE goals. By leveraging technological and digital solutions, the hub will enhance traceability, optimise raw material use, and develop innovative recycling and reuse processes. This will not only reduce greenhouse gas emissions but also create new employment opportunities and support the local economy. As a forerunner in this domain, the Circular Hub sets a precedent for future regulatory production models, showcasing a sustainable pathway that integrates economic, environmental, and social benefits within the luxury fashion industry. These case studies collectively illustrate the diverse approaches within the Italian T&C industry, showcasing successful implementations of CE principles that not only address environmental concerns but also pave the way for a more sustainable fashion ecosystem.

5.3.2 The Textile Industrial District of Prato

Prato is well known as one of the preeminent IDs in Italy, and it holds the distinction of being the most extensive textile centre in Europe, as well as one of the foremost global hubs to produce recycled woolen yarns and fabric (Weibel-Orlando, 2012). The origins of Prato's textile industry can be traced back to the late nineteenth century when the Italian city emerged as a global hub for textile manufacturing, excelling notably in regenerating wool from discarded rags. Prato's specialisation in the textile industry, however, dates back to the twelfth century, but the shift from artisanal to industrial production occurred in the late nineteenth century during the Industrial Revolution. Mechanisation was driven by technologies imported from France and England, especially wool-recycling equipment from Yorkshire, and significant capital investment. Local companies gained a cost advantage by using recycled wool from old clothes and industrial scraps, leading to specialisation in recycled wool production. Simple items like plaids and military blankets were produced on a large scale by highly vertically integrated firms. Post-World War II, the demand for lighter, more sophisticated products led to the decline of these vertically integrated companies. This resulted in the fragmentation of the production system into many small and medium enterprises, located in the same area, each specialising in different phases of the textile process (spinning, weaving, finishing, etc.), i.e., a textile district. Such a configuration has remained largely unchanged since then. However, major crises occurred in the late 1980s and in the period 2001–2008 (Bressanelli et al., 2022) which led to a significant downsising of the district, losing approximately 50% of its companies and 50% of its employees in the last 10 years.

This decline was further compounded by the relocation of manufacturing processes outside the district and an influx of semi-finished products from low-cost countries. The textile district now comprises approximately 7,000 companies operating within the T&C sector, with over 2,000 of these firms specifically dedicated to textiles (Table 5.3). The textile companies in Prato are renowned for their diverse product offerings, which include fabrics for the apparel industry, textile furnishings, yarns for knitwear, knitted garments, non-woven textiles, and specialised fabrics for industrial applications.

Notably, Prato is an important area for the ready-to-wear fashion industry and the crafting of "made-in-Italy" garments (ITMA Services, 2022). Prato's textile district

Table 5.3 Textile industry in Prato, Florence

Category	Number of active companies	Average revenue (2022, million €)	Average number of employees (2022)
Yarn producers	80	11	27
Fabric producers	150	9	17
Other textile producers	90	9	25
Wet processing	130	4	26
Other third-party textiles	2,050	0.5	4
Total third-party services[a]	2,180	0.7	5

Source Confindustria Toscana Nord (2024)
[a] Spinning mills, knitting mills, winding, twisting, warping, printing mills, control and darning centres, quilting mills, laminating, steaming, carbonising, gauzing, etc.

has not only been a manufacturing centre but has also influenced the city's urban landscape and cultural identity. The district's development, characterised by a polycentric production structure and family entrepreneurship, has contributed to the creation of a modern heritage in Prato (Del Bianco, 2024). Additionally, the influx of Chinese migrants into the Prato area has played a critical role in the district's evolution, sparking discussions on labour dynamics and the city's social fabric (Hamilton & Fels, 2014). The district's historical success in the textile sector, coupled with its commitment to sustainability, provides the context for examining how circularity can be effectively implemented and scaled within a well-established industrial setting. The industrial cluster of Prato is embracing CE and eco-innovation practices, leading to environmental, economic, and societal benefits. Prato's T&C industry is undergoing a paradigm shift towards sustainable circularity, driven by the necessity to reduce environmental impacts and business risks (Palm et al., 2021). The district has faced significant economic pressures over the years. Factors such as rising labour costs, competition from newly industrialised countries, and shifts in the global market and regulatory regime have contributed to the challenges faced by the Prato textile district (Lan, 2014). The Chinese, taking advantage of the crisis in the district and the wide availability of industrial facilities, have bought or rented these facilities once used for textile activities and converted them for the production of low-cost garments (Fig. 5.2). These garments are produced by sourcing yarn and fabrics from outside the district (i.e., from China or other low-labour-cost countries) by Chinese or (increasingly) other migrant labour force (Hamilton & Fels, 2014).

The large representation of this community in Prato reflects the impact of globalisation and immigration on the industrial landscape of the city (Bellini & Leonardi, 2020). The growth of household businesses engaged in textile production and the trend towards emigration for economic success has reshaped the entrepreneurial landscape in Prato (Guercini et al., 2017). Additionally, the impact of the COVID-19 pandemic, coupled with competitive pressures and changes in technology and

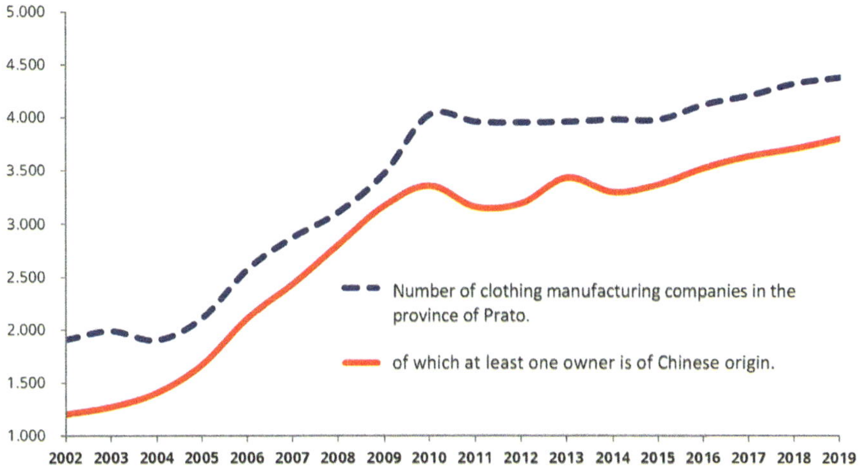

Fig. 5.2 Active clothing manufacturing companies registered with the Prato Business Registration Office. *Source* Confindustria Toscana Nord (2024)

consumer preferences, has created volatility in the economic sustainability of textile industries globally (Santos, 2021). This heightened competition poses a serious threat to the profitability and long-term sustainability of local businesses. While the ID maintained flexibility and service standards for the fashion industry, it struggled to ensure adequate remuneration for all stakeholders, especially weak suppliers. The district also faced the risk of losing specialised textile expertise, as many skilled workers approached retirement without sufficient generational turnover to pass on their knowledge. As a result, the industry has faced the need to adapt and innovate to remain competitive in the global market (Lan, 2014).

In this difficult environment, economic and institutional actors in the district have invested significant resources in CE initiatives to project Prato as a "circular" city and textile district. Annually, about 22,000 tonnes of rags are regenerated in Prato, saving approximately 60 million kWh of energy, 500,000 m^3 of water, and 18,000 tonnes of CO_2 (Prato Chamber of Commerce, 2022). Additionally, an increasing number of companies are investing in Life Cycle Assessment analyses to objectively measure the environmental performance of their production processes and are certifying their products (yarns, fabrics) with "green" labels. These companies are also emphasising their circular initiatives in marketing plans. Local industrial associations and other institutions are lobbying for recognition of Prato's prominent role at the European level in textile waste treatment. Finally, the municipality of Prato has launched several projects focused on sustainability and circularity to promote its image as a "circular" city and district.

5.3.3 *Case Studies*

Company A stands as a premier manufacturer of high-quality fabrics and jerseys. It sells them to leading apparel and fashion brands. The company manages its operations from design to sales, marketing, and stringent quality control, while strategically outsourcing the actual production to specialised suppliers. Over the years, Company A has demonstrated remarkable financial growth and stability within the Prato ID. The revenue has grown from approximately €37.61 million in 2013 to approximately €96.85 million in 2022. This represents a cumulative growth of approximately 157% over the decade. From 2021 to 2022 it has increased 24.8% from €77.62 million to €96.85 million. At the heart of the business model is its dedication to the production of regenerated wool, an area where it has established itself as a pioneer. The company has also started two projects called Project 43 and Project 53, both initiatives aimed at promoting sustainability and CE practices within the textile industry. Project 43 focuses on developing high-quality, environmentally friendly textiles by utilising recycled materials, emphasising the reduction of waste and the use of sustainable fibres. This project aims to create luxurious, sustainable fabrics while maintaining a low environmental impact. Project 53 is dedicated to minimising textile waste by recycling pre-consumer garment offcuts and unsold or second-class knits into new, high-end fabrics. This project leverages Company A's expertise in fibre recycling to produce premium textiles, ensuring that materials are reused efficiently and sustainably, thus contributing to a zero-waste approach in the fashion industry.

These materials are then repurposed as secondary raw materials in the creation of high-quality regenerated wool. To enhance the sustainability of its operations, Company A has meticulously coordinated a closed-loop supply chain. The relationships between Company A and its suppliers are based on long-term agreements that provide mutual benefits. These agreements ensure that suppliers have capacity saturation, even during periods of low demand. They also receive financial support for the upgrade of production equipment to meet the increasing quality standards of Company A. To facilitate coordination with these suppliers and provide visibility to final customers, Company A has recently implemented an IT platform to exchange production, accounting, and environmental data, such as energy and water consumption, with suppliers. This advanced system not only facilitates the recycling of materials but also enables garment makers to monitor the life cycle of their products.

In its commitment to environmental stewardship, Company A has successfully developed and patented processes that eliminate the use of chemicals and dyes. Moreover, the company has crafted an extensive collection featuring over 100 unique colour recipes that require no dye, significantly reducing the environmental impact of traditional colouring methods. The commitment to sustainability is further exemplified by its rigorous Life Cycle Assessment (LCA), which evaluates the environmental impact of its fabrics. Each product is marketed with detailed information on water consumption, CO_2 emissions, and energy usage, providing transparency and fostering consumer trust. Additionally, the wastewater generated through Company

A's production processes is carefully treated and recycled in collaboration with the local district wastewater treatment plant, reinforcing the company's commitment to sustainable practices. In addition, Company A is deeply investing in the education and development of future textile professionals. It has established a corporate academy that serves as a centre for learning and innovation. Here, professionals and students from around the world are invited to immerse themselves in the rich tradition of textile production, while also engaging with contemporary issues such as sustainability and the principles of the CE. This educational initiative not only enhances professional skills but also propagates the values that the company upholds in its quest to redefine industry standards.

Company B is a leading manufacturer of carded yarns and fabrics. Founded in the 1950s, the company has made significant strides over the years in the textile recycling industry, regenerating both pre- and post-consumer textile waste and transforming approximately five million kilograms of textile waste each year into valuable products. In recent years, Company B has embarked on several ambitious circular initiatives aimed at supporting the sustainability of the textile industry. One such initiative is Company B's active participation in the Italian Recycled Textile Association, where it plays a proactive role in preserving the ecosystem of small and micro enterprises that are integral to the collection, sorting, and production of regenerated fibres. This involvement underscores the company's commitment to fostering a supportive network for these crucial industry segments. Additionally, to enhance its environmental impact and operational efficiency, Company B was a founding member of GIDA (Gestione Impianti Depurazione Acque), a consortium dedicated to the reuse of wastewater produced by textile companies in Prato. This consortium is a pivotal part of the local strategy for managing water resources sustainably, ensuring that wastewater is treated and repurposed effectively, thereby reducing the environmental footprint of textile production processes (Gestione Impianti Depurazione Acque, 2024).

Currently, all wet processing companies are part of the consortium. This initiative not only aids in regulatory compliance but also promotes more efficient recycling processes. The company owns and operates specialised carbonisation and water-shredding plants, enabling it to manage the complete transformation process from discarded rags to high-quality regenerated wool fibres. This capability allows Company B to offer a palette of over 250 colours in regenerated fibres, which can be blended to create numerous "no-dye" yarns, offering a wide range of hues and shades without the need for additional dyeing processes. Company B's products are distinguished by a unique system that provides customers with detailed information about the product's composition, the suppliers involved in its production, and an independent third-party assessment of the product's environmental impact, including water, energy, and CO_2 consumption. This transparent approach ensures that customers are fully informed of the environmental footprint of their purchases, reflecting Company B's deep commitment to sustainability and ethical production practices.

Company C is a pioneering apparel and fashion brand start-up that has quickly carved a niche in the sustainable fashion industry. Based in Prato and founded in 2017, the company specialises in designing and selling eco-compatible and fully recyclable garments such as sweaters, gloves, socks, towels, and shawls, primarily made from regenerated fibres. In 2023, Company C produced around 72,000 circular fashion items and participated in substantive social impact projects, advocating for a sustainable model within the fashion industry. The firm extensively utilises recycled fibres such as wool, cashmere, and cotton. This method supports resource conservation by reducing the demand for virgin materials and significantly lowering water and energy consumption. In 2023, 85% of the fibres used were recycled, highlighting C's dedication to minimising its environmental footprint. This commitment to sustainability is not confined to C's product line but extends deeply into its innovative circular business model. Company C actively involves its customers to extend the life cycle of its garments. This engagement is facilitated through various initiatives aimed at extending the life of the garments. Its take-back service, augmented by blockchain for improved traceability, successfully reclaimed 14% of post-consumer textile waste. Furthermore, it offers free repairs and recycling options. Customers are encouraged to return their old cashmere sweaters after a minimum of two years from the purchase date in exchange for a voucher, promoting the idea of longevity rather than fast fashion. Additionally, to reduce packaging waste, it offers reusable and returnable packaging options. Customers who choose to send their packaging back to a designated third-party provider receive a discount voucher, further incentivising sustainable practices.

Company C's products carry significant extra-economic value, recognised by consumers who are willing to pay a premium for garments that not only meet their aesthetic and functional needs, but also contribute positively to the community and environment. It reinforces its community ties through its social responsibility programmes, where for every purchase made, a small amount is donated to a local social project or charity chosen by the customer. The company's design philosophy exemplifies its commitment to eco-design principles, incorporating at least 50% recycled materials into each product and adopting a timeless design to counteract overconsumption. Company C has implemented a pre-order model to synchronise production more closely with market demand, thus reducing waste. Since September 2023, the company has collaborated with "BCome" to incorporate traceability information and transparent data on the environmental impact of its 100 best-selling products into their respective product sheets on the website. This initiative involved testing a Digital Product Passport (DPP) solution at trade fairs, where QR codes were added to the labels of these top-selling products. BCome has developed a comprehensive methodology, regularly updated to comply with European Regulations, for the LCA, Benchmark, and Traceability modules. This enables the research team to rely on a validated Life Cycle Assessment (LCA) study to monitor, quantity, and provide information on the environmental impact of each product. It allows us to benchmark ourselves against industry standards, facilitating continuous improvement and adherence to best practices within our sector (Rifo Lab, 2023).

Company C's success can be attributed to its effective utilisation of local skills from the Prato ID and its ability to tap into the growing market demand for circular products. Company C is committed to producing within a 30 km radius of Prato, Tuscany. This practice significantly reduces transportation emissions and supports the local economy. The company adheres to a production philosophy centred on high-quality craftsmanship using recycled and recyclable materials, ensuring each item's durability. The brand's image is closely linked to its network of local, artisan suppliers. These artisans are not only skilled but are also committed to the values of sustainability, ethics, and transparency, as evidenced by their adherence to a signed code of conduct with Company C. In recognition of their commitment and craftsmanship, these artisans gain visibility and acknowledgement on Company C's website, enhancing their reputational equity. Moreover, Company C is dedicated to reducing inventory levels across the supply chain and improving capacity planning for its suppliers. This is exemplified by its refusal to participate in traditional mass marketing events like Black Friday, opting instead for a pre-sale model that allows for better production scheduling and longer lead times for its craftsmen.

5.4 Analysis

This section synthesises the key findings and directly addresses the research questions, examining the specific obstacles and potential advancements within the realms of sustainability, the CE, and digitalisation, as well as their combined impacts.

5.4.1 RQ1. How Are Italian Textile and Clothing Companies Addressing Sustainability in Their Corporate Strategies and Activities?

The journey towards sustainability is fraught with several critical challenges and substantial opportunities. This discussion aims to highlight the main issues and potential solutions, focusing on the role of digital technologies and specific examples from the Italian T&C industry. One of the foremost issues in the journey towards sustainability T&C industry is the high environmental impact, particularly concerning water and energy usage. Industries globally consume vast amounts of natural resources, leading to significant environmental degradation. Furthermore, the implementation of sustainable technologies often demands substantial financial investments, posing a barrier for many organisations. Companies also struggle with balancing profitability and sustainable practices, as eco-friendly initiatives can sometimes lead to increased costs. Additionally, consumer resistance to higher prices for sustainable products poses a challenge, as many consumers are reluctant to pay a premium for environmentally friendly options.

Despite these challenges, there are substantial opportunities in the sustainability landscape. The implementation of LCA offers a robust method for measuring and reducing environmental impact by analysing the entire life cycle of products. The adoption of green supply chain management and eco-design practices can optimise resource use and minimise waste. Marketing products as green-labelled can significantly enhance a brand's image and build consumer trust. Moreover, there is a growing consumer demand for sustainable products, driven by increased environmental awareness, which can drive market growth. Digital technologies play a crucial role in overcoming sustainability challenges and maximising opportunities. Digital tools for LCA provide detailed data on environmental impact, enabling more accurate assessments and improvements. Digital platforms enhance supply chain management by increasing coordination and efficiency, thereby reducing resource wastage. Blockchain technology provides additional transparency and traceability, making it easier to verify the accuracy of sustainability claims. These claims often include information on a company's use of eco-friendly materials, efforts to reduce environmental impact, and commitment to ethical practices. Furthermore, digital marketing and e-commerce platforms enable businesses to effectively reach and engage sustainability-conscious consumers.

Sustainability in the Italian T&C industry is evident in the integration of eco-friendly practices into corporate strategies and daily operations. Companies like WRÅD are at the forefront, utilising 80% recycled materials in their products and engaging in educational programmes to promote sustainable fashion practices. Their take-back programmes, which have collected over 5,000 garments, demonstrate a commitment to reducing resource consumption and encouraging responsible consumer behaviour. Furthermore, the Prato textile district regenerates approximately 22,000 tonnes of rags annually, saving significant amounts of energy, water, and CO_2 emissions. The district's emphasis on LCA analyses and green certifications further underscores its dedication to measuring and improving environmental performance.

Company A, for example, has eliminated the use of chemicals and dyes, developing over 100 unique colour recipes that significantly reduce environmental impacts. Their rigorous LCA provides transparency on water consumption, CO_2 emissions, and energy usage for each product. Additionally, they collaborate with the local district wastewater treatment plant to recycle wastewater generated during production. Company B's founding membership in GIDA, a consortium dedicated to reusing wastewater, underscores their commitment to sustainable water resource management. GIDA, currently, is a joint-stock company with mixed public and private capital. The consortium is composed of three main partners: the Municipal Administration of Prato, Confindustria Toscana Nord Lucca Pistoia Prato, and Alia Servizi Ambientali S.p.A. Their collaboration with a local Art Institute to produce garments designed for easy repair and regeneration reflects their dedication to eco-design principles. Company C's extensive use of recycled fibres, constituting 85% of their material usage, highlights their commitment to resource conservation and reduced environmental footprint. Their social responsibility programmes and the adoption of a pre-order model to reduce waste further demonstrate their holistic approach to sustainability.

In summary, Italian T&C companies are integrating sustainability into their strategies by addressing environmental challenges, such as resource consumption and waste. They use LCA to measure and reduce impacts, adopt eco-friendly supply chain practices, and leverage digital technologies for transparency and efficiency. Notable initiatives include the use of recycled materials, the Prato district's recycling projects, and focus on improving production methods. These actions demonstrate the industry's commitment to sustainability and its role in promoting sustainable practices.

5.4.2 RQ2. What Strategies and Activities Relating to the CE Are Being Pursued in the Italian Textile and Clothing Industry?

The transition to a CE encounters significant hurdles. Fragmented supply chains with low margins and poor coordination lead to inefficiencies and increased costs. The recycling processes are often complex and expensive, adding to the operational challenges. Additionally, limited infrastructure for textile waste collection and processing hampers effective recycling efforts. Consumer reluctance to participate in recycling or take-back programmes further exacerbates these challenges. However, the CE presents numerous opportunities for innovation and efficiency. Developing closed-loop supply chains can enhance resource efficiency and reduce waste by reusing materials within the supply chain. Establishing consortia for specialised sorting and recycling can optimise recycling processes. Innovative business models, such as rental services and take-back programmes, can extend product life cycles. Engaging consumers through education and incentives can increase participation in recycling programmes, fostering a more sustainable consumption pattern.

Digitalisation significantly aids the CE by providing advanced solutions to its challenges. Advanced data analytics optimise recycling processes by improving efficiency and reducing costs. Digital platforms manage rental services and take-back programmes, enhancing consumer participation and logistical efficiency. Mobile apps and online portals facilitate consumer education and engagement in recycling programmes, making sustainable practices more accessible and appealing.

For instance, Candiani Denim has implemented closed-loop systems, recycling initiatives, and water recovery efforts, significantly reducing its environmental footprint. By reprocessing over 20,000 tonnes of denim waste and cutting water usage by 50% per pair of jeans, Candiani exemplifies effective waste management and resource utilisation. The implementation of CE principles is prominently seen across multiple companies within the Prato ID. Company A has established itself as a pioneer in producing regenerated wool by setting up a dedicated internal unit for collecting and repurposing textile waste. This unit gathers scraps from various production stages and converts them into secondary raw materials, effectively closing the loop in its supply chain. This initiative not only reduces waste but also supports a sustainable production cycle. Similarly, Company B transforms approximately five million kilograms of textile waste annually into valuable products, emphasising its role in the textile recycling industry. Their ownership of specialised carbonisation and water-shredding plants allows them to manage the entire process of transforming discarded textiles into high-quality regenerated fibres, offering a palette of over 250 no-dye yarns. Company C's circular business model includes a take-back service enhanced by blockchain technology, enabling the reclamation of post-consumer textile waste and offering free repairs to extend the life cycle of its garments. Their approach to reusable packaging and the pre-order model further reduces waste and promotes sustainable consumption patterns. While the transition to a CE presents significant challenges, the strategic adoption of innovative practices and digital technologies offers substantial opportunities for improving resource efficiency, reducing waste, and promoting sustainable consumption. The examples from Candiani Denim, Gucci, and the Prato ID illustrate how these principles can be effectively implemented to achieve a more sustainable future.

Overall, Italian T&C companies are pursuing various strategies and activities related to the CE to enhance sustainability and resource efficiency. Key strategies include developing closed-loop supply chains, as demonstrated by Candiani Denim, which recycles denim waste and reduces water usage. The Prato industrial district exemplifies effective CE implementation with companies like Company A and B, focusing on regenerating wool and transforming textile waste into high-quality fibres. These efforts include specialised facilities for processing waste and innovative business models like take-back services and reusable packaging. Digital technologies, such as blockchain, are also utilised to track and manage these processes, ensuring transparency and efficiency. These initiatives highlight a strong commitment to CE principles, promoting waste reduction, resource reuse, and sustainable consumption patterns.

5.4.3 *RQ3. What Role Are Digital Technologies Playing in the Transition to Sustainability and the CE in the Italian Textile and Clothing Industry?*

Digital transformation itself is not without its own set of challenges. Limited digital literacy and technological adoption among small and micro-sized enterprises hinder the widespread implementation of digital tools. The costs associated with digital transformation can be prohibitive for many organisations. Resistance to change within traditional business models poses a significant barrier, as many businesses are reluctant to move away from established practices. Additionally, data privacy and security concerns are critical challenges that need to be addressed to protect sensitive information. Despite these obstacles, digitalisation offers numerous opportunities for enhancing industrial processes. Improved traceability and transparency across supply chains can lead to better oversight and management. Enhanced coordination and data sharing among suppliers can streamline operations. The implementation of DPP provides comprehensive product information, enhancing transparency and consumer trust. The use of advanced data analytics drives sustainability and efficiency improvements by analysing complex data sets to inform decision-making. To address these challenges and leverage opportunities, digital solutions are crucial. Training programmes and user-friendly digital tools can increase adoption among small enterprises. Cloud-based solutions and scalable digital tools reduce costs and improve accessibility, making digital transformation more feasible for businesses of all sizes. Blockchain and IoT create secure, immutable records of product data, ensuring data integrity and transparency. Robust cybersecurity measures and data governance frameworks are essential to protect sensitive information and ensure compliance with data privacy regulations. The role of digitalisation in advancing the CE and sustainability in the T&C industry cannot be overstated.

The Circular Hub by Gucci, supported by Kering, represents a significant step towards integrating digital solutions for sustainable practices (Kering Group, 2022). This hub focuses on optimising raw material use and enhancing product durability, repairability, and recyclability through advanced technological and digital innovations. By leveraging tools such as blockchain technology and the Internet of Things (IoT), the hub aims to improve traceability, optimise resource use, and develop innovative recycling processes. This initiative aligns with broader environmental goals, such as Italy's National Recovery and Resilience Plan and the EU's 2030 emissions reduction targets, highlighting the synergy between digital innovation and sustainability. Digitalisation plays a crucial role in enhancing sustainability and transparency within these companies. Company A has implemented an IT platform to exchange production, accounting, and environmental data with suppliers, facilitating better coordination and visibility in the supply chain. This platform allows garment makers to monitor the life cycle of their products, contributing to more informed and sustainable production decisions. Company C leverages blockchain technology in its take-back service, enhancing traceability and transparency. Their deployment of the DPP offers consumers insights into the product's origins and environmental impacts, promoting informed purchasing decisions. While digital transformation poses significant challenges, it also offers substantial opportunities for enhancing industrial processes and advancing sustainability. Through strategic adoption of digital tools and technologies, businesses can improve traceability, transparency, and efficiency, driving progress towards a more sustainable future. The examples from these companies illustrate the critical role of digitalisation in achieving these goals.

5.5 Discussion: Integrating Sustainability, the CE, and Digitalisation.

The integration of sustainability and CE principles into existing digital infrastructures presents a complex challenge. Ensuring that all stakeholders are aligned with sustainability and CE goals requires effective communication and consensus-building.

Managing the transition period where old and new systems coexist can be challenging, as it requires careful planning and execution to minimise disruptions. Additionally, ensuring regulatory compliance and keeping up with evolving standards adds another layer of complexity to this integration process. However, the integration of these principles with digitalisation also opens up significant opportunities.

Leveraging digitalisation can scale sustainable and CE practices across industries, leading to more widespread and impactful changes. Enhanced collaboration through digital networks and platforms fosters innovation and collective problem-solving. There are numerous opportunities for innovation in product design, supply chain management, and consumer engagement that arise from this integration. Streamlining compliance and reporting processes through digital solutions can reduce administrative burdens and improve efficiency. Digital ecosystems that support sustainable and circular practices across the value chain are essential for successful integration. Digital collaboration tools facilitate stakeholder engagement and collective decision-making, ensuring that all parties are aligned with the goals. Real-time data monitoring and analytics optimise the transition process, minimising disruption and ensuring a smooth transition. Automated compliance management systems and real-time regulatory updates ensure effortless compliance, allowing businesses to focus on innovation and growth.

Table 5.4 summarises the multifaceted challenges and opportunities presented within sustainability, CE, and digitalisation, emphasising the crucial role that digital technologies play in transforming these areas. By leveraging digital tools and platforms, industries can overcome significant barriers, optimise processes, and drive towards a more sustainable and efficient future. The integration of these principles not only enhances environmental and economic performance but also fosters innovation, collaboration, and resilience in an ever-evolving industrial landscape.

Nevertheless, companies also have been investing in education and community development, further promoting sustainable practices. Company A has established a corporate academy that serves as a centre for learning and innovation in textile production, engaging professionals and students in sustainability and CE principles. This initiative not only enhances professional skills but also propagates the company's values. Company C's commitment to social responsibility is evident in its donation programmes linked to each purchase, supporting local social projects and charities. Their refusal to participate in mass marketing events like Black Friday, opting instead for a pre-sale model, aligns with their sustainable production philosophy and supports better capacity planning for local artisans. Companies within the Prato ID exemplify the successful integration of CE, sustainability, and digitalisation practices. Their innovative approaches to waste management, eco-friendly production processes, transparent operations, and community engagement set a benchmark for the textile and fashion industry's transition towards more sustainable and responsible practices. The Italian T&C industry's adoption of CE, sustainability, and digitalisation practices is driving significant advancements in environmental responsibility and economic resilience. Through innovative strategies, collaborative efforts, and technological advancements, the industry is setting a benchmark for sustainable fashion practices globally.

Table 5.4 Sustainability, circular economy, and digitalisation challenges and opportunities for the Italian Textile and Clothing industry

Aspect	Challenges	Opportunities	Digitalisation assistance
Sustainability	High environmental impact (water, energy usage)	Implementation of LCA to measure and reduce environmental impact	Digital tools for LCA providing detailed environmental impact data
	Significant investment required for sustainable technologies	Enhance relationship with stakeholders	Digital platforms enhancing supply chain management coordination and efficiency
	Balancing profitability with sustainable practices	Marketing green-labelled products to enhance brand image and consumer trust	Blockchain for transparency and traceability, ensuring authenticity of sustainability claims
	Consumer resistance to higher costs of sustainable products	Growing consumer demand for sustainable products driving market growth	Digital marketing and e-commerce platforms to reach sustainability-conscious consumers
Circular economy (CE)	Fragmented supply chains with low margins and poor coordination	Development of closed-loop supply chains to enhance resource efficiency and reduce waste	IoT for tracking and managing material flows within closed-loop systems
	High costs and complexity in recycling processes	Establishment of consortia for specialised sorting and recycling	Advanced data analytics for optimising recycling processes and improving efficiency
	Limited infrastructure for textile waste collection and processing	Creation of innovative business models like rental services and take-back programmes to extend product life cycles	Digital platforms for managing rental services and take-back programmes, enhancing consumer participation and logistical efficiency
	Consumer reluctance to participate in recycling and take-back programmes	Engaging consumers in sustainable practices through education and incentives	Mobile apps and online portals for consumer education and engagement in recycling programmes
Digitalisation	Limited digital literacy and technological adoption among small and micro-sized enterprises	Improved traceability and transparency across supply chains	Training programmes and user-friendly digital solutions to increase adoption among small enterprises

(continued)

Table 5.4 (continued)

Aspect	Challenges	Opportunities	Digitalisation assistance
	High costs associated with digital transformation	Enhanced coordination and data sharing among suppliers	Cloud-based solutions and scalable digital tools reducing costs and improving accessibility
	Resistance to change within traditional business models	Implementation of DPP to provide comprehensive product information	Blockchain and IoT for creating secure, immutable records of product data
	Data privacy and security concerns	Use of advanced data analytics to drive sustainability and efficiency improvements	Cybersecurity measures and robust data governance frameworks to protect sensitive information
Combined sustainability, CE, and digitalisation	Integrating sustainability and CE principles into existing digital infrastructures	Leveraging digitalisation to scale sustainable and CE practices across the industry	Development of integrated digital ecosystems supporting sustainable and circular practices across the value chain
	Ensuring all stakeholders are aligned with sustainability and CE goals	Enhanced collaboration and innovation through digital networks and platforms	Digital collaboration tools facilitating stakeholder engagement and collective decision-making
	Managing the transition period where old and new systems coexist	Opportunities for innovation in product design, supply chain management, and consumer engagement	Real-time data monitoring and analytics to optimise the transition process, ensuring minimal disruption
	Ensuring regulatory compliance and keeping up with evolving standards	Streamlining compliance and reporting processes, reducing administrative burdens	Automated compliance management systems and real-time regulatory updates for minimal effort compliance

5.6 Conclusions

The Italian T&C industry stands at a critical juncture, balancing its rich artisanal legacy with the pressing need to adopt sustainable, circular, and digital practices. This chapter has explored the multifaceted efforts of companies and IDs within Italy, particularly focusing on the integration of sustainability and CE principles and the role of digitalisation in facilitating these transitions. Through an examination of empirical evidence from grey literature and case studies, this chapter underscored the ongoing transition towards sustainability and circularity within the Italian T&C industry. The research highlighted the critical challenges and opportunities in this transition, such as high environmental impact, significant investment requirements, fragmented supply chains, and consumer resistance to sustainable products (Arpizio et al., 2022). Despite these challenges, opportunities for innovation, enhanced transparency, and improved supply chain efficiency through digital technologies were identified. Digital technologies play a crucial role in aiding this transition. By leveraging digital tools to facilitate sustainability and CE practices, the Italian T&C industry can overcome challenges and capitalise on opportunities for growth and innovation. Additionally, blockchain and digital platforms enhance supply chain transparency and efficiency, while DPP and advanced data analytics are promising in supporting sustainability and CE practices.

The integration of sustainability, CE, and digitalisation is exemplified by successful case studies from companies like Candiani Denim, WRÅD, Gucci, and the Prato ID. These examples highlight how digital tools and innovative practices can drive sustainable and circular economies, setting benchmarks for the industry. Prato, known for its extensive textile production, has embraced eco-innovation practices to mitigate environmental impacts and business risks. The district's initiatives, such as the regeneration of textile waste and investment in green labels, showcase a proactive approach to sustainability. Additionally, the involvement of local industrial associations and the municipality in promoting Prato as a "circular" city and district demonstrates a collaborative effort to achieve environmental goals at a regional level.

For managers in the Italian T&C industry, the findings of this chapter have several important implications. Managers should prioritise investments in digital technologies that enhance supply chain transparency and efficiency. Tools like blockchain, digital platforms, and advanced analytics can drive significant improvements in sustainability practices. Integrating CE principles requires a strategic approach to supply chain management, emphasising closed-loop systems and recycling initiatives. Managers should collaborate with industry stakeholders to build a robust recycling infrastructure and promote consumer education on sustainability. Effective implementation of sustainability and CE practices necessitates collaboration among various stakeholders, including local industrial associations, municipalities, and educational institutions. Managers should foster partnerships that support eco-innovation and sustainable practices. To mitigate consumer resistance to sustainable products, managers should focus on marketing strategies that highlight the long-term benefits of sustainability. Transparent communication about the environmental impact of products and the use of green labels can build consumer trust and drive demand. The dynamic nature of digital technologies and sustainability practices requires continuous learning and adaptation. Managers should invest in training programmes to enhance digital literacy and keep abreast of the latest advancements in sustainability and CE.

Lastly, it is important to acknowledge the limitations of this study. The research relied heavily on the District of Prato in the Tuscany region as a case study, analysing three cases from a scientific perspective in conjunction with grey literature. Future research should investigate other districts or conduct comparative analyses of different districts and their strategies related to sustainability and CE. The journey towards establishing a sustainable CE in the Italian T&C industry is rife with challenges. However, the integration of sustainability, CE, and digitalisation holds the promise of substantial opportunities. By leveraging innovative practices and digital technologies, the industry can achieve environmental and economic progress, creating a more sustainable and inclusive future. Insights from the Prato industrial cluster's initiatives, digital mapping projects, and diverse community experiences can guide stakeholders in fostering a thriving ecosystem that values diversity, promotes innovation, and upholds sustainability principles.

Acknowledgements This study was carried out within the MICS (Made in Italy—Circular and Sustainable) Extended Partnership and received funding from the European Union Next-Generation EU (PIANO NAZIONALE DI RIPRESA E RESILIENZA (PNRR)—MISSIONE 4 COMPONENTE 2, INVESTIMENTO 1.3—D.D. 1551.11-10-2022, PE00000004). This manuscript reflects only the authors' views and opinions, neither the European Union nor the European Commission can be considered responsible for them.

References

Alves, L., Cruz, E. F., Lopes, S. I., Faria, P. M., & da Cruz, A. M. R. (2021). Towards circular economy in the textiles and clothing value chain through blockchain technology and IoT: A review. *Waste Management & Research the Journal for a Sustainable Circular Economy, 40*(1), 3–23. https://doi.org/10.1177/0734242x211052858

Arpizio, C. D., Levato, F., Capellini, M., Flammini, B., Luthra, P., & Improta, G. (2022). Brands can accelerate and capitalize on this opportunity as more fashion consumers become champions for sustainability. How brands can embrace the sustainable fashion opportunity.

Becattini, G. (2017). The Marshallian industrial district as a socio-economic notion. In *Revue d'Economie Industrielle* (Vol. 157, Issue 1, pp. 13–32). Editions Techniques et Economiques. https://doi.org/10.4000/rei.6507

Belfanti, C. M. (2015). Renaissance and made in Italy: Marketing Italian fashion through history (1949–1952). *Journal of Modern Italian Studies, 20*(1), 53–66. https://doi.org/10.1080/1354571X.2014.973154

Bellini, A., & Leonardi, L. (2020). Prato: the social construction of an industrial city facing processes of cultural hybridization. 547–570. https://doi.org/10.1007/978-3-658-30956-5_30

Bressanelli, G., Visintin, F., & Saccani, N. (2022). Circular economy and the evolution of industrial districts: A supply chain perspective. *International Journal of Production Economics, 243*. https://doi.org/10.1016/j.ijpe.2021.108348

Candiani Denim. (2024, June 12). *Candiani Denim - Sustenibilitá*. https://www.candianidenim.com/it/sostenibilita

Ceccagno, A. (2017). *City making and global labor regimes: Chinese immigrants and Italy's fast fashion industry*. Springer.

Chae, Y.-M., & Hinestroza, J. P. (2020). Building circular economy for smart textiles, smart clothing, and future wearables. *Materials Circular Economy, 2*(1). https://doi.org/10.1007/s42824-020-00002-2

Chen, X., Memon, H. A., Wang, Y., Marriam, I., & Tebyetekerwa, M. (2021). Circular economy and sustainability of the clothing and textile industry. *Materials Circular Economy, 3*(1). https://doi.org/10.1007/s42824-021-00026-2

Colombi, C. (2015). Made in Italy: Rethinking a century of Italian design. *Journal of Modern Italian Studies, 20*(1), 140–142. https://doi.org/10.1080/1354571x.2015.974912

Colucci, M., & Vecchi, A. (2021). Close the loop: Evidence on the implementation of the circular economy from the Italian fashion industry. *Business Strategy and the Environment, 30*(2), 856–873. https://doi.org/10.1002/bse.2658

Comune di Prato. (2023). *Next Generation Prato Documento di indirizzo per il Recovery Fund.*

Confindustria Moda, & Sistema Moda Italia. (2023). *L'industria tessile-abbigliamento in Italia - il quadro generale.* www.sistemamodaitalia.it

Confindustria Toscana Nord. (2024). *Il distretto tessile-moda pratese in pillole.* https://www.confindustriatoscananord.it/media/UFFICIO_STUDI/2024_04_Distretto_PO_pillole.pdf

Coppola, C., Vollero, A., & Siano, A. (2023). Developing dynamic capabilities for the circular economy in the textile and clothing industry in Italy: A natural-resource-based view. *Business Strategy and the Environment, 32*(7), 4798–4820. https://doi.org/10.1002/bse.3394

Del Bianco, C. (2024). Generating modern heritage through changing urban environments and identities: A case study from Prato's (Italy) industrial district, history, and multiculturalism in a polycentric urban setting. *Curator, 67*(1), 309–328. https://doi.org/10.1111/cura.12591

Dominidiato, M., Guercini, S., Milanesi, M., & Tunisini, A. (2023). Supplier-customer relationships for sustainability-led innovation in the textile industry. *Journal of Business and Industrial Marketing, 39*(13), 15–26. https://doi.org/10.1108/JBIM-01-2023-0060

Dursun, E., Ulker, Y., & Gunalay, Y. (2022). Blockchain's potential for waste management in textile industry. *Management of Environmental Quality an International Journal, 34*(4), 1174–1197. https://doi.org/10.1108/meq-03-2022-0085

EURATEX. (2024). *Facts & key figures 2024.* www.euratex.eu

European Commission & Centre for Industrial Studies (CSIL). (2021). *Data on the EU textile ecosystem and its competitiveness—Final report.*

Gestione Impianti Depurazione Acque. (2024, June 12). *GIDA consortium.* https://www.gida-spa.it

Gucci, & Kering Group. (2023). *Gucci, supported by Kering, invests in first circular hub to power a "circular made in Italy."* https://www.kering.com/en/news/gucci-supported-by-kering-invests-in-first-circular-hub-to-power-a-circular-made-in-italy/

Guercini, S., Milanesi, M., & Ottati, G. D. (2017). Paths of evolution for the Chinese migrant entrepreneurship: A multiple case analysis in Italy. *Journal of International Entrepreneurship, 15*(3), 266–294. https://doi.org/10.1007/s10843-017-0209-0

Hamilton, G. G., & Fels, D. (2014). The social sources of migration and enterprise: Italian peasants and Chinese migrants in Prato. *East Asia, 31*(4), 269–287. https://doi.org/10.1007/s12140-014-9219-x

Islam, M. T., Jahan, R., Jahan, M., Howlader, Md. S., Islam, R., Islam, Md. M., Hossen, Md. S., Kumar, A., & Robin, A. H. (2022). Sustainable textile industry: An overview. *Non-Metallic Material Science, 4*(2), 15–32. https://doi.org/10.30564/nmms.v4i2.4707

ITMA Services. (2022). *The Italian textile industry a key economic sector.*

Kering Group. (2022). *Sustainability progress report 2020–2023—Kering Group.*

Kreci, M., & Enegbuma, S. A. (2023). *Sustainability transition in the fashion industry a multi-case study of manufacturers in the Prato textile Industrial District (Italy).*

Lan, T. (2014). Industrial district and the multiplication of labour: The Chinese apparel industry in Prato, Italy. *Antipode, 47*(1), 158–178. https://doi.org/10.1111/antl.12104

Lazerson, M. H., & Lorenzoni, G. (1999). *The firms that feed industrial districts: A return to the Italian source.*

Mazzoni, F. (2021). Insights into regional development. *Hal, 2020*(3), 661–676. https://doi.org/10.9770/ird.2020.2.3(4)

Mckinsey & Company, & The Business of Fashion. (2024). *The state of fashion 2024.*

Montagna, G., Antonietta Sbordone, M., Esposito, A., Marques, A., Amato, I., & Gaspar, R. (2022). New territories of sustainable fashion proposals for the transition towards sustainability and circularity. *Human Factors for Apparel and Textile Engineering, 32.* https://doi.org/10.54941/ahfe1001546

Motta, M., Conti, G. M., Lo Scocco, G., & Didero, R. (2024). Craftmanship and digitalization in the Italian Knitwear Industry. A paradigm shift for the narrative of made in Italy. In F. Zanella,

G. Bosoni, E. Di Stefano, G. L. Iannilli, G. Matteucci, R. Messori, & R. Trocchianesi (Eds.), *Multidisciplinary aspects of design* (pp. 705–713). Springer Nature Switzerland.
Naidoo, D., Nhamo, L., Lottering, S., Mpandeli, S., Liphadzi, S., Modi, A. T., Trois, C., & Mabhaudhi, T. (2021). Transitional pathways towards achieving a circular economy in the water, energy, and food sectors. *Sustainability, 13*(17), 9978. https://doi.org/10.3390/su13179978
Palm, C., Cornell, S. E., & Häyhä, T. (2021). Making resilient decisions for sustainable circularity of fashion. *Circular Economy and Sustainability, 1*(2), 651–670. https://doi.org/10.1007/s43615-021-00040-1
Peters, J., & Simaens, A. (2020). Integrating sustainability into corporate strategy: A case study of the textile and clothing industry. *Sustainability, 12*(15), 6125. https://doi.org/10.3390/su12156125
Prato Chamber of Commerce. (2022). The economic situation in the provinces of Pistoia and Prato.
Rifo Lab. (2023). Sustainability report rifo. https://rifo-lab.com/en/pages/moda-sostenibile
Saccani, N., Bressanelli, G., & Visintin, F. (2023). Circular supply chain orchestration to overcome circular economy challenges: An empirical investigation in the textile and fashion industries. *Sustainable Production and Consumption, 35*, 469–482. https://doi.org/10.1016/j.spc.2022.11.020
Sandvik, I. M., & Stubbs, W. (2019). Circular fashion supply chain through textile-to-textile recycling. *Journal of Fashion Marketing and Management, 23*(3), 366–381. https://doi.org/10.1108/JFMM-04-2018-0058
Santos, E. (2021). How does the pandemic has affected the economic sustainability of textile industry in France? *SHS Web of Conferences, 129*, 09017. https://doi.org/10.1051/shsconf/202112909017
The European House—Ambrosetti & Confindustria Venezia. (2022). *Just fashion transition.*
Weibel-Orlando, J. (2012). *The utility of employing integrated research analysis paradigms: A case study of the textile industry in Prato, Italy. 41*, 1–42.
Wrad. (2024, June 12). *Wrad living*. https://www.wradliving.com

Guilherme Hörner Bussolo (guilherme.hornerbussolo@unifi.it) is a Ph.D. candidate in Industrial Engineering at the University of Florence, Italy, where he conducts research at the IBIS Lab, specialising in the Circular Economy within the textile industry. His research involves analysing EU legislation on Circular Economy and textiles, assessing the current state of Italian textile companies, and developing Circular business models to enhance supply chain competitiveness. Additionally, he collaborates with the Technical University of Denmark in Copenhagen on research focused on Product/Service Systems in the textile and clothing industry. With over eight years of global experience spanning three continents, Guilherme has held key roles in operations and procurement. He has a strong background in Lean management and Green Supply Chain Management, with expertise in optimising processes and integrating sustainable practices into industrial operations. Guilherme holds a Master's degree in Logistics Engineering from the Budapest University of Technology and Economics, Hungary, a Green Belt certification from the Pontifical Catholic University of Paraná, Brazil, and a Bachelor's degree in Logistics from the University of the Itajaí Valley, Brazil. He is dedicated to advancing sustainability in industry through both research and practical application.

Gianmarco Bressanelli (gianmarco.bressanelli@unibs.it) is a Post-Doctoral researcher who works at the RISE Laboratory (Research and Innovation for Smart Enterprises), at the University of Brescia (Italy). Gianmarco does research in Circular Economy, Sustainability, and Digitalisation, focusing on their implications on Operations and Supply Chain Management, combining case study research with simulation and assessment methodologies. He holds a Ph.D. in Mechanical and Industrial Engineering. His research has been presented at national and international conferences, as well as published in national and international journals. He teaches in the courses of Production Systems, Industrial Plants Design, and Supply Chain Management at the University of Brescia. He also manages and carries out several innovation projects with companies and dissemination activities in the fields of Circular Economy, Operation, and Supply Chain Management.

Filippo Visintin (visintin.filippo@unifi.it) is an Associate Professor of Service Design and Management at the School of Engineering of Florence University. He is the coordinator of the International Master's Degree Programme in Management Engineering of the University of Florence and the Scientific Director of the Information-Based Industrial Services Laboratory. He is the national coordinator of the REsilient, SusTainable, and circulAr leatheR and Textile supply chains (RESTART) project. He is co-founder and co-owner of SmartOperations srl, a University of Florence's spin-off now part of the Digital360 group. His research interests include servitisation of manufacturing, healthcare operations management, discrete-event simulation, and circular economy. He is the author of several research papers published in prestigious international journals such as the European Journal of Operational Research, OMEGA, Industrial Marketing Management, International Journal of Production Economics, Computers in Industry, Computers and Industrial Engineering, Flexible Service and Manufacturing Journal, Management Decision, Journal of Intelligent Manufacturing, Production Planning and Control, Sustainable Production and Consumption, and IMA Journal of Management Mathematics. He served as guest editor for Management Decision and Flexible Service and Manufacturing Journal.

Nicola Saccani (nicola.saccani@unibs.it) is an Associate Professor of Operations Management at the Department of Mechanical and Industrial Engineering of the University of Brescia, Italy. He is a senior member of the Research and Innovation for Smart Enterprises (RISE) Laboratory. He holds a Ph.D. in Design and Management of Logistics and Production Systems. His research activity concerns supply chain management and Service Management. He coordinates the ASAP Service Management Forum Initiative (http://www.asapsmf.org).

Chapter 6
Romania

Mădălina Dumitru and Voicu D. Dragomir

Abstract This chapter investigates the sustainability strategies formulated in relation to circular economy (CE) practices, and the digital technologies supporting them, in the clothing retail sector in Romania. The study is motivated by the high impact on consumer preferences of large clothing retailers through their physical stores. The research adopts a triangulated methodology, with data being collected through a combination of field research, content analysis, and website/app functionality investigation, in 15 retailers that have a significant presence in Romania and worldwide. The data was analysed using an original evaluation sheet including 23 criteria. The findings indicate that major clothing retailers in Romania apply the CE principles in a very selective way. Some practices (such as labelling recycled fibers/materials for individual items) are present in most stores, while others are absent in most cases (e.g., returning used/old clothes for shopping credit). There are significant differences between companies' CE practices in clothing retail in Romania. Regarding the use of digital technologies to support CE practices, the findings indicate certain aspects that can be improved upon by the brands included in the sample, such as the availability of targeted loyalty programs, apps, surveys, and social media messages.

Keywords Circular economy practices · Clothing retail sector · Digital technologies · Sustainability strategies · Field research · Content analysis

6.1 Introduction

The textile industry is the second largest user of natural resources of all sectors, after the food industry (Paras et al. 2019). As such, adopting circularity principles across the entire value chain, from raw materials producers to product buyers, is imperative

M. Dumitru · V. D. Dragomir (✉)
Bucharest University of Economic Studies, Bucharest, Romania
e-mail: voicu.dragomir@cig.ase.ro

M. Dumitru
e-mail: madalina.dumitru@cig.ase.ro

M. Wynn and T. Wiegand (eds.), *Sustainability, the Circular Economy and Digitalisation in the European Textile and Clothing Industry*, SDGs and Textiles,
https://doi.org/10.1007/978-981-97-9116-3_6

to reduce the environmental footprint (Dragomir & Dumitru, 2024). One of the least discussed aspects in this domain is the implementation of circular economy (CE) practices in the retail and consumer use stages. This chapter focuses on the sustainability, CE, and digitalisation options adopted by clothing stores in Romania.

Romania, one of the youngest members of the European Union (EU), is a former communist country, and its history still influences the sustainability actions and practices of its companies. After the fall of communism, the business environment was characterised by a lack of interest in sustainability aspects in the 1990s (Albu et al., 2021). Then, in the new millennium, the first sustainability reports were issued by local subsidiaries of multinational companies (UNDP et al., 2007). The most important driver of sustainability in Romania is regulation—or more specifically the law implementing the EU Non-Financial Reporting Directive (Directive 2014/95/EU, 2014), which became effective in 2017. After the implementation of the Directive, companies started focusing more on the information needs of a larger number of stakeholders, implementing social responsibility practices, and reporting on their internal operations (Albu et al., 2021).

One dimension of corporate responsibility is the CE, which is defined as "an industrial system that is restorative or regenerative by intention and design, [that] replaces the *end-of-life* concept with reintegration, eliminates the use of toxic chemicals, which impair reuse, and aims for the elimination of waste through the superior design of materials, products, systems, and, within this, business models" (Ellen MacArthur Foundation, 2013, p. 7). The challenges related to the CE in the textile and clothing sector in Romania may be grouped into the following categories: legislation, waste management, workforce, and local raw materials production (Staicu & Pop, 2018). Romania represents a promising market for used clothes, due to the fact that, in this country, "the at-risk-of-poverty rate remains among the highest in the EU, reaching 22.6% in 2020" (World Bank, 2023, p. 1). Romania is also the largest importer of used clothes in Europe (Paras et al., 2019), which means that most of the second-hand clothing is not collected in the country. Moreover, the recycling and reprocessing of textiles in Romania are low due to the lack of infrastructure, technologies, and awareness of the public (Paras et al., 2019).

Digital technologies are of relevance to the CE because they "can track the journeys of products, components and materials and make the resulting data securely accessible" (European Commission, 2020, p. 17). Digitalisation is one of the "trends emerging towards the CE" (Koszewska, 2018, p. 346). The European Commission (2020, p. 2) considers that the CE can be "powered by digital technologies, such as the internet of things, big data, blockchain and artificial intelligence". In the textile and clothing sector, CE-related digital technologies are product passports (Centobelli et al., 2022), digital receipts (Sandvik & Stubbs, 2019), 3D technologies (Bürklin & Wynants, 2020), customisation process technologies (Dissanayake, 2020), QR codes for smartphone apps (Wiegand & Wynn, 2023), tagging and watermarks (European Commission, 2020), and blockchain and the internet of things (Alves et al., 2022). Retailers operating in Romania indicate their willingness to focus on digitising their business processes to enable new, sustainable business models (Purcărea et al., 2022). For instance, Zara considers that its website is "supporting sustainability as a continuous project" (Gheorghe & Matefi, 2021, p. 1).

Facing coercive institutional pressures, Romanian companies have become more concerned with their sustainability-related actions. Retailers, however, still fail to send coherent and consistent messages regarding their sustainability practices to customers (Purcărea et al., 2022), even though Romanian consumers have a positive attitude towards environmental protection (Lakatos et al., 2016; Orzan et al., 2018). The aim of this chapter is to provide an analysis of the CE practices adopted by clothing stores present in Romania. The chapter covers three main areas—sustainability strategies, CE practices, and digitalisation—and addresses the following research questions (RQs):

RQ1 What sustainability strategies and CE practices are implemented in the clothing retail sector in Romania?

RQ2 How are CE practices signaled on the labels available in stores for customers in Romania?

RQ3 How is digitalisation impacting the clothing retail sector in Romania and how are these technologies supporting the transition to the CE?

This chapter is divided into five sections. Following this introductory section, the research methodology is outlined: the investigative method is supported by content analysis. Next, the findings are set out, addressing the three RQs. A discussion section then assesses the sustainability and CE practices identified in the Romanian clothing stores. A final concluding section summarises the main themes discussed in the chapter and identifies possible areas for future research.

6.2 Research Method

The aim of the chapter is to provide an analysis of the sustainability strategies, CE practices, and digitalisation adopted by clothing stores in Romania. The sample includes 15 stores that have a significant presence in Romania and worldwide. They were selected following initial visits to the biggest malls in Bucharest. The capital city of Romania has a total mall retail space of 1.24 million square meters and this type of retail is expanding (Ungureanu, 2023). For each parent company only one brand (store) was selected. For instance, for Inditex (Spain), the main brand (Zara) was selected and analysed in the store located in AFI Cotroceni Mall (the largest in Romania). All the sample stores sell their own brand, except for Peek & Cloppenburg, which sells both its own brand and multi-brand. Premium or luxury clothing retail was not analysed. Stores were selected after a visual inspection of the map of each mall. The largest stores were identified and included in the sample. Data regarding the sample is presented in Table 6.1. The reliability of the data collected is assured by the fact that the authors visited eight of the stores together and the information on digitalisation was cross-checked.

Table 6.1 Brands and stores in the research sample

Brand	Parent	Brand stores worldwide	Store in Bucharest
Adidas	Adidas (Germany)	13,000	Băneasa
C&A	C&A (Germany)	1300	AFI Cotroceni
Coccodrillo	CDRL SA (Poland)	500	Băneasa
Desigual	Desigual (Spain)	425	Băneasa
H&M	H&M (Sweden)	4465	Unirii
Mango	Mango (Spain)	2600	AFI Cotroceni
Marks & Spencer	Marks & Spencer (UK)	1487	Plaza Romania
Mothercare	Mothercare (UK)	700	Băneasa
New Yorker	New Yorker (Germany)	1200	AFI Cotroceni
Next	Next plc (UK)	700	Băneasa
Nike	Nike (USA)	1032	Băneasa
North Face	VF Corp. (USA)	370	Băneasa
Peek & Cloppenburg	Peek & Cloppenburg (Germany)	140	AFI Cotroceni
Sinsay	LPP (Poland)	737	Drumul Taberei
Zara	Inditex (Spain)	2047	AFI Cotroceni

The methodology is a combination of field research, content analysis, and website/app functionality investigation, covering the three themes of sustainability strategy, CE practices, and digitalisation. This mixed method also allowed the authors to validate the information from one environment (the physical store) with other media (apps and websites).

Based on an analysis of the CE literature with a focus on fashion retail (Centobelli et al., 2022; Dragomir & Dumitru, 2022; Geegamage et al., 2023; Hultberg & Pal, 2021; Pal & Jayarathne, 2022; Sandberg, 2023; Sugg, 2022), the authors created an evaluation sheet (Table 6.2). The S (sustainability) section includes eight criteria regarding the sustainability and circularity information visible in physical stores (on posters and window ads) or websites. The L (labels) section includes five criteria which were validated by checking the products labels. The D (digitalisation) section includes seven criteria related to the CE and digitalisation. The R (Romanian) section is related to the availability of the retailer's labels, app, and website in Romanian (3 criteria). Information translated to Romanian has the highest impact on the Romanian consumer, especially when the information goes beyond the size and price of the item sold. The overall instrument includes 23 criteria. Thus, the data was collected from a variety of sources relevant to the research questions. All items included in the evaluation sheet were quantified as 1 or 0. For some items, additional information was obtained by the authors (for example, the organisations with which the brands had partnerships or collaborations for CE practices). The data were collected in early January 2024.

To answer RQ1, the authors visited the physical stores. When visiting the stores, the authors analysed the information available, for instance, on billboards posted on the walls. Next, they checked the products and the labels. Afterwards, the publicly available information on posters or via other written materials and information sheets, such as the ones regarding the environmental impact, were analysed. In addition, they asked the shop assistants questions for further clarifications.

As regards RQ2, practices include, for example, the return policies, repairability of the products, and the use of recycled fibers. The information needed to answer this RQ was collected from the physical stores, especially from the labels attached to each item, but also from more general announcements available in stores (at the check-out desk).

To answer RQ3, the authors collected data from the physical stores, for instance, by accessing the QR codes available on the products' labels and searching for digital information. Additionally, the brands' mobile apps were installed and investigated for evidence regarding customer incentives linked to CE practices. The authors searched for information regarding the transition to the CE on the website. This method allowed the authors to see how the strategies are implemented at the national level. Instagram is the most influential social media platform for young people who are fashion-oriented. The most recent 100 posts on the official Instagram account of each company were checked (approximately three months of social media activity) to see if the platform is used to promote CE practices. Further information was collected on whether apps and websites are in Romanian. The methodology and research questions are visually depicted in Fig. 6.1.

6.3 Findings

6.3.1 Sustainability and the CE in the Corporate Strategy and Activities of Clothing Retail Stores in Romania (RQ1)

In general, the CE mainly addresses the environmental aspects of sustainability, especially materials consumption and waste generated (Geissdoerfer et al., 2017). A synthesis of the references to sustainability and CE found in clothing retail stores in Romania is presented in Table 6.3.

The first contact of customers with the notion of sustainability is through advertisements placed in the physical environment. Billboards, flyers, window stickers, and labels are just a few of the tools that can be utilised in physical stores to draw the consumer towards the idea of "sustainability in fashion". Although this form of advertising is convenient and inexpensive, only six of 15 companies (40%) included

Table 6.2 The evaluation sheet (research instrument and information source)

Physical/in-store retail indications of sustainability and CE practices (only for stores in Bucharest, Romania) (RQ1)
S1. Reference to sustainability (in physical stores) (Centobelli et al., 2022)
S2. Reference to circularity or any of the 3Rs - reduce, reuse, recycle - (in physical stores)—educational aspects (flyers, videos) (Centobelli et al., 2022)
S3. Reference to circular design practices to facilitate recycling and extend product life (in physical stores) (Hultberg & Pal, 2021; Münster et al., 2022)
S4. Reference to repair and alteration services/repair kits (in physical stores) (Centobelli et al., 2022; Dragomir & Dumitru, 2022; Hultberg & Pal, 2021)
S5. Existence of in-store donation/recycle bin (for used garment returns) (Centobelli et al., 2022; Dragomir & Dumitru, 2022; Hedegård et al., 2020)
S6. Returning used/old clothes for shopping credit (check in-store information or website) (Dragomir & Dumitru, 2022; Jia et al., 2020)
S7. Reference (flyers, links) to return policy and take-back services (in physical stores) (Centobelli et al., 2022)
S8. Partnerships/collaborations for CE practices (mentioned in physical stores) (Poldner et al., 2022; Sandberg, 2023)
Labels (physical stores in Bucharest, Romania) (RQ2)
L1. Labelling of eco-friendly fibres (e.g., mix of cotton, bio-based polyester) for individual items (Centobelli et al., 2022; Sugg, 2022)
L2. Labelling recycled fibres/materials for individual items (Dragomir & Dumitru, 2022; Hultberg & Pal, 2021)
L3. Label detailing the environmental impact of individual items (for in-store physical items) (Dragomir & Dumitru, 2022; Geegamage et al., 2023)
L4. Labelling of Fair Trade certification of individual items or packaging (in the physical stores) (Centobelli et al., 2022)
L5. Use of digital product passport (DPP) for customers to have information about individual items in physical stores (Centobelli et al., 2022; Dragomir & Dumitru, 2022; Pal & Jayarathne, 2022)
Digitalisation (what Romanian customers have access to in terms of corporate information) (RQ3)
D1. Customer loyalty program having incentives for customers to engage in CE practices (on website) (Sandberg, 2023)
D2. Publishing clothing use guides on the retailer's website (on website) (Dragomir & Dumitru, 2022)
D3. Use of social media (Instagram) to promote CE practices (on social media) (Dragomir & Dumitru, 2022)
D4. Use of retailer's mobile app to promote CE practices (on mobile app, if available) (Dragomir & Dumitru, 2022)
D5. In-store donation/recycle bins are localised in the app or on the website in Romanian (Hellström & Olsson, 2024)
D6. Customer survey on website that refers to sustainability or CE practices (on website) (Hellström & Olsson, 2024)

(continued)

Table 6.2 (continued)

D7. Disclosure of standards to assess circularity and environmental impact of products (mentioned in physical stores, website, or app) (Dragomir & Dumitru, 2022)
Localisation (information in Romanian)
R1. In-store labels describing sustainability and CE practices are in Romanian
R2. App in Romanian
R3. Website in Romanian

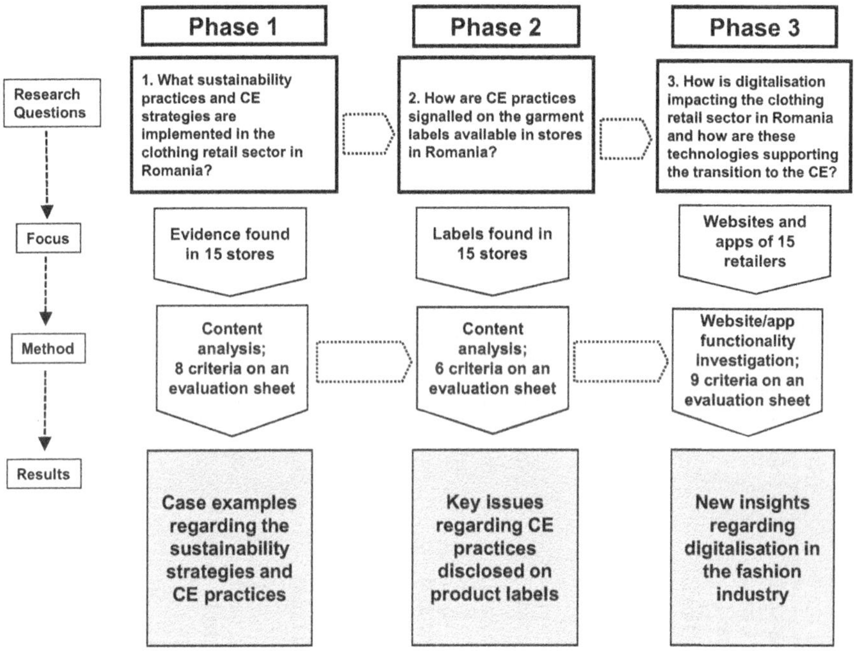

Fig. 6.1 Phases in the research process

references to sustainability in their physical stores (criterion S1). In four cases, "sustainability" was limited to urging people to reuse an old bag or purchase a paper bag instead of a plastic bag, as recommended at the check-out. In contrast, C&A featured large billboards with the text mentioning "sustainable actions for a better future" (in Romanian) and in-store advertising for the company's "most sustainable collection" (with a dedicated logo). On the other hand, Nike chose a more subtle approach by placing the sustainability references on labels that describe their program "Move to Zero" in relation to climate change adaptation.

References to circularity in physical stores (criterion S2) are even less frequent, appearing in only five locations (33.33%). The walls of the stores are not decorated in most of the stores analysed. Among the brands that chose to advertise their circularity initiatives, Zara has implemented a program called "#joinlife", which targets a more

Table 6.3 Evidence of sustainability and CE criteria in clothing retail stores in Romania

Brand	S1	S2	S3	S4	S5	S6	S7	S8
Adidas	Yes, paper bag purchase	Yes	Yes, Stella McCartney (store poster)	Not found	Not found	Not found	Yes	Yes, Stella McCartney
C&A	Yes, billboards	Yes	Not found	Not found	Not found	Not found	Not found	Not found
Coccodrillo	Not found	Not found	Not found	Yes	Not found	Not found	Yes	Not found
Desigual	Not found	Not found	Not found	Yes	Not found	Not found	Yes	Not found
H&M	Not found	Not found	Not found	Not found	Yes	Yes, voucher in app	Not found	Yes, Zlatan
M&S	Not found	Not found	Not found	Not found	Not found	Not found	Yes	Not found
Mango	Not found	Not found	Not found	Yes, additional buttons	Not found	Not found	Not found	Not found
Mothercare	Yes, paper bag purchase	Yes	Not found	Not found	Not found	Not found	Yes	Yes, 3M
New Yorker	Not found	Not found	Not found	Yes, repair kits	Not found	Not found	Yes	Not found
Next	Not found	Not found	Not found	Not found	Not found	Not found	Yes	Not found
Nike	Yes, move to zero	Yes	Not found	Not found	Not found	Not found	Yes	Not found
North Face	Not found	Not found	Yes, label “Circular design”	Not found	Not found	Not found, registration not available in Romania	Yes	Yes, Smartwool
P&C	Yes, paper bag purchase	Not found	Not found	Not found	Not found	Not found	Yes	Not found

(continued)

Table 6.3 (continued)

Brand	S1	S2	S3	S4	S5	S6	S7	S8
Sinsay	Not found	Not found	Not found	Not found	Not found	Not found	Not found, only underwear return policy	Not found
Zara	Yes, paper bag	Yes, Join Life	Not found	Not found	Yes	Not found	Not found	Yes, Red Cross

Notes The abbreviations in the header indicate CE practices: *S1* Reference to sustainability (in physical stores, as advertisements or brochures); *S2* Reference to circularity or any of the 3Rs (in physical stores) + educational aspects (flyers, videos); *S3* Reference to circular design practices to facilitate recycling and extend product life; *S4* Reference to repair and alteration services/repair kits; *S5* Existence of in-store donation/recycle bin (for used garment returns); *S6* Returning used/old clothes for shopping credit; *S7* Reference (flyers, links) to return policy and take-back services; *S8* Partnerships/collaborations for CE practices.

sustainable business model. This includes a holistic approach to circular initiatives in the value chain, from design to a second life of the products (Zara, 2023). While this program is mentioned on some labels, there is no explanation of what it means for retail consumers in Romania. C&A displays a large billboard in store stating that "we offer your old clothes a new life" (in Romanian) and has a dedicated section for products made with recycled fibers.

In the fashion industry, circular design refers primarily to incorporating parts of used products into new garments (Dragomir & Dumitru, 2022). References to circular design (criterion S3) were found in only two stores in the sample studied. For instance, a North Face product label related to circular design states: "This product can keep on going. Explore hard, repair as needed, leave the labels attached and then get it back to us to recycle when you're done" (in English). However, Romanian customers do not actually have the option to return the products after using them because there are no appropriate programs implemented by the state or the clothing retailers.

Options for repair and alteration (criterion S4) were identified in four stores. Such options include repair kits, extra buttons, staples, or other consumables (for example, in Coccodrillo). In some cases, the spare parts are visibly attached to the price tag, which is a useful indication that small repairs are easy to perform. In-store donation/ recycle bins for used garment returns (criterion S5) exist in only two stores included in the sample. This finding is particularly important for Romania, because more than 20% (INS, 2023) of the population lives in poverty or is at risk of poverty. Also, as referenced in criterion D5, there is no option for a separate collection of used textiles in Romania. However, according to the European Union Waste Framework Directive (Directive 2018/851, 2018), the EU requires the implementation of separate waste collection systems for textiles by 2025 (Article 11(b)). In relation to garment returns, only one company offers monetary compensation for such practice (criterion S6). H&M offers a five lei (one euro) voucher for a bag of used textiles. However, customers may need to explicitly ask for the voucher at the counter when handing in the bag (based on the authors' personal experience).

The criterion with the most relevant evidence in relation to Q1 is S7 (return policies and take-back services). Significant information appears in 10 out of 15 stores (66.67%). Romanian Law 449/2003 regarding selling conditions to individual consumers specifies that, for in-person sales, the consumer can demand repairs or replacements only if the product has major flaws. Reimbursements are applicable only if the product cannot be repaired or replaced. Any other consumer benefits (like replacements or reimbursements for products without defects) are left to each company to decide upon. Most sample companies have strict rules for product returns. For instance, regarding a product from Coccodrillo, one must change it within 30 days from purchasing it (or 15 days in the case of Nike) for a product with the same or higher price, keep the receipt, original label, and packaging, and change the product only in the shop from which it was purchased. The value of the product cannot be reimbursed to the customer. These conditions are an obstacle to implementing CE principles in Romania. H&M has a more permissive policy in this regard, and accepts returns in any store.

Partnerships or collaborations for CE practices (criterion S8) may help companies solve technical recycling problems or offer alternatives to buying products (Dragomir & Dumitru, 2022). In the sample, 20% of the stores have such partnerships. For example, Adidas has a partnership with Stella McCartney for shoes inspired by nature and made with natural and renewable raw materials. This information is featured on a billboard in the physical store, showing the company's commitment to sustainability in general.

6.3.2 CE Practices Disclosed on the Labels of Products Present in Clothing Retail Stores in Romania (RQ2)

Labels are a generally accepted source of information that assists buyers in making informed decisions. European consumers appreciate more the environmental information linked directly to products than what is published in company annual reports (Koszewska, 2011). A summary of the information related to CE practices disclosed on the labels of in-store products is presented in Table 6.4.

The first criterion (L1) is met by products present in 80% of the sampled stores. Labels indicate the use of eco-friendly fibers such as certified cotton, cashmere, or the percentage in which eco-friendly fibers are used for specific products. In very few cases, the positive environmental effect is also mentioned on the label ("reduces waste and our carbon footprint"—Nike). For instance, an Adidas garment product label maintains that it is "made with at least 95% regenerative cotton", explaining that "regenerative cotton is grown using farming methods that seek to improve soil health, watersheds and biodiversity".

Reference to L2 (labelling recycled fibres/materials for individual items) can be found on some product labels in 13 of the 15 stores (86.67%). An example is a Coccodrillo product which, according to the label, "contains padding made of 100% recycled raw material with GRS [Global Recycled Standard] certification". A GRS Certificate is issued based on the supply chain control system at every production stage, starting from recycling of raw material through processing of the materials up to the retail stage (SCS, 2024). The customers are urged to "give clothes a second life!" In one case, reference to recycled materials used for packaging was found: "this bag contains 30% recycled material. This item can be recycled" (Mothercare). Thus, two steps in the CE were addressed: the production and end-of-life of this packaging item, which was made of plastic.

The environmental impact of individual in-store items (L3) is detailed in 30% of the cases analysed. In some instances, labels indicate several CE practices. For example, Desigual has a campaign called "Love the World" that encompasses several facets of the CE, including recycling and lower resource use. According to a label, the "product is created with recycled fibers. This process uses less water and energy, so it has a smaller impact on the environment". Nike refers to its environmental impact in connection to its "Move to Zero" strategy (showing the company's commitment

Table 6.4 CE practices disclosed on the labels of the products in physical stores

Brand	L1	L2	L3	L4	L5	R1[a]
Adidas	Yes	Yes	Not found	Yes	Not found	No (EN, FR, DE, ES etc.)
C&A	Yes	Yes	Yes	Yes, FSC	Yes	Yes (RO)
Coccodrillo	Yes	Yes	Not found	Yes, FSC, Oeko Tex	Not found	No (EN)
Desigual	Yes	Yes	Yes	Yes	Yes	No (EN, FR, ES)
H&M	Yes	Yes	Yes, denim with lower impact	Yes	Yes	Yes (RO)
M&S	Yes	Yes	Not found	Yes, responsibly managed forest	Not found	No (EN)
Mango	Not found	Yes	Not found	Yes	Yes	No (EN)
Mothercare	Yes	Yes	Yes	Yes, responsibly sourced cotton	Not found	No (EN)
New Yorker	Not found	Not found	Not found	Not found	Not found	–
Next	Yes	Yes	Not found	Yes, FSC, Global Recycled Standard	Not found	No (EN)
Nike	Yes	Yes	Yes	Not found	Not found, QR code does not point to products	No (EN, FR, ES)
North Face	Yes	Yes	Not found	Yes, Oeko Tex	Not found	No (EN, DE, FR, ES)
P&C	Yes	Yes	Not found	Yes	Not found	No (EN, DE, ES)
Sinsay	Yes	Yes	Not found	Yes	Yes	No (EN)
Zara	Not found	Not found	Not found	Yes	Yes	No (EN)

Notes The abbreviations in the header indicate CE practices: *L1* Labelling of eco-friendly fibres (e.g., mix of cotton, bio-based polyester) for individual items; *L2* Labelling recycled fibres/materials for individual items; *L3* Label detailing the environmental impact of individual items; *L4* Labelling of certification of individual items or packaging (FSC, fair trade, or any other relevant certification); *L5* Use of digital product passport (DPP) for customers to have information about individual items; *R1* In-store labels describing sustainability and CE practices are in Romanian.

[a] Where *DE* German; *EN* English; *ES* Spanish; *FR* French; *RO* Romanian

to achieve zero carbon and zero waste). It relies on combining recycled fibers with organic materials and discloses the actual and assumed percentages of such materials.

Nike presents the reduction in carbon footprint using special materials compared to traditional materials: "Nike *forward* material reduces the carbon footprint by an average of 75% (due to the use of recycled materials, the lower process energy use, and the lower material density)". A lower impact on the environment is claimed on an H&M label of "next generation denim". First, it states that recycled fibers are used for the denim fabric, sewing thread, and labels. Second, the company does not rely on harmful manufacturing processes ("all metal trims are non-electroplated") or substances ("no potassium permanganate") for the denim. Third, "lower impact garment finishes were used", a fact certified by a low-impact score in "environmental impact measurement by JEANOLOGIA". Fourth, cleaning instructions are presented on the label ("wash only when needed, wash at low temperature, and avoid tumble drying"). Finally, the same label refers to options other than landfill ("repair, resell, and eventually recycle").

In almost all physical stores analysed (86.67%), labels showing certification of individual items or packaging (L4) were found. Certification was obtained from FSC, Oeko Tex, Global Recycled Standard, and others. This corporate responsibility practice is important for consumers who want to have an indication of product quality and manufacturing integrity. Shopping malls not only offer functional purchasing (i.e., when a product is strictly needed for its purpose) but also aim to create "an experience" for the consumer, with a particular focus on product quality and responsible use. Linking CE practices with product quality is an important step in advancing customer education. However, this is hampered by the fact that only two of the 15 companies translated their in-store labels to Romanian (criterion R1).

Although most physical stores displayed products with labels containing barcodes and QR codes, in only 40% of the cases were digital product passports used (criterion L5). The digital passport is a link printed on the label which can be accessed using the customer's mobile phone and points to the description of the product on the seller's website. The product page should (ideally) contain information on sizes, materials, weight, cleaning and maintenance, environmental impact, certifications, and disposal. In some instances, available inventories are also indicated on the respective webpage, referring to physical stores or online shopping. The use of QR codes is more prevalent, but some retailers allow the customer to scan the barcode using the retailer's mobile app. However, even if digital passports were functional, not all information was in Romanian on the product webpage (criteria R2 and R3, in the next section). For example, Desigual provides digital product passports only in English (or Spanish).

6.3.3 Digitalisation Supporting CE Practices of the Largest Clothing Retailers in Romania (RQ3)

The company's loyalty program is an important tool for promoting CE practices (criterion D1). Customers who are enrolled in the program can receive notifications and suggestions through multiple channels, such as newsletters, Instagram ads, and mobile app notifications, as well as advertisements in physical stores. If the company intends to push certain CE practices towards the consumers, it can use any of these advertising means and the direct connection through the loyalty program. In this study, only three companies (20%) use their loyalty programs for CE purposes. Desigual and H&M extend the return period for customers enrolled in their loyalty program, while Sinsay allows customers to return without the receipt if the purchased item was scanned together with the loyalty program code.

Clothing use guides on the retailer's website (criterion D2) can be accessed in two ways: through the DPP or in the general "use guides" section of the websites. A typical use guide would contain instructions on how to wash and care for a piece of clothing. For example, one of Zara's winter coats has these instructions on the website (retrieved using the digital passport): "By taking care of your clothes, you are also taking care of the environment. To keep jackets and coats clean, simply air them out and clean them with a cloth or clothes brush. This process is gentler on fabrics and, in addition, avoids water and energy consumption during washing processes. Clothes care guide: Machine wash at max. 30 °C, delicate. Do not use bleach. Do not iron. Do not dry clean. Do not tumble dry". As indicated in Table 6.5, this is a common practice for most clothing retailers. Only three retailers do not provide relevant information in this regard: Marks and Spencer (M&S) does not offer this type of information to Romanian customers, while the websites of Next (children's clothing) and New Yorker (men's and women's fashion) contain only information regarding the sizes and materials of clothing items.

Instagram is a powerful tool that can be used to promote CE practices (criterion D3). The results obtained through searching for the most recent 100 posts on the official Instagram account of each company are less relevant in terms of contributing to the CE. Only Adidas and North Face have mentioned CE practices in at least one post. In the post flagged on the North Face's Instagram feed, the text reads: "Honoured to see our Circular Designs selected by @fastcompany for the 2023 Innovation by Design Awards", and the associated images pointed to in-store recycling options. However, North Face does not offer any CE options for Romanian customers (such as recycling or returning used items). In the case of Adidas, the flagged post contained the following text: "Because of you, 250,000 kg/500,000 lbs of plastic waste will be cleaned up by @parley.tv's Global Clean Up Network to help end plastic waste".

The use of the retailer's mobile app to promote CE practices (criterion D4) is an important way of engaging customers in the CE paradigm. Any mention of CE principles or practices within the retailers' Android app was investigated. Only five companies have configured their apps to promote CE principles to consumers through general statements (H&M, Zara), clothing use guides (Mango), products in dedicated

Table 6.5 CE practices supported by digitalisation

Brand	D1	D2	D3	D4	D5	D6	D7	R2	R3
Adidas	Not found[a]	Yes	Yes	No app[a]	No	Not found	Not found	No app[a]	Yes (RO)
C&A	Not found	Yes	Not found	Not found	No	Not found	Yes, ZDHC	Yes (RO)	Yes (RO)
Coccodrillo	Not found	Yes	Not found	Not found	No	Not found	Yes, Oeko Tex	No app	Yes (RO)
Desigual	Yes, extended returns	Yes	Not found	Not found	No	Not found	Yes, better cotton	No app	No (EN)
H&M	Yes, voucher for recycling	Yes	Not found	Yes	Yes	Not found	Yes, Jeanologia	Yes (RO)	Yes (RO)
M&S	Not found	Not found	Not found	Not found	No	Not found	Not found	No app	No (EN)
Mango	Not found[a]	Yes	Not found	Yes	No	Not found	Not found	Yes (RO)	Yes (RO)
Mothercare	Not found	Yes	Not found	Not found	No	Not found	Not found	No app	Yes (RO)
New Yorker	Not found	Not found	Not found	Not found	No	Not found	Not found	No (EN)	No (EN)
Next	Not found	Not found	Not found	Not found	No	Not found	Yes, better cotton	Yes (RO)	Yes (RO)
Nike	Not found	Yes	Not found	Yes	No	Not found	Yes, PRe sustainability	No (EN)	No (EN)
North Face	Not found[a]	Yes	Yes	No app[a]	No	Not found	Yes, responsible down	No app[a]	Yes (RO)
P&C	Not found	Yes	Not found	Not found	No	Not found	Not found	No app	Yes (RO)
Sinsay	Yes, returns without receipt	Yes	Not found	Yes	No	Not found	Yes, CmiA	Yes (RO)	Yes (RO)
Zara	Not found	Yes	Not found	Yes	Yes	Yes, asks for suggestions	Yes, CanopyStyle	Yes (RO)	Yes (RO)

Notes The abbreviations in the header indicate CE practices: *D1* Customer loyalty program having incentives for customers to engage in CE practices; *D2* Publishing clothing use guides on the retailer's website; *D3* Use of social media (Instagram) to promote CE practices; *D4* Use of retailer's mobile app to promote CE practices; *D5* In-store donation/recycle bins are localised in the app or on website in Romanian; *D6* Customer survey on website that refers to sustainability or CE practices; *D7* Disclosure of standards to assess circularity and environmental impact of products; *R1* App in Romanian; *R2* Website in Romanian.

[a] Not available in Romania. Where *DE* German; *EN* English; *ES* Spanish; *FR* French; *RO* Romanian

categories (Nike), or the extended return period (Sinsay). In most cases (except for Nike), the apps are also localised in Romanian, which is a bonus for the consumer.

In Romania, textiles are not included in government-sponsored recycling programs (which are limited to appliances, electronics, batteries, plastics, metal cans, and glass bottles). Therefore, it is difficult for Romanian consumers to know where to dispose of their used textiles and other clothing items. The location of in-store donations/recycle bins in the app or on retailers' websites (D5) is a needed facility for anyone who wants to donate or recycle their used clothing items. Only two retailers indicate on their websites the locations of recycling bins: Zara and H&M. These are also the only two retailers that provide recycling bins. Inditex (Zara's parent company) has installed recycling bins in other stores owned by the company, such as Bershka, Pull & Bear, and Zara Home.

Customer surveys on retailers' websites that refer to sustainability and CE practices (D6) are an important instrument to get feedback from (potential) customers and promote CE practices. However, only Zara has implemented such a survey, which is advertised near the check-out desk. Even in this case, the survey does not mention CE aspects, but asks customers for feedback regarding in-store facilities. This is a missed opportunity for the clothing retailers in Romania to engage with customers and collect their opinions on recycling, donations, returns, product quality, and any other relevant aspects.

The disclosure of standards to assess circularity and environmental impact of products (criterion D7) refers to more general disclosures, found in physical stores, on websites, or in apps, and not specifically tied to any one item. Criterion D7 does not refer to the annual reports of the respective companies. Where available, the Romanian version of the website was checked, specifically the "Sustainability" section. In the sample, six out of 15 companies chose to ignore any circularity or sustainability standards in their general disclosures (as of January 2024). It is surprising that these companies ignored aspects related to circularity and sustainability on any communication channel accessible to consumers. This result is consistent with the findings on criterion D3 regarding the presence of CE topics on the retailers' Instagram accounts. Overall, the results suggest that some of the companies are not interested in educating and engaging with Romanian consumers on the topic of the CE.

6.4 Discussion

In Romania, the largest clothing retailers have opened stores almost exclusively in shopping malls. In some cases, a parent company like Inditex has stores for each of its brands in the same mall. In Bucharest, specifically, shopping malls have extensive commercial space that is highly appreciated by consumers. Bucharest malls do not have unoccupied retail space, which is proof to their commercial success. In this context, clothing stores have a significant impact on consumer preferences, shaping the target market. The present research was motivated by the high impact of large clothing retailers through their physical stores found in the largest shopping malls in Bucharest, Romania. The sample was selected based on the size of store space in the respective malls, which may not be perfectly correlated to sales.

A visual summary of the CE information and practices found in physical stores and on brands' websites is provided in Fig. 6.2. There is significant variation in the types of CE options available to Romanian consumers. What is apparent is that none of the analysed practices is present at all retailers, which is a sign that there is considerable scope for progress in this regard. The sample study includes only international retailers because Romanian brands have not penetrated the local market in terms of large retail space and sale volumes. In this sense, the investigation shows what CE practices have been "imported" into Romania by store owners, which, in some cases, are franchisees. There is a gap between the CE practices of the international chain and the options available to Romanian consumers. For example, product labels may contain advice on how to recycle garments while the store does not offer any recycling facility. In most cases, labels pointing to CE aspects are not translated into Romanian, suggesting that the retailer is not interested in promoting CE principles on the local market.

Figure 6.2 indicates that most clothing retailers in Romania are not promoting sustainability principles and practices in their physical stores. References to sustainability are poorly represented in the entire sample. Store managers may believe that consumers are not interested in sustainability aspects while shopping for clothes, although C&A (Germany) advertised its sustainability practices on almost every shelf in the store. In contrast, New Yorker (Germany) does not have any indication of sustainability practices. These two brands offer similar types of clothing at comparable prices, but C&A's retail space is considerably larger. From this perspective, C&A's impact on consumer preferences could be more significant, but very few retailers in Romania promote their sustainability practices directly to consumers, either in store or on the website. For example, Zara (having a commercial space comparable to C&A's) does not have any type of in-store advertising referring to sustainability or circularity practices (except for the paper bags at check-out). The Romanian mall shopper has very little exposure to sustainability or circularity principles and practices while shopping in the physical stores of clothing retailers.

Digitalisation in the fast fashion industry can take multiple forms and can have different purposes. In relation to the consumer, several aspects of digitalisation were identified that could drive the customer to adopt CE practices. Firstly, DPPs (starting from the QR codes printed on labels) are a very convenient way for prospective buyers to assess the quality of a garment but also to learn more about responsibly using and discarding the product. For a DPP to be useful, it must point to a comprehensive product page, written in the language of the user. This existed in only 40% of the brands analysed. However, clothing use guidelines are published on the website of most companies, helping the consumer to maintain the product in good condition. On the other hand, it was surprising to find very poor engagement on social media (as of December 2023) on the topic of the CE. For example, only one company offers recycling incentives paired with vouchers linked to the loyalty program. The study indicates that the sample companies do not leverage the combined power of the loyalty program and social media in advancing CE principles. Finally, companies

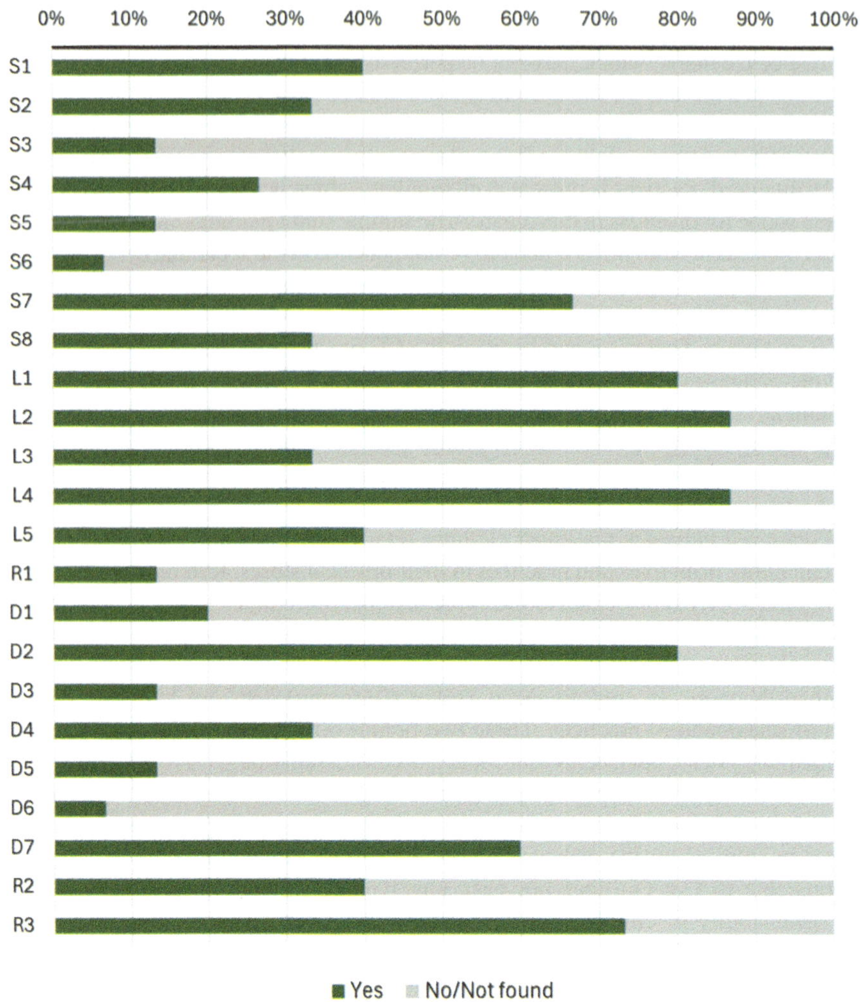

Fig. 6.2 A visual overview of CE and digitalisation practices implemented by the largest clothing retailers in Romania. *Source* Chapter authors. *Note* On the *X*-axis: the % figures indicate the scores obtained by all retailers on each criterion relative to the maximum score of the respective criterion. On the *Y*-axis: the codes are the criteria for evaluation shown in Table 6.2

do not engage with consumers through satisfaction surveys. Only one company has created a feedback webpage which is available to anyone, irrespective of buying or not from the store.

The retailers' overall scores (Table 6.6 and Fig. 6.3) ranged between the highest score of 70% (H&M) and the lowest score of 9% (New Yorker), in relation to the maximum score possible. The distribution of the overall scores is close to the normal distribution. In the present sample, some retailers were aware of the sustainability issues surrounding the fast fashion industry (Brewer, 2019), and engaged with the consumers through explicit labels and recycling options (Papadopoulou et al., 2022). Other retailers seem to disregard sustainability principles and offer no relevant information in connection with sustainable materials, product care, and garment end-of-life.

The scores granted to in-store evidence of sustainability and CE practices for the retailers included in the sample are the lowest, at 30% (RQ1, Table 6.6, criterion S). Most of the stores do not have information related to these aspects at all. This is a significant aspect to be improved. The highest scores are granted to the information presented on the labels in relation to the CE (RQ2, Table 6.6, criterion L), proving the importance of the label in the shopping process. The scores for digitalisation (criterion D) are low, 32.38%. Even when they exist, digital technologies may not offer the buyer the expected experience. For instance, the apps would not work properly, have information in another language, and contain limited information related to the product. These practices need to be improved, as they are significant from a CE point of view: a person can use the digital technologies before deciding to buy a product, but also after buying it, throughout his entire life.

Table 6.6 The cumulative scores of analysed clothing retailers (by research category)

Brand	Criterion S (max. 8)	Criterion L (max. 5)	Criterion D (max. 7)	Localization R (max. 3)	Total (max. 23)
Adidas	5	3	2	1	11
C&A	2	5	2	3	12
Coccodrillo	2	3	2	1	8
Desigual	2	5	3	0	10
H&M	3	5	5	3	16
M&S	1	3	0	0	4
Mango	1	3	2	2	8
Mothercare	4	4	1	1	10
New Yorker	2	0	0	0	2
Next	1	3	1	2	7
Nike	3	3	3	0	9
North Face	3	3	3	1	10
P&C	2	3	1	1	7
Sinsay	1	4	4	2	11
Zara	4	2	5	2	13
Total points	36	49	34	19	138
Average points per company	2.4	3.27	2.27	1.27	9.2
Percentage (average points accomplished/maximum possible) (%)	30	65.33	32.38	42.22	40

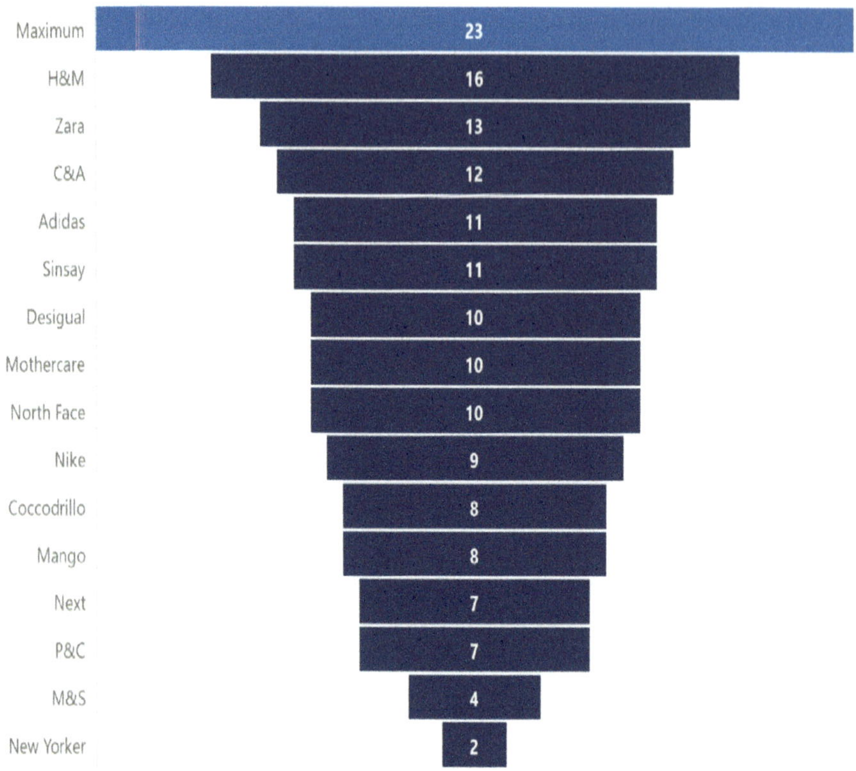

Fig. 6.3 A funnel chart of best to worst scores on CE practices in physical stores and online, for Romanian customers. *Source* Chapter authors. *Note* The numbers on the horizontal bars indicate the overall scores of each retailer, in comparison to the maximum score of 23 (top of the funnel chart)

Even when clothing companies invest in sustainable production and disclose their circularity initiatives, it is not clear how Romanian consumers perceive these practices or if they shape their shopping and disposal habits according to sustainability principles (Purcărea et al., 2022). In the end, the responsible consumer will have to ask the following questions (Gheorghe & Matefi, 2021; Papasolomou et al., 2023): 1. Do I need a new item of clothing? 2. What is this garment made of? 3. Where was it produced? 4. How long will it last? 5. How can I dispose of my old clothes in a responsible manner?

6.5 Conclusion

In this research, the investigative methodology and content analysis were applied to observe the extent to which clothing retailers in Romania promote and apply CE practices in physical stores. The selected sample consists of 15 large stores located in malls in Bucharest, and the data collected were recorded as regards 23 items in four categories: sustainability, labels, digitalisation, and localisation. The results indicate that major clothing retailers in Romania apply the principles of CE in a very selective way, meaning that some practices are present in most stores while others are absent in most cases. Moreover, there are significant differences between companies, indicating that there is no isomorphism within CE practices in clothing retail in Romania. The main contribution of this research is to assess the stage of implementation of the CE principles in fast fashion retail in Romania, as well as the degree of exposure of the Romanian consumer to CE principles in the field of garment retail. Regarding the use of digital technologies to support CE practices, the findings indicate certain aspects which can be improved upon by the brands included in the sample (such as the availability of the targeted loyalty programs, the apps, the surveys, and the social media messages).

This research has implications for various actors within the clothing retail value chain in Romania. The European Commission has proposed a strategy to reduce waste in the fast fashion value chain by returning and recycling used garments, but large retailers in Romania are not ready to receive and recover used textiles. At the same time, large clothing stores (i.e., physical retail) do not communicate adequately with the consumer on options to return used clothes or how to care for and repair clothes to extend their useful life. In this regard, fast fashion retail in Romania is not ready to achieve the European Commission's objectives regarding the reduction of waste in the value chain. The research suggests that both retailers and state authorities could play a key role in providing financial incentives and ethical motivation for consumers to adopt CE principles: the purchase of products made of ecological materials, the extension of the life of purchased products, and the responsible return of used products. It is important for large retailers to give consumers the opportunity to express their wishes and provide feedback on CE practices by creating a method of engagement with consumers on this topic. This type of engagement can be achieved by digitalisation to support the CE.

There are several limitations to this research. Firstly, there was no hard evidence regarding the value of sales per brand so that it was not possible to precisely assess the market share of each brand. This is because some of these brands are part of franchise contracts held by non-listed companies. Secondly, a decomposition of physical versus online products sold per brand was not available. Even at a global level, the major retailers do not provide this type of financial information, as it is not mandatory. In this respect, it was not possible to assess the impact of physical stores on the preferences of consumers, in comparison to online shopping. Thirdly, comprehensive quantitative data on the retail spaces available to each of these brands, in Bucharest or in other cities in Romania, was not available. For example, the impact of Inditex is much larger than just through the sales of the Zara brand, because its other brands also have large commercial spaces (Bershka, Pull & Bear, Stradivarius, Massimo Dutti). From this perspective, Inditex has by far the largest commercial presence in Romania in the domain of physical retail. Finally, there is no data on final sales, i.e., sales less returns. Most clothing companies, especially through their online stores, advertise their permissive return policies to boost sales. However, quantitative information on how many clothing items remain in use or reach the point at which they are discarded by consumers was not in evidence. As regards digitalisation, it is also worth noting that only the technologies available for the retailer's relation with the customer were analysed, and not the entire supply chain or the enterprise resource planning software.

There are numerous avenues of future research in the context of the clothing retail industry and the CE in Romania. Mirroring this investigation from the consumer's perspective, future studies could provide insights into how much customers are concerned with any of these CE practices. Secondly, it would be of value to ascertain how many consumers would use any CE options, if given the opportunity and necessary information. Thirdly, experimental research could be conducted to discover the minimum financial incentives necessary to activate consumer behavior towards CE practices. These avenues of investigation are merely illustrative of the significant potential for future research in the domain of the CE, either in physical or online retail, in the Romanian, and wider, national contexts.

References

Albu, N., Albu, C. N., Apostol, O., & Cho, C. H. (2021). The past is never dead: The role of imprints in shaping social and environmental reporting in a post-communist context. *Accounting, Auditing & Accountability Journal, 34*(5), 1109–1136. https://doi.org/10.1108/AAAJ-08-2019-4131

Alves, L., Ferreira Cruz, E., Lopes, S. I., Faria, P. M., & Rosado Da Cruz, A. M. (2022). Towards CE in the textiles and clothing value chain through blockchain technology and IoT: A review. *Waste Management & Research: The Journal for a Sustainable CE, 40*(1), 3–23. https://doi.org/10.1177/0734242X211052858

Brewer, M. K. (2019). Slow fashion in a fast fashion world: Promoting sustainability and responsibility. *Laws, 8*(4), 24. https://doi.org/10.3390/laws8040024

Bürklin, N., & Wynants, J. (2020). Opening new opportunities to close the loop: How technology influences the CE. In G. Vignali, L. F. Reid, D. Ryding, & C. E. Henninger (Eds.), *Technology-driven sustainability* (pp. 219–240). Springer International Publishing. https://doi.org/10.1007/978-3-030-15483-7_12

Centobelli, P., Abbate, S., Nadeem, S. P., & Garza-Reyes, J. A. (2022). Slowing the fast fashion industry: An all-round perspective. *Current Opinion in Green and Sustainable Chemistry, 38*, 100684. https://doi.org/10.1016/j.cogsc.2022.100684

Directive 2014/95/EU of the European Parliament and of the Council of 22 October 2014 amending Directive 2013/34/EU as regards disclosure of non-financial and diversity information by certain large undertakings and groups (2014). https://eur-lex.europa.eu/legal-content/EN/TXT/PDF/?uri=OJ:L:2014:330:FULL&from=EN

Directive (EU) 2018/851 of the European Parliament and of the Council of 30 May 2018 amending Directive 2008/98/EC on waste (Text with EEA relevance) (2018). http://data.europa.eu/eli/dir/2018/851/oj

Dissanayake, D. G. K. (2020). Does mass customization enable sustainability in the fashion industry? In R. Beltramo, A. Romani, & P. Cantore (Eds.), *Fashion industry—An itinerary between feelings and technology*. IntechOpen. https://doi.org/10.5772/intechopen.88281

Dragomir, V. D., & Dumitru, M. (2022). Practical solutions for circular business models in the fashion industry. *Cleaner Logistics and Supply Chain, 4*, 100040. https://doi.org/10.1016/j.clscn.2022.100040

Dragomir, V. D., & Dumitru, M. (2024). The state of the research on CE in the European Union: A bibliometric review. *Cleaner Waste Systems, 7*, 100127. https://doi.org/10.1016/j.clwas.2023.100127

Ellen MacArthur Foundation. (2013). *Towards the CE Vol. 1: An economic and business rationale for an accelerated transition*. Ellen MacArthur Foundation. https://www.ellenmacarthurfoundation.org/towards-the-circular-economy-vol-1-an-economic-and-business-rationale-for-an

European Commission. (2020). *A new CE action plan. For a cleaner and more competitive Europe* (COM/2020/98 final). European Commission. https://eur-lex.europa.eu/legal-content/EN/TXT/?uri=COM:2020:98:FIN

Geegamage, T., Ranaweera, A., & Halwatura, R. (2023). Pre-loved or hatred? Consumers' perception of value towards second-hand fashion consumption in Sri Lanka. *Research Journal of Textile and Apparel*. https://doi.org/10.1108/RJTA-02-2023-0022

Geissdoerfer, M., Savaget, P., Bocken, N. M. P., & Hultink, E. J. (2017). The CE—A new sustainability paradigm? *Journal of Cleaner Production, 143*, 757–768. https://doi.org/10.1016/j.jclepro.2016.12.048

Gheorghe, C. A., & Matefi, R. (2021). Sustainability and transparency—Necessary conditions for the transition from fast to slow fashion: Zara Join Life collection's analysis. *Sustainability, 13*(19), 11013. https://doi.org/10.3390/su131911013

Hedegård, L., Gustafsson, E., & Paras, M. K. (2020). Management of sustainable fashion retail based on reuse—A struggle with multiple logics. *The International Review of Retail, Distribution and Consumer Research, 30*(3), 311–330. https://doi.org/10.1080/09593969.2019.1667855

Hellström, D., & Olsson, J. (2024). Let's go thrift shopping: Exploring circular business model innovation in fashion retail. *Technological Forecasting and Social Change, 198*, 123000. https://doi.org/10.1016/j.techfore.2023.123000

Hultberg, E., & Pal, R. (2021). Lessons on business model scalability for CE in the fashion retail value chain: Towards a conceptual model. *Sustainable Production and Consumption, 28*, 686–698. https://doi.org/10.1016/j.spc.2021.06.033

INS. (2023). *În anul 2022, unul din cinci români era afectat de sărăcie*. Institutul Național de Statistică. https://insse.ro/cms/sites/default/files/com_presa/com_pdf/saracia_si_excluziunea_sociala_r2022.pdf

Jia, F., Yin, S., Chen, L., & Chen, X. (2020). The CE in the textile and apparel industry: A systematic literature review. *Journal of Cleaner Production, 259*, 120728. https://doi.org/10.1016/j.jclepro.2020.120728

Koszewska, M. (2011). Social and eco-labelling of textile and clothing goods as a means of communication and product differentiation. *Fibres & Textiles in Eastern Europe, 19*(4), 20–26.

Koszewska, M. (2018). CE—Challenges for the textile and clothing industry. *Autex Research Journal, 18*(4), 337–347. https://doi.org/10.1515/aut-2018-0023

Lakatos, E., Dan, V., Cioca, L., Bacali, L., & Ciobanu, A. (2016). How supportive are Romanian consumers of the CE concept: A survey. *Sustainability, 8*(8), 789. https://doi.org/10.3390/su8080789

Münster, M. B., Sönnichsen, S. D., & Clement, J. (2022). Retail design in the transition to CE: A study of barriers and drivers. *Journal of Cleaner Production, 362*, 132310. https://doi.org/10.1016/j.jclepro.2022.132310

Orzan, G., Cruceru, A., Bălăceanu, C., & Chivu, R.-G. (2018). Consumers' behavior concerning sustainable packaging: An exploratory study on Romanian consumers. *Sustainability, 10*(6), 1787. https://doi.org/10.3390/su10061787

Pal, R., & Jayarathne, A. (2022). Digitalisation in the textiles and clothing sector. In *The digital supply chain* (pp. 255–271). Elsevier. https://doi.org/10.1016/B978-0-323-91614-1.00015-0
Papadopoulou, M., Papasolomou, I., & Thrassou, A. (2022). Exploring the level of sustainability awareness among consumers within the fast-fashion clothing industry: A dual business and consumer perspective. *Competitiveness Review: An International Business Journal, 32*(3), 350–375. https://doi.org/10.1108/CR-04-2021-0061
Papasolomou, I., Melanthiou, Y., & Tsamouridis, A. (2023). The fast fashion vs environment debate: Consumers' level of awareness, feelings, and behaviour towards sustainability within the fast-fashion sector. *Journal of Marketing Communications, 29*(2), 191–209. https://doi.org/10.1080/13527266.2022.2154059
Paras, M. K., Curteza, A., & Varshneya, G. (2019). Identification of best reverse value chain alternatives: A study of Romanian used clothing industry. *Journal of Fashion Marketing and Management: An International Journal, 23*(3), 396–412. https://doi.org/10.1108/JFMM-04-2018-0060
Poldner, K., Overdiek, A., & Evangelista, A. (2022). Fashion-as-a-service: Circular business model innovation in retail. *Sustainability, 14*(20), 13273. https://doi.org/10.3390/su142013273
Purcărea, T., Ioan-Franc, V., Ionescu, Ş.-A., Purcărea, I. M., Purcărea, V. L., Purcărea, I., Mateescu-Soare, M. C., Platon, O.-E., & Orzan, A.-O. (2022). Major shifts in sustainable consumer behavior in Romania and retailers' priorities in agilely adapting to it. *Sustainability, 14*(3), 1627. https://doi.org/10.3390/su14031627
Sandberg, E. (2023). Orchestration capabilities in circular supply chains of post-consumer used clothes—A case study of a Swedish fashion retailer. *Journal of Cleaner Production, 387*, 135935. https://doi.org/10.1016/j.jclepro.2023.135935
Sandvik, I. M., & Stubbs, W. (2019). Circular fashion supply chain through textile-to-textile recycling. *Journal of Fashion Marketing and Management: An International Journal, 23*(3), 366–381. https://doi.org/10.1108/JFMM-04-2018-0058
SCS. (2024). *Global recycled standard. Recycled content for both finished and intermediate products*. SCS Global Services. Retrieved April 23, 2024, from https://www.scsglobalservices.com/services/global-recycled-standard
Staicu, D., & Pop, O. (2018). Mapping the interactions between the stakeholders of the CE ecosystem applied to the textile and apparel sector in Romania. *Management & Marketing, 13*(4), 1190–1209. https://doi.org/10.2478/mmcks-2018-0031
Sugg, B. (2022). Circular textiles innovation during COVID-19: Not the silver lining some had hoped for. *Journal of Fashion Marketing and Management: An International Journal*, 1–16. https://doi.org/10.1108/JFMM-07-2021-0180
UNDP, Line, M., & Braun, R. (2007). *Baseline study on CSR practices in the new EU member states and candidate countries*. United Nations Development Programme. http://www.cbcsr.mk/Upload/Content/Documents/BASELINE_STUDY_ON.pdf
Ungurcanu, A. (2023, August 29). *Invazie de mall-uri și centre comerciale în România! Care sunt cele mai importante deschideri și ce branduri noi vin în țară*. Wall-Street.ro. https://www.wall-street.ro/articol/Real-Estate/300254/invazie-de-mall-uri-si-centre-comerciale-in-romania-care-sunt-cele-mai-importante-deschideri-si-ce-branduri-noi-vin-in-tara.html
Wiegand, T., & Wynn, M. (2023). Sustainability, the CE and digitalisation in the German textile and clothing industry. *Sustainability, 15*(11), 9111. https://doi.org/10.3390/su15119111
World Bank. (2023). *Poverty & equity brief. Romania*. World Bank Group. https://databankfiles.worldbank.org/public/ddpext_download/poverty/987B9C90-CB9F-4D93-AE8C-750588BF00QA/current/Global_POVEQ_ROU.pdf
Zara. (2023). *Join Life | Zara Romania*. https://www.zara.com/ro/ro/z-join-life-mkt1399.html

Mădălina Dumitru (madalina.dumitru@cig.ase.ro) is Professor of Accounting at the Bucharest University of Economic Studies, Romania. She was employed by this university in 2001, and obtained her Ph.D. in 2007 from the same university. Her research and teaching areas are management accounting, sustainability reporting, and accounting education, in an international or Romanian context. A significant part of her work analyses the impact of the European regulations on company reporting. During the last few years, she has published papers on the circular economy in the fashion industry and in agriculture.

Voicu D. Dragomir (voicu.dragomir@cig.ase.ro) is Professor of Accounting at the Bucharest University of Economic Studies and a Ph.D. supervisor. He earned his Ph.D. in Accounting in 2010 and holds a BA in Psychology. His research and teaching areas include corporate governance, circular economy models, research methodologies, and corporate strategy. He has published in prestigious journals published by Elsevier, Taylor and Francis, and Emerald. He has published a book with Springer on the topic of corporate environmental strategy. He serves as reviewer for several high-profile journals and has coordinated multiple research projects.

Chapter 7
Denmark

Cathryn Anneka Hall, Karen Marie Hasling, Kerli Kant Hvass, and Nikola Kiørboe

Abstract This chapter examines both the textile manufacturing and textile recovery sectors in Denmark. It investigates the sustainability measures and circular strategies currently being actioned in the Danish textile manufacturing field and explores, across both sectors (manufacture and recovery), the role of digital technologies towards a circular economy. Using Danish Industrial Classification (DB07) codes and the Danish Central Business Register (CVR) database alongside expert knowledge, a list of 115 textile manufacturers and eleven recovery actors were found and included in the study. Websites from each company were studied and the data was analysed for circular strategies across materials, production, use, recovery and social activities. The findings demonstrate a focus on production and material related strategies and less emphasis on recovery within manufacture. Overall, there was a very limited uptake of digital technologies. The chapter concludes by exploring opportunities for new business models within repair, upgrade and recycling towards a Danish circular textile economy.

Keywords Textile manufacture · Textile sorting · Textile recycling · Circular business models

C. A. Hall (✉) · K. M. Hasling
Kolding School of Design, Kolding, Denmark
e-mail: cah@dskd.dk

K. M. Hasling
e-mail: kmh@dskd.dk

K. Kant Hvass
Aaborg University, Copenhagen, Denmark
e-mail: kerlikh@plan.aau.dk

K. Kant Hvass · N. Kiørboe
Revaluate, Hellerup, Denmark
e-mail: nikola@revaluate.dk

M. Wynn and T. Wiegand (eds.), *Sustainability, the Circular Economy and Digitalisation in the European Textile and Clothing Industry*, SDGs and Textiles,
https://doi.org/10.1007/978-981-97-9116-3_7

7.1 Introduction

To transition to a circular economy (CE), many actors, strategies, models and pieces of legislation need to combine to progress forward. For the textile industry, like many others, this forms a complex globalised picture, with complex supply chains and unresolved recovery routes (EMF, 2017). In order to explore solutions, it can be useful to consider a more localised position before expanding outward. Thus, this chapter first examines both textile manufacturing and textile recovery sectors in Denmark. Secondly, it investigates the sustainability measures and circular strategies currently being actioned in the Danish textile manufacturing field. Thirdly, across both sectors (manufacture and recovery), the role of digital technologies towards a CE is explored. The chapter concludes by drawing on all the findings to discuss the circular opportunities that present themselves from a Danish perspective. To do this, the chapter addresses the following four research questions (RQs):

RQ1: What is the scope of a) the textile manufacturing and b) the textile recovery industries in Denmark?
RQ2: What sustainability measures and circular strategies are being employed by the textile manufacturers in Denmark?
RQ3: What is the role of digital technologies in supporting sustainability and circularity within textile manufacture and recovery in Denmark?
RQ4: What circular opportunities are there for manufacturing and recovery in Denmark?

This chapter is structured around five sections. Following this brief introduction, the research method is outlined (Sect. 7.2). This is followed by the findings (Sect. 7.3), which comprises four sub-sections that address the first three RQs. Sections 7.3.1 and 7.3.2 explore the scope of textile manufacture and recovery respectively (RQ1), Sect. 7.3.3 addresses sustainability and circular strategies employed by textile manufacturers in Denmark (RQ2) and Sect. 7.3.4 explores digitalisation across manufacture and recovery (RQ3). Then, Sect. 7.4 presents a discussion of the findings, in which the fourth and final RQ is addressed, regarding circular opportunities in Denmark. Finally, the concluding Sect. 7.5 draws together the main issues covered in the chapter and points to possible future areas for follow-on research studies.

7.2 Research Method

This research aims to explore both the textile manufacturing and textile recovery sectors in Denmark including the sustainability measures and circular strategies being carried out and the role of digital technologies. To do this, a list of Danish textile manufacturing and recovery firms was complied, and their websites were analysed and coded to draw out findings.

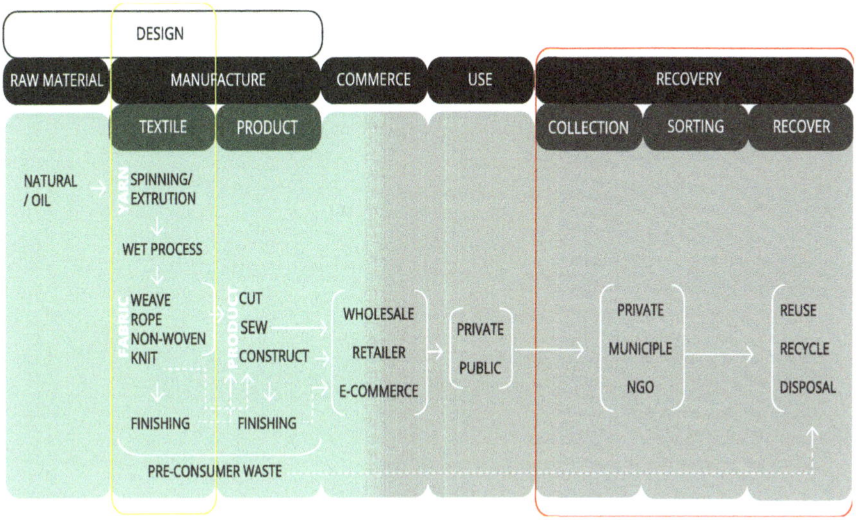

Fig. 7.1 The textile value chain highlighting the focus areas of this study based on Stamm et al. (2019) and Arponen et al. (2018)

For the first part (manufacturing), the research scope was focused on textile manufacture and excluded those companies that made only "finished goods" (such as garments or interior products) rather than the construction of the textile materials themselves (Fig. 7.1). The exception was knitwear and carpets in which the textile process and the construction of the finished product cannot always be easily separated.

The companies within these criteria were found using the relevant Danish Industry 2007 Codes (DB07). A DB07 code is a "6-digit industry nomenclature that is primarily prepared for statistical use" (Danmarks Statistik, 2014, para. 1) similar to the European NACE codes. The research focused on Code 13 "Manufacture of textiles" where nine of the eleven sub-codes were included in the study (Fig. 7.2). There were two sub-codes that were excluded, as it was assumed that companies under "13.92: Manufacture of finished textile goods, except articles of clothing" would construct finished textile goods and thus fall out of scope (Fig. 7.2).

Whilst the DB07 Code 14 "Manufacture of articles of clothing" would suggest the construction of textile products and thus be excluded, the inclusion of two of its sub-codes were allowed due to the nature of knitted textiles production methods. Unlike weaving, in which fabric is produced before products are constructed, knitted textiles and the construction of the products using this method are often combined. Therefore, it was considered relevant to include the two sub-codes within "14.3: Manufacture of knitted and crocheted articles of clothing" (Fig. 7.2).

The Danish CVR (The Central Business Register) was used to search for the list of companies under the eleven DB07 codes used in the research. From the 470 companies listed under these codes as active, exclusions were only made if the

- — 13: Manufacture of textiles
  - — 13.1: Pre-treatment and spinning of textile fibres
    - — 13.10: Pre-treatment and spinning of textile fibres
      - 131000: Pre-treatment and spinning of textile fibres
  - — 13.2: Weaving textiles
    - — 13.20: Weaving of textiles
      - 132000: Weaving of textiles
  - — 13.3: Finishing of textiles
    - — 13.30: Finishing of textiles
      - 133000: Finishing of textiles
  - — 13.9: Manufacture of other textiles
    - — 13.91: Manufacture of knitted fabrics
      - 139100: Manufacture of knitted fabrics
    - — 13.92: Manufacture of finished textile goods, except articles of clothing
      - 139210: Manufacture of home textiles
      - 139220: Manufacture of finished textile goods except home textiles and clothing
    - — 13.93: Making carpets
      - 139300: Manufacture of carpets
    - — 13.94: Manufacture of ropes, ropes, twine and net fabrics
      - 139400: Manufacture of ropes, ropes, twine and net fabrics
    - — 13.95: Manufacture of non-woven fabric and goods made of non-woven fabric, except articles of clothing
      - 139500: Manufacture of non-woven fabric and non-woven goods except clothing
    - — 13.96: Manufacture of other technical and industrial textiles
      - 139600: Manufacture of other technical and industrial textiles
    - — 13.99: Manufacture of other textiles ian
      - 139900: Manufacture of other textiles ian

- — 14: Manufacture of articles of clothing
  - + 14.1: Manufacture of articles of clothing, except articles of clothing made of fur skins
  - + 14.2: Manufacture of goods from fur skins
  - — 14.3: Manufacture of knitted and crocheted articles of clothing
    - — 14.31: Production of knitted and crocheted hosiery
      - 143100: Manufacture of knitted and crocheted hosiery
    - — 14.39: Manufacture of other knitted and crocheted articles of clothing
      - 143900: Manufacture of other knitted and crocheted articles of clothing

Fig. 7.2 Danish Industry Codes (DB07) 13 and 14 that have been included (light grey) and excluded (dark grey) in this study

company: did not have a website; did not have any textile manufacturing within Denmark (partial manufacturing abroad was accepted); was based in Greenland; or was classified by the researchers as a "hobby"/"artistic" production of textiles.

This left 112 companies. Two further manufacturers were added through a "snowball" method (Naderifar et al., 2017) in which the companies appeared during a Google search for another company or were mentioned on another company's website. In addition, the list was given to Betina Simonsen, CEO of the Lifestyle and Design Cluster that works to promote innovation and sustainable growth within interior, fashion and creative industries throughout Denmark. She and her team were asked to check the list and establish if there were any manufacturers, pertinent to the RQs, missing. This led to the addition of one manufacturer. Therefore, the final list consisted of 115 firms which were then mapped (including the basic data used

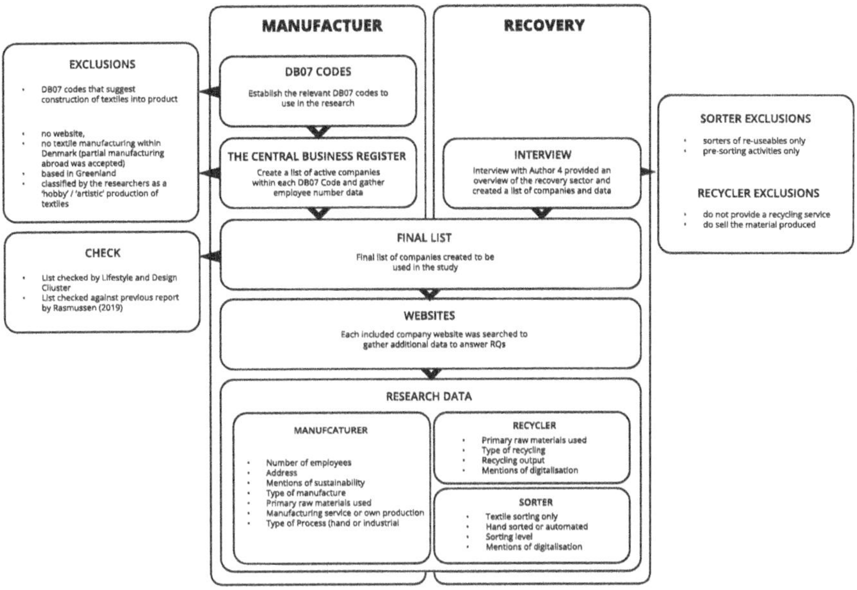

Fig. 7.3 Research process, exclusions and types of data gathered

in the study for each company) by location onto a Google map by Hall and Hasling (2024), and the information provided on their websites was used to answer the RQs. Figure 7.3 provides an overview of the research process.

Whilst using DB07 codes was an effective way to compile the list of manufacturers based in Denmark, there were also limitations. Although companies in Denmark are given the most appropriate DB07 code, it may not be wholly representative of the nature of their business. For example, most companies that provide embroidery services are listed under code "139900: Manufacture of other textiles" and many of these also provide printing services. Yet other manufacturers that provide primarily print but also embroidery services were listed under "133000: Finishing of textiles". Therefore, there will be companies that would be classed as "textile manufacturers" that are classified under DB07 codes not considered for this study. Nevertheless, even though using the DB07 codes was not the perfect method, it was chosen as the most appropriate within the time constraints.

For the second part of the research (recovery), the relevant DB07 code "383200 Recycling of sorted materials" was not an effective method to find the textile recovery companies. This was mainly because the companies categorised under this code were fairly diverse, including those that deal with non-textile waste material such as metal, electronics and plastics. The number of textile waste companies was expected to be a small proportion of the 240 in total. Furthermore, the recovery companies may be categorised under other DB07 codes and thus, an alternative method was chosen. Building upon the authors' previous experience in the field of Danish textile collection, sorting and recovery, a semi-structured interview (Brinkmann, 2014) method between author1 and author4 was used to discuss and map the Danish recovery sector. From this, after some exclusions (see Fig. 7.3 and Sect. 7.3.2), a list of thirteen textile recovery actors was collated from previous work and knowledge and data points noted. After the list had been created, it was observed that the recovery companies were categorised across a range of different DB07 codes reaffirming the validity of the method used.

The final list of 115 manufacturing companies (see Appendix, Table 7.5) and thirteen recovery companies was cross referenced with a previous research report of the Danish textile industry using DB07 codes from DAKOFA (Danish network for waste and resources) conducted by Rasmussen (2019). Once the final list had been compiled (April, 2024), each of the 115 manufacturing companies' websites was searched for different types of data (see Fig. 7.3) including mentions of sustainability/circularity and digitalisation. Corporate Social Responsibility (CSR) reports were also used to gather information, but most companies were too small to publish these, and therefore website text was the main source of information. Recovery company websites were also searched for appropriate data and information and together with the authors' previous experience in this field, the data was collected. It should be noted that one recovery actor did not have a website (the website is still in the process of being built). In this case, information from the interview and other sources of information (such as an online article about their recycled yarn) was drawn upon to gain the data required. Google translate was used to convert the Danish text into English for this research.

Whilst the study of company websites was the best method available, there could be room for error because each website is constructed differently, and information written on an unexpected webpage could be missed. Data from the websites was collected and allocated to a number of categories (for specific categories, see research data in Fig. 7.3). For two of these categories, "Primary raw materials used" and "Mentions of sustainability", the data gathered was first recorded using keywords. For example, "wool" or "natural materials" was used based on material information available. Furthermore, "UN SDGs" (United Nations Sustainable Development Goals) indicated a reference to the goals on company websites and keywords "efficiency in production" indicated if manufacturers were making efforts (rather than just setting out ambitions) to reduce environmental impacts in their manufacture (i.e. reducing wastewater, etc.).

Within the sustainability category, five themes emerged: "materials", "production", "use", "recovery" and "social activities". When collecting data for raw materials a limitation arose; not all the websites specified specific fibre types. Therefore, three overarching fibre categories were utilised based on the information available: protein (mainly fibres from animals, such as wool), cellulose (fibres from plants, such as cotton) and oil-based fibres (such as polyester and nylon). The same categories were applied to describe the recyclers.

The data compiled for the study was used alongside relevant literature to create a round discussion and answer the RQs set out above. It is recognised that data collected from company websites provides limited insight into a company's activities, especially those that are not legally required to produce CSR reports until 2025/2026 (The

Danish Business Authority, 2024). However, the purpose of this study, in answering the RQs, is to present an overview of the Danish situation and open a discussion for future scenarios based on the findings. Therefore, the methods selected where deemed acceptable. If the study was to be expanded upon in the future, other methods could be applied, such as questionnaires and/or interviews with the companies to gain a deeper insight.

7.3 Findings

This section will answer the first three RQs of this study, as set out in Sect. 7.1. Data from all 115 manufacturing companies and thirteen recovery company websites was gathered following the categories previously outlined regarding types of manufacture/recovery, location, mentions of sustainability/circularity, mentions of digitalisation and so on. This is summarised in the sections below.

7.3.1 *What Is the Scope of Textile Manufacturing in Denmark? (RQ1a)*

As 99% of all businesses in the EU are classified as micro, small or medium sized enterprises—SMEs (European Commission, 2024), it is unsurprising that nearly all manufacturers in this study fall under the SME definition of less than 250 employees (European Commission, 2003). The majority of companies were micro size (75% of total 115 companies), and over half of the total companies in the study (62%) or two-thirds of the 87 micro businesses (83%) were run by one employee. The remaining manufacturers were small enterprises (17%) and medium sized enterprises (7%). Only one company's employee numbers exceeded the SME definition (Table 7.1).

Across the manufacturers, nine different types of manufacture were found (Fig. 7.4), namely, yarn, weave, knit, print, embroidery, dye, carpet, rope and non-woven. Whilst most companies concentrated their efforts in one area of manufacture, others offered two different manufacturing types and only one example offered three. Typically, companies offered complimentary manufacture, such as two textile finishing processes—print and embroidery.

Table 7.1 Danish textile manufacturers in relation to their employee numbers

SIZE OF COMPANY	MICRO			SMALL			MEDIUM		LARGE
EMPLOYEES	1	2-4	5-9	10-19	20-29	30-49	50-99	100-250	250+
NUMBER OF COMPANIES	72	7	8	12	4	3	6	2	1
%	75%			17%			7%		1%

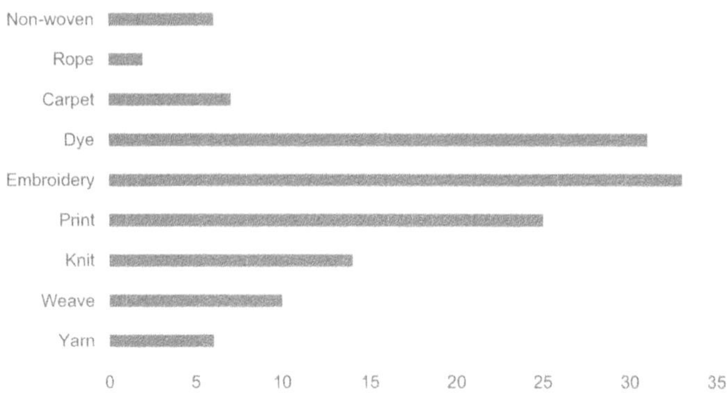

Fig. 7.4 Danish textile manufacturers in relation to the type of textiles produced

Surprisingly, the most common manufacturing process across Denmark was embroidery (33 companies), closely followed by dye (31 companies) and print (25 companies). The least common manufacturing was the creation of rope (only 2 companies) whereas the rest (yarn, weave, knit, non-woven and carpet) all had within 6–14 manufacturers.

Studying the embroidery companies more closely (Table 7.2), it is evident that the majority of firms are micro in size and mainly run by individuals. Only three of these companies conducted embroidery by hand and the rest used digital embroidery machines. Unlike traditional textile disciplines, such as weave and knit, which require large equipment, digital embroidery machines can be small and therefore easy to purchase and store. Furthermore, the primary service offered by these manufacturers was stitching company logos to garments and products (such as corporate workwear or sport teams) whilst others offered personalised embroideries (such as adding names and birthdates to soft toys).

Of the three hand embroiderers, two provided services for new embroidery work, repair and restoration, while the third was based on a social business model creating work (embroidery of a pre-designed emblem onto a t-shirt) for women on the run from war.

Across all the manufacturers in Denmark, 61% provided a service rather than just manufacturing their own products, and there was an even spread of manufacturing types across both these (service/own products). 69% of all manufacturers used industrial processes, leaving 31% using hand techniques. Overwhelmingly, dying was the most common hand produced process (26 of 35 hand manufacturers). The rest of

Table 7.2 Danish embroidery manufacturers in relation to their size

SIZE OF COMPANY	MICRO			SMALL
NUMBER OF EMPLOYEES	1	2-4	5-9	10-19
NUMBER OF EMBROIDERY COMPANIES	27	2	2	2

the hand manufacturing skills were made up of three embroiderers, two knitters, two weavers, one printer and one carpet manufacturer. Specifically, the dyers were producing hand dyed yarns, whose popularity seems to be growing. Many of the companies have grown from a hobby project into a business and there are examples of manufacturers that not only sell yarns themselves but supply their hand dyed yarns to other retailers. There was only one example of a dyer that has more than one employee, but this aligns with the view that this is an expanding manufacturing niche.

Growth within manufacturing in a European context is not the norm (Marschinski & Martinez Turegano, 2019). More notably, the export of Denmark's manufacturing skills has progressively increased over the last few decades. This is evident within the "history" pages on some of the older companies' websites, explaining how now some or all of their manufacturing was relocated/expanded abroad. Consequently, in the initial search, many companies registered in Denmark had to be excluded from the study, as they had no remaining manufacture in Denmark. The companies included in the study that had both manufacturing within Denmark and abroad tended to be non-woven production.

Finally, Danish manufacturing was analysed geographically. All the companies were placed on a map and layers were added to visually establish the location in relation to manufacturing type. The map has been published by authors Hall and Hasling (2024). Figure 7.5 shows all manufacturers, in which no clear pattern is established, as the spread is fairly even across the country.

When divided by manufacturing type, there were only two examples that provided any sort of location clustering; knit and carpet manufacture (Fig. 7.6). This is interesting as these are the only two categories that often combine textile and construction

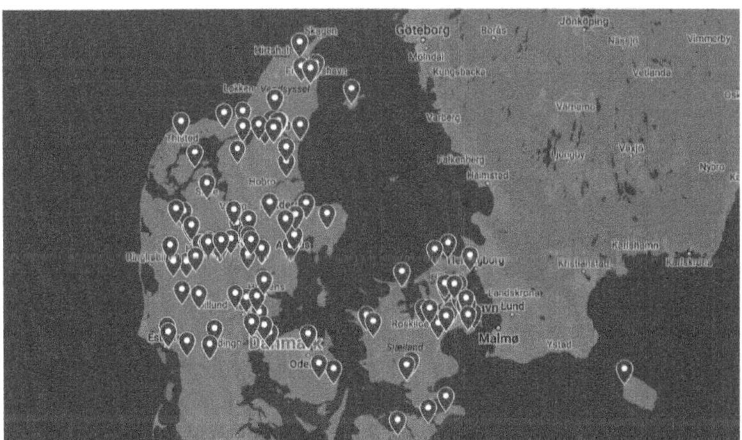

Fig. 7.5 Danish textile manufacturers' locations. *Source* Hall and Hasling (2024)

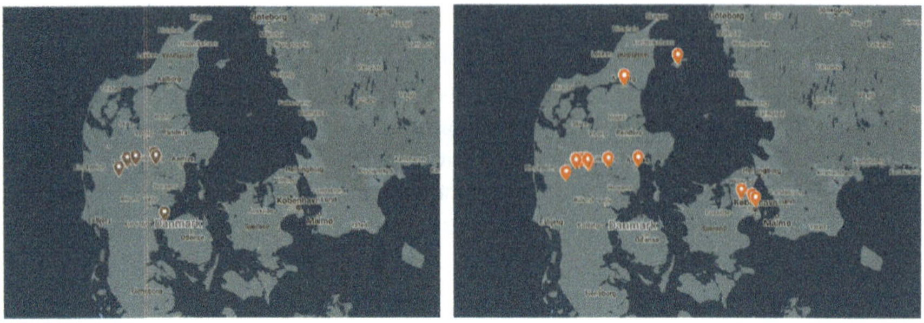

Fig. 7.6 Danish knit (left) and carpet (right) manufacturers' locations. *Source* Hall and Hasling (2024)

in manufacture (although it should be noted that there are examples of knit manufacturers creating knitted fabric without creating a product). The clustering location of these manufacturers is not surprising, falling across the middle of Jutland which holds a rich history of textile production (The Agency for Culture & Palaces, 2023). This was the only notable pattern when the data was explored in relation to location, including when location was compared to hand/industrial, service/own products and sustainability.

7.3.2 What Is the Scope of Textile Recovery Industries in Denmark? (RQ1b)

Like many other European countries, the Danish textile recovery network consists of a mixture of private, public and NGO actors. The textile waste collected within this network is predominantly post-consumer.

Pre-consumer textile waste is generated by the manufacturer (before reaching the consumer), sometimes referred to as post-industrial waste. This can be easily sorted and collected in fibre categories, as evidenced by Spanish recycler Recover, who uses pre-consumer waste from factories globally to create colour specific recycled yarns (Recover, 2024). However, more generally there is lack of transparency surrounding pre-consumer waste as the responsibility of what happens to it often falls to the manufacturer rather than the client (Runnel et al., 2017). Even if this textile waste is captured, Runnel et al. (2017) found that there is lack of data on how it is passed on, such as selling for reuse/recycling or disposing of it. In contrast, post-consumer has a more unified collection system. In Denmark, with the small number of manufacturers, there is no system or study that outlines these pre-consumer waste routes.

Since the publication of the EU waste legislation requiring member states to establish separate collection of post-consumer textile waste by 1 January 2025, a spotlight has been shone on the current system (European Commission, 2022b). Denmark has paved the way by implementing this post-consumer textile waste collection system in advance of the 2025 deadline, and in doing so the network is in a transition. Where previously the focus was on collecting reusable textiles, there are now two major streams of post-consumer textile waste: reusable and recyclable textiles. Recyclable clothing is defined by The Danish Environmental Protection Agency as clothing that others cannot benefit from, e.g., damaged, containing holes or stained clothes and textiles (excluding oil and paint stains, which should be placed in general waste), whereas the opposite is true of reusable clothing (Miljøministeriet, 2024).

In Denmark the strategy is to collect these two streams separately. However, both types of textiles inevitably leak into the opposing streams and thus sorting is required. Reusable textiles are collected by private actors and NGOs through charity shops, general donation bins and through donation bins within recycling centres. Recyclable textiles are collected by the municipality either at people's doors, placed in dedicated collection bins nearby or at recycling centres.

Denmark has three main types of actors within the recovery network: municipal waste management companies; NGOs and the private companies. Municipal waste management companies (MWMCs) are those formed by a single or multiple municipalities to collect different waste fractions, for example, paper, plastic, metal amongst others. Some of these also collect textiles. NGOs are typically charities or associations that collect and sell clothing to raise money for a certain cause. Finally, the private companies are specialists in textile collection and/or sorting, which they conduct to create a profit.

Sorting of textile waste comes in many different levels. Here we will discuss three types of sorting that occur in Denmark: pre-sorting, simple sorting and detailed sorting. The first step is to remove non-textile items and dirty/wet textiles (removed as waste and often sent for incineration) from the collection, but leaving the rest of the textiles unsorted. This is described as pre-sorting. Companies that soley undertake this type of task, and do not continue to conduct simple and/or detailed sorting, send the textiles on to private sorters abroad (these "pre-sorters" have been excluded from the study).

Other companies, in addition to pre-sorting, provide "simple" sorting. This is the removal of reusable textiles from recyclable streams and vice versa. This results in three categories: waste, recyclable and reusable textiles. In some cases, this type of sorting is not viewed as a "sorting" activity, but described as "optimisation of logistics", meaning the company is optimising how to transport textiles to the next step. In Denmark, MWMCs are legally not allowed to sort textiles, and if they do conduct these "optimisation" activities, they sit on the edge of this legislation (and have been excluded from the study). Finally, detailed sorting: this is often conducted by private companies and NGOs by either first conducting simple sorting themselves, and then sorting the resulting categories further into more detailed groups (for example, by garment for reuse or fibre type for recycling). This enables these specific textile groupings to be directed to the appropriate reuse or recycling routes.

It is not easy to separate sorting actors by focus area (re-use or recycling) as both types of textiles leak into the opposing streams. This means that sorters that collect re-usable textiles are also tasked with sorting out the recyclable qualities and Vise

Table 7.3 Sorting companies that sort recyclable textiles in Denmark used within the study

Name	Company type	Method	Sorting Type
Trasborg	Private	Hand	Detailed
UFF Humana	NGO	Hand	Simple
NeoTex	Private	Hand	Simple
NewRetex	Private	Automated	Detailed
Røde Kors	NGO	Hand	Detailed
Frelsens Hær	NGO	Hand	Detailed
Blå Kors	NGO	Hand	Detailed

verse. Therefore, all sorters, excluding those that only pre-sort were collated into a list. This resulted in seven companies that sort textiles in Denmark (Table 7.3).

The findings show that there is a mix of private companies and NGOs that sort recyclable and reusable textiles in Denmark, and there are two sorters that conduct simple sorting while the rest sort in a more detailed way. The majority conduct sorting by hand but the exception is NewRetex, which has developed an automated textile sorting technology. The landscape of automated textile sorting is rapidly changing. While NewRetex is the only automated textile sorter in Denmark, they form part of a growing sector across Europe and the globe (WRAP, n.d.).

The study found six textile recycling companies based in Denmark (Table 7.4). To be included in the study, the companies needed to either recycle textiles in Denmark as a service, or recycle textiles in collaboration with others as a service to produce new textile materials to sell to clients. It should be noted that while Textile Change is a key player in the Danish recycling landscape, who is developing technology to separate and recycle polyester and cotton to produce new yarns, they have been excluded from the study as they are the developers rather than users of the technology.

Much of Denmark's textile waste is not recycled by Danish companies, but rather these textiles are exported, often to another sorting facility to be sorted again, before being sent to recyclers. Of the six recyclers, the findings show that two conduct their

Table 7.4 Recycling companies in Denmark

Name	Level	Waste	Materials	Output	Location
Textile Pioneers	Fibre	Pre- and post-consumer	Cotton	Yarn	Asia
Haack Recycling	Fibre	Post-consumer	All/mixed	Non-woven	Denmark
Convert	Fibre	Post-consumer	All/mixed	Non-woven	Denmark
Hviid.nu	Fibre	Pre- and post-consumer	Cotton	Yarn	Europe/Asia
NewRetex	Fibre	Post-consumer	Cotton (and poly/cotton)	Yarn	Europe
Retextil Global	Fibre	Pre- and post-consumer	All/mixed	Non-woven	Europe

activities locally and two collaborate with recycling companies in Europe, one in Asia, and one both in Europe and Asia, to recycle and produce textile materials/products to sell to clients.

This study is focused on the recycling of textiles via shredding or other means into fibres, and excludes "recycling" activities such as cutting textiles into squares for wiper clothes (Hall, 2021). There are three levels of recycling that can be conducted: fibre, polymer and monomer. These three levels describe what remains intact after the recycling process. At fibre level, the fibres are left, whereas recycling at polymer level breaks the fibres into polymer chains. Finally, recycling at monomer level reduces the polymer chains to the original building blocks (monomers or oligomers). After recycling—regardless at which level—to create a textile again, monomers must be built up to polymers and to fibres, before they can be turned into a non-woven material or spun into yarn (Harmsen et al., 2021).

Across Denmark, the recycling companies are focused on fibre level recycling. Overall, recycling companies concentrate on cotton fibre and there is an even mix of yarn and non-woven output. If the four companies which recycle textiles outside of Demark are excluded, mixed fibres are most prevalent (used by Convert and Haack Recycling), although mono-materials could also be used within their processes and turned into a mono-non-woven product.

All the companies utilise post-consumer textile waste, except for Textile Pioneers and Hviid.nu, who also recycle pre-consumer factory offcuts. While Hviid.nu is a Danish company, they produce their textiles in various countries across Europe and in Asia. They have produced recycled aprons for Danish company Jespers Torvekøkken, who rent these out through Elis' laundry service. In addition to working with factory offcuts, they both offer a service to recycle textile waste produced by companies in Denmark (and elsewhere), but it is not clear exactly what types of textiles these are.

Producing recycled yarns for textile fabrics is undertaken by three of the six companies, and the remaining three create non-woven production. However, these non-woven materials are not for the more typical hidden applications, such as insulation and the inside of mattresses (Hall, 2021), but rather a designed product. For example, Haack Recycling utilises print to create a new aesthetic of the recycled boards they make, which are used as wall panels; Retextil Global produces composite profiles/materials for visible outdoor applications such as furniture; and Convert works with the company Really, which sorts by colour to control the aesthetics of their recycled hard boards used for table tops and other applications (Hall & Larsen, 2023).

Overall, there are only a few companies in the recovery sector for textiles within Denmark (unsurprising considering Denmark's size). However, this snapshot should not be viewed without considering the wider Nordic, European or global textile recovery network. For example, across Europe, Prato in Italy has a well-established wool recycling industry (Hall, 2018) and outside Europe, wool is recycled in India and Pakistan (Hall, 2021). Until these established wool recovery networks reach capacity, it does not make sense for countries such as Denmark to create their own wool recycling industry to compete, unless wool consumption dramatically increases. This balance between supply and demand was explored in the SATIN project by

Nordic Innovation (2020). They found that Denmark is set to have an oversupply of cellulose-rich, polyester and mixed fibres waste but other countries such as Sweden and Norway have excess capacity. They concluded that collaboration is required across the Nordic region to overcome the imbalance in supply and demand.

7.3.3 What Sustainability Measures and Circular Strategies Are Being Employed by the Textile Manufacturers in Denmark? (RQ2)

67 (or 58%) of all textile manufacturers in the study and across all types of manufactures mentioned sustainability measures or circular strategies (S&C) of some kind on their websites. It should be noted that manufacturers' "ambitions" for S&C were not included, as there is no certainty that these would be actioned or achieved. Notably, within manufacturers of yarn, weave, knit, non-woven and carpet, the majority of the manufacturers mentioned S&C (Fig. 7.7), whereas only half of the print and rope manufacturers mentioned S&C. Most notably, within the embroidery companies, over two-thirds of the manufacturers did not mention S&C at all (23 of 33). Here the focus of their websites was less on the processes used to manufacture, but more on the quality and speed of the service that they could provide. As discussed previously, the majority of these embroidery companies are run by individuals, who are mainly focused on maintaining a thriving enterprise.

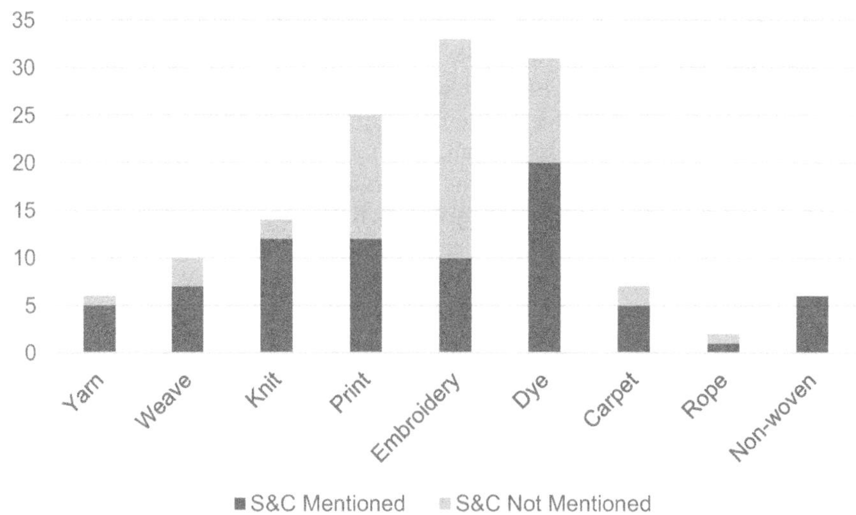

Fig. 7.7 Danish manufacturers that mention/do not mention sustainability and circularity (S&C) on their websites

The communication of S&C can be confusing, and imminent legislation is set to make reporting on the topic more uniform across Europe. Reporting will be required from 2024 for companies with over 500 employees, and from 2026 SMEs will also be included. Micro-sized companies will be exempt, but can report voluntarily (The Danish Business Authority, 2024). The current situation means transparency over S&C initiatives can be lacking. One way companies can clarify activities is by using pre-defined themes, such as the UN SDGs. Within Denmark, 11 of the 68 manufacturers that mentioned S&C, referred to the UN SDGs. It is interesting to note that these 11 companies were not represented by a specific manufacturing type. They included nearly all types of manufacture (only embroidery was not represented), and all employee number groupings (except companies run by a single employee).

Due to the lack of uniform methods used by the companies, S&C mentions were noted using keywords which were categorised across five themes: materials, production, use, recovery and social activities. The findings across these themes are discussed below.

7.3.3.1 Materials

46% of the manufacturers that mentioned S&C referred directly to materials in their communication. Many companies referred directly to the types of materials they used or broader material groups, such as "natural materials", but there was no explicit information explaining why this was a more sustainable material choice. For example, H.Hjelholts Uldspinderi (2021a) state that within a sustainability context, "we only use natural materials in our production. That is primarily wool from sheep-…acrylic and plastic do not belong in our yarns" (para. 5). While this has sustainable connotations, they do not go as far as explaining why plastic materials do not "belong".

The only fibre type highlighted for its sustainable benefits was wool. For example, yarn manufacturer Danspin (n.d.) states "pure wool is 100% natural, 100% renewable and 100% biodegradable" (para. 1) and Gabriel (n.d.b) connected the benefits of wool and polyester with longevity: "we only want to use fibres that age with beauty and offer long-lasting functional and aesthetic value" (para. 3).

By far the most common method used by manufacturers to describe sustainable material efforts was through certifications. 17% of all the companies in the study, and 30% of those that mentioned S&C, used material certifications. Overwhelmingly, OEKO-TEX®, which indicates that a textile has been tested for harmful substances (OEKO-TEX, 2024), was the most common certification. Other certifications included GOTS (Global Organic Textile Standard), which, as the name suggests, certifies organic materials plus environmental and responsible production and finally, Cradle2Cradle, which assesses safety, circularity and responsibility of materials. However, almost a third of the 31 manufacturers within the material S&C theme referred to organic materials without certifications. Similarly, almost a third mentioned recycled materials in their production. Of all the manufacturers that mention S&C, only 13% used recycled material. Although this might seem low, many of the manufacturers only offer a production service, and in these cases the materials used (within the manufacturer's capabilities) would be dictated by the client.

Overall, it was not possible to establish the exact materials used by the manufacturers. However, based on the information available, an estimate of materials used by each of 115 manufacturers (all in the study) was created. The materials were broadly categorised across three types: protein fibres (such as wool), cellulose fibres (such as cotton) and oil-based fibres (such as polyester). While many manufacturers (19%) used materials from multiple categories, 37% of all manufacturers could not be categorised by their materials because they either performed a textile finishing process (28 dyers and 12 printers), or rent access to equipment (1 quilt embroidery studio and 1 weaving studio), where the material responsibilities are placed on the client.

Although the method used is an estimate based on the information available, Fig. 7.8 provides an overview of the fibre types used by manufacturers that could potentially turn into pre-consumer waste in their production. Many manufacturers are counted in two or all three of the fibre type categories, as they use a range of material types. The results clearly show a higher number of oil-based fibres being used. 67% of the manufacturers using oil-based fibres used only these types, leaving 33% to manufacturer oil-based as well as protein and/or cellulose fibre. When the oil-based materials were divided into an estimation of polyester, nylon or other oil-based fibres, polyester was the most prevalent fibre type. However, a significant number of the manufacturers using polyester were embroidery companies, who used polyester sewing thread rather than a textile fabric. When excluded from the data, the remaining manufacturers only used slightly more polyester than nylon.

Studying the data further, those manufacturers that used cellulose fibres nearly always combined these with protein and/or oil-based fibres. Only 3% of all 115

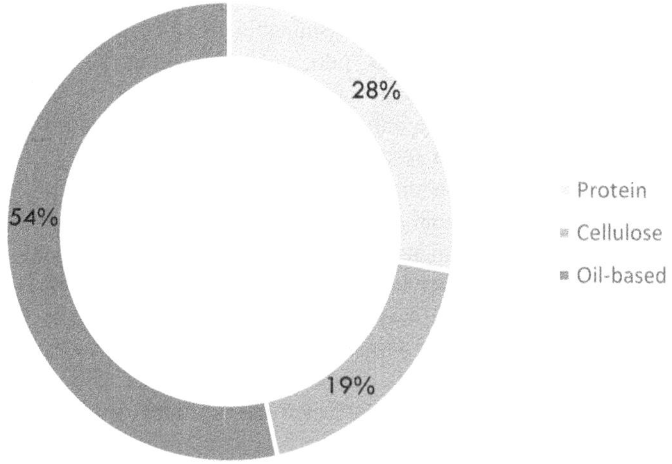

Fig. 7.8 Estimated material types used by Danish manufacturers (excluding 42 manufacturers which could not be categorised by material type)

manufacturers in the study used solely cellulose fibres, compared to 10% using protein only, and 31% oil-based only.

7.3.3.2 Production

61% of the manufacturers that mentioned S&C communicated improvements to production. The keywords used can be divided into initiatives or projects involving “energy”, “chemicals”, “reducing waste” and “increasing supply chain transparency”.

The initiatives surrounding energy were divided between switching to green/ renewable energy sources or implementing initiatives to reduce energy consumption. For example, Fletco Carpets (n.d.a, para. 2) state, “in 2016 we made the conscious decision to only utilize 100% Danish and CO_2-neutral wind energy in our manufacturing plant”. In addition to wind energy, other manufacturers were using solar panels to generate electricity. While investments like these were normally taken on by companies with more employees (Fletco Carpets have 50–99), manufacturer Handdyed by Charlotte Spagner (n.d., para. 4), which has only three employees, also sourced electricity with solar panels.

Other companies approached this through energy efficiency measures, such as Herning Fabrics A/S (2023) who have introduced “low nox burners for the steam boilers, which reduce the CO_2 emissions from the supplied natural gas” (para. 6), as well as increased use of dope dyed yarns (where dyeing occurs in the yarn and not after weaving) to “reduce the number of process baths from 16 to 2–4” (para. 7). These types of energy efficiency initiatives are evident across the different textile manufacturing types, such as Fibertex Nonwovens (2022), which has created a more efficient transmission of heat from their rollers to material. They explain “with this efficiency upgrade, an energy-intensive part of the production process can now be left out without compromising product properties” (p. 12). Finally, this was also demonstrated by hand dyer By Mehlsen (n.d., para. 3) who has removed the need to use energy in their process. They removed heat entirely, by utilising the natural fluctuating air temperatures in Denmark across the seasons to produce different outcomes.

However, efficiency in production was not only linked to energy but also related to the reduction of waste. Rodinia Generation (2023) is a micro factory concept that provides on-demand production services with digital printing technology to create “ultra-fast and sustainable” production. This model is echoed by the brand Nadia Wire (2024), which does not hold product stock, but rather knits its products on-demand. Two yarn manufacturers sold waste yarns to customers. For example, Cepra (2024) sells their hand dyed yarns that do not pass quality control checks at a discounted price, and Skive Garn ApS (known as Henrichsens Uldspinderi (n.d.) on its weblink) holds a factory sale for leftover yarns every Tuesday and Thursday.

Textile dyers, specifically, focused on the reduction of water and waste dye. Handdyed by Charlotte Spagner (n.d.) explains “the recipes for the various colours are calculated in a way that the yarn absorbs almost all of the dye” (para. 5) and Fru

Krogh (2024) observes "it has become a sport to use the same water over and over and refine the process so much that now many weeks pass between when I change the water in a pot" (para. 9). H.Hjelholts Uldspinderi (2021b) specialises in the spinning of yarns but has chosen, because of limited washing facilities at their own location, to export the washing, drying and even some dying processes to partners in England. They explain:

> We have chosen our partners in England because they have a long tradition of treating wool - in addition, they have significantly more efficient systems where they can recirculate the water much better than we can, so the process in England is much greener than we can perform ourselves (para. 11).

The most prevalent method of sustainable production was the use of environmentally friendly or natural dyes. Many of the hand dyers did this, such as Weeds and Wool (n.d.), who "mainly colour with plants I either grow in my garden or pick from the edge of the ditch where I live" (para. 2). In addition, the methods of collecting plants for dying were also considered, as Vildspire (2024) illuminates:

> When I collect, I always make sure to only cut from the plant, not uproot it and to collect widely so as not to empty an entire area - that way the plants can survive well and come back again and again, even if I use a little for my colouring. My aim is always that you should not be able to see that I have gone over with the scissors (para. 11).

For printing, more environmentally friendly inks were also cited. Unicorn Screen Print ApS (n.d.) explains how their "printing inks are…of course free of PVC, lead and phthalates. The bit of chemistry that is used here is made from malic or citric acid, and can in principle be drunk" (para. 4).

Finally, many of the manufacturers drew on their location within Denmark to promote responsibility and traceability. Danfloor (n.d.) provides an example of this:

> Both design and production take place in Denmark. This means that we have control over all stages of the production process and have associated various environmental management systems, which ensure that everything takes place with the least possible impact on the environment. In purely practical terms, this also means that when the carpet is to be delivered, we have control over the logistics and ensure the shortest transport time with the smallest possible CO2 footprint (para. 3).

7.3.3.3 Use

36% of manufacturers that mentioned S&C discussed the use of the products they produce. Perhaps it is not surprising that this percentage is less than the mentions of materials and production, because the use phase for manufacturers is mainly outside of their control. However, one way of doing this was by referring to the quality of the product produced. References to quality were only included if it was related to S&C in some way, rather than a promotion of their workmanship. For example, LogoNålen (2023) explains "we never compromise on quality. We use advanced technology to embroider or print your logo onto the garment with precision and durability in mind" (para. 2). Here the mention of durability emphasises their desire for a longer lasting product, which is a vital part of the CE. Similarly, JV Knitwear ApS (n.d.) explains "with our many years of experience, we have become specialists in knitting quality products with a long shelf life" (para. 3).

Furthermore, there were more explicit mentions of the use phase. In particular, Det Vilde Spinderi (n.d.) explains "we believe that local textile designs and involvement of the end user can help create a more sustainable textile production" (para. 3), and Herning Fabrics A/S (2023) states that by increasing durability of their textiles they hope to "contribute to reducing overconsumption and its negative environmental impact" (para. 2). Finally, Vildspire (2024) spells out the importance of the use phase clarifying "sustainability for me is also about how and for how long the product is used" (para. 14). In another way, the extended use of the manufacturing equipment was also promoted by H.Hjelholts Uldspinderi (2021a). They describe with pride, that, although old, they have kept their machinery going rather than replace it, in their efforts towards sustainability.

Examples were also found within carpet manufacturing of care and maintenance guidelines for customers to follow, and extend the life of the product. Carpet manufacturer Ege Carpets (n.d.a) even went as far as producing mini films to demonstrate how to remove common substances from the textile, such as blood, chewing gum, pen stains and nail varnish.

Finally, two hand embroidery individually run companies offered repair services to extend the life of textiles. Berthe Bramsen Broderi (n.d.) describes their workshop as "well-stocked with all kinds of materials and tools to carry out restoration and repairs of older embroidery, lace, carpets and other textiles" (para. 5). Likewise, Lille Klode (2024) explains "I also do private commissions, so if you have a favourite shirt with a hole…or something else entirely, I could be the right person to help you" (para. 2).

7.3.3.4 Recovery

Only 12% of manufacturers that mentioned S&C discussed the recovery of their textiles. While it is lower on the waste hierarchy priority list, for a CE, recovery is a vital final step. Whilst recycling of the final product by a manufacturer's customers can be difficult to influence, the recycling of waste produced by the manufacturer themselves was only mentioned by a few companies. For example, JP Air Tech (n.d.c) explains that they "seek to increase the recyclable fraction of our waste load, and we are searching for ways to enhance pre-sorting" (para. 6). Henrik Ellerbæk A/S (2023) states that all their "waste at the factory is sorted and sent for recycling" (para. 10).

Recovery initiatives were predominantly taken on by larger companies, with all but two having more than 30 employees. While some only mention their efforts in recycling, others provide more detail. For example, Foamtex (n.d.) describes the benefits of a mono-material design approach. They explain, "if the carpet yarn is PP [polypropylene], the whole carpet can be manufactured in PP, and this gives the possibility of recycling the carpet" (para. 1). Gabriel (n.d.a) has launched a fabric line called "Gabriel Loop" which is "a new category of products based on recycled textile waste—also called textile-to-textile recycling" (para. 2). However, it is unclear whether this process uses their own waste or that of others. Dansk Wilton (n.d.a) explains the challenges of recycling carpet waste for new carpets (closed-loop), when the recycling process reduces the quality of the materials. Instead their "Re:Shape" pilot project creates new products from the recycled carpets (open loop). They explain "we have designed various interior design solutions, from acoustic elements to small furniture. And our brand book is covered in this Colour Line carpet waste" (para. 15).

Furthermore, Ege Carpets (n.d.b) has designed their CircleBack carpets to be more easily disassembled, and launched a take-back system so they can "recycle 98–100% of the used carpet's components" (para. 2). While their carpet offcuts are used in "riding arenas and sports gear", the CircleBack initiative means that "the yarn fibres are reborn into new yarn of which some is turned into new carpet" (closed loop) (para. 9).

7.3.3.5 Social

36% of manufacturers that mentioned S&C discussed social initiatives towards sustainability. One example, UMM (n.d.) uses a social business model providing employment in the form of hand embroidery for "mothers on the run from war", creating "empowerment, strength, community and hope for a better future" (para. 1). Others focused their efforts on smaller changes for their employees in the day-to-day. For example, JP Air Tech (n.d.a) explains how they have given their workers a social space that is "like being at home" (para. 5) and Textil Print (n.d.) has proudly sponsored a local handball and football club for decades. They explain the benefit of this:

> The majority of our employees come from the local area - and our young, committed printers have often become acquainted with Textil Print through the sports clubs or the local community. It provides security and stability - both for you and us (para. 7).

However, most of the bigger companies describe social efforts through their policies on anti-corruption and human rights. Thygesen & Birk Fabrics (n.d.), for example, outlines this by clearly stating that "if the supplier is involved in violation of human rights, either directly or indirectly, he must cease immediately" (para. 4). Furthermore, Thygesen & Birk Fabrics is not allowed to "offer, promise, receive, allow, encourage, or use any form of corruption, bribery, or extortion" (para. 9) and "cannot use or take advantage of any kind of forced or involuntary work. It is forbidden to use child labour" (para. 11). Some manufacturers took this one step further by introducing a whistle blowing policy. For example, Dansk Wilton (n.d.b) explains, "should you have any concerns of a breach or suspected breach of any laws, company policies, standards, or behaviours, you are able to report it confidentially" (para. 1).

Yet, the most prevalent mention within the social category was animal rights (60% of all social mentions). Specifically, the manufacturers who used wool, and clearly communicated sourcing fibres or yarns that were mulesing-free. H.Hjelholts Uldspinderi (2021a) explains that "mulesing is a reprehensible process that does not align with our values of animal welfare" (para. 8). Galaxy Glow Yarn (2024) summarises the feeling from the manufacturers by saying "the yarn is carefully selected so that quality and animal welfare go hand in hand" (para. 1).

7.3.3.6 Summary

Overall, a wide range of approaches are being taken by Danish textile manufacturers to address sustainability and circularity. The focus of these approaches relates to the reduction of impacts in the manufacturing process. This result was expected, as production is the main activity for these companies, and therefore S&C in this area would be investigated first. The next priority for manufacturers was sustainable materials, mainly achieved through certifications.

While much of the sustainable discussion surrounded wool (material choice and animal welfare in the social category) the most predominant fibres used within textile manufacture across Denmark were found to be oil-based fibres, such as polyester and nylon. While recovery of materials is a key issue for a CE, initiatives in this area were evidenced in only a limited number of manufacturers. While this could be put down to the fact that many of the manufacturers had no control over what their customers did with the textiles, this was not the case for the use phase (which could also be considered beyond the manufacturer's reach). Compared with recovery, use was considered by more than double the number of companies. This was due to the manufacturers' ability to design textiles to be more durable, and by providing knowledge to the customer on how to care and thus extend the life of the textiles produced. While all these findings are encouraging, there is still plenty of work to be done within textile manufacture to reach true sustainability in the field and to transition to a CE.

7.3.4 What Is the Role of Digital Technologies in Supporting Sustainability and Circularity Within Textile Manufacture and Recovery in Denmark? (RQ3)

In order to answer this RQ, the findings are addressed in two sub-sections concerning digitalisation within manufacture, and then digitalisation within recovery.

7.3.4.1 Digitalisation Within Manufacture

Only 3% (4 of 115) of manufacturers mentioned digital technologies in relation to S&C. Whilst this is a very low number, increased mentions might have been gained if the study had included digitalisation not specifically promoting S&C. In this context, Denmark has been shown to be an EU frontrunner in digitalisation within Europe (European Commission, 2022a).

Whilst it could be concluded from these numbers that digitalisation is not prevalent in the communication of S&C by manufacturers of textile materials in Denmark, this might be explained by the fact that they do not discuss it on their websites. For example, Scan HTP, a printing company, explicitly explains the environmental benefits of digital printing over traditional screen printing:

> Digital print meets an increasing demand for sustainable production. The entire process saves resources and has a significantly lower impact on the environment than any of the conventional printing techniques (Scan HTP, 2024, para. 1).

This point is true of all the print manufacturers providing a digital printing service, even if not all of them point this out. Rather, their websites focus on other selling points, such as how reliable and cost effective their manufacturing service is. This might change if the companies in question are legally required to publish CSR reports by 2026, or do so voluntarily (The Danish Business Authority, 2024).

Whilst the research carried out shows that digital tools are not a focus area within textile manufacture, this overlooks the prior digitalisation of the industry with the introduction of automated machinery, such as digital knitting and weaving machines. While this has speeded up production, the energy consumption within the manufacture of textiles (yarn, fabric and wet treatment) has become very high (Sandin et al., 2019). This contributes to these areas forming a large proportion of a garment's climate impact (Wennberg & Östlund, 2019), and it therefore becomes more important for manufacturers to consider greener or regenerative energy sources for their production.

Nevertheless, some digital initiatives have been implemented. For example, JP Air Tech (n.d.b) has installed energy metres to create (what was assumed to be digital) data "to evaluate and compare measurements, identify energy-demanding processes, and ultimately reduce energy use through optimal maintenance" (para. 1). At the same time, Herning Fabrics A/S (2023, para. 8) encourages customers to take digital meetings to reduce increasing carbon footprints. These serve as examples of companies taking measures to reduce the environmental impacts of their businesses.

Furthermore, Herning Fabrics A/S (2023) has started using new administration software (partly AI-based) to "report precise facts (e.g., about chemical purchases) in relation to the indirect environmental impacts" (para. 13). This software, they explain, has been implemented to help validate "environmental data towards the upcoming digital product passport" (para. 13). Here Herning Fabrics A/S is preparing for the upcoming Ecodesign for Sustainable Products Regulation (ESPR) in which "digital product passports will be the norm for all products regulated under the ESPR, enabling products to be tagged, identified and linked to data relevant to their circularity and sustainability" (EU Commission, 2022a, 2022b, 2022c, p. 5). If manufacturers are not currently adopting digital tools, there is set to be an increase in this area once the legislation is passed.

Finally, Fletco Carpets (n.d.b) communicates their link to the digital "Mindful Materials" product library for sustainability. While it was unclear if this was a resource for others to use, or something the company's products were listed on, it emphasises how digital libraries are used to help customers access information in a single location to make more informed decisions about materials and production processes, without having to visit a manufacturer and ask questions. Other examples include the Higg Index. This also points to a wider discussion on digital technologies that allow companies that manufacturer outside of Denmark to communicate and make environmental reductions, but this falls outside the scope of this study.

7.3.4.2 Digitalisation Within Recovery

The recovery sector is, itself, part of a circular and sustainable ecosystem and therefore any mention of digitalisation found on websites was included in the study. Again, the research method relied upon the company mentioning initiatives they may be undertaking.

The findings show that only one recovery actor promoted digital technology: NewRetex. NewRetex's whole business is centred around automated sorting and therefore digital tools are at the very core. NewRetex (n.d.a) explains that they use "sensor, robot and automation technology… to carry out intelligent and automated sorting of discarded textiles" (para. 3). Today, NewRetex (n.d.b) has two robots that sort the textile waste, near-infrared (NIR) sensors and cameras. "Machine learning", they explain, means that each textile is scanned and compared to the known compositions within the software and then sorted accordingly. The benefit of this is that it "ensures efficient and documentable sorting according to colour and material composition" (para. 4).

The use of digital technology for sorting textile waste has long since been a discussion, but there are still many challenges to overcome. Sunde et al. (2023) explain that by integrating advanced machine learning and Internet of Things (IoT) technologies, the manpower and costs required for sorting could be substantially decreased. Furthermore, they assert that:

> The emergent Digital Product Passports (DPP) technology facilitates the effortless acquisition of data about a garment's origin and constituent fibers. Harnessing machine vision and learning techniques enables the categorization of these items for subsequent processes (Sunde et al., 2023, p. 9).

However, as Pehrsson et al.'s (2023) assessment of textile sorting technology points out, there are two important challenges still to overcome. Firstly, the databases behind the technologies require more development, and secondly, the accuracy and precision of the scanning are not yet at the level required. To unlock the potential of data sorting in the future, they call for digitisation of the entire sorting process, including pre-sorting.

The NewRetex digital sorting process is amongst a small list of companies that are utilising digital technology. Other examples include: SipTex and Fibresort. Thus, while the recovery network across Denmark might be small (and by and large utilising non-digital methods), NewRetex's innovations represent a key initiative that should allow Denmark to contribute to the next digital generation of textile sorting.

7.4 Discussion: What Circular Opportunities Are There for Manufacturing and Recovery in Denmark? (RQ4)

The findings from this research provide the basis for further discussion of opportunities for the transition to a CE in Denmark. This section addresses the final RQ: what circular opportunities are there for manufacturing and recovery in Denmark?

Manufacturing and recovery are only two parts of the wider circular system and each of these represents a method to add value to a product. As Achterberg et al. (2016) point out, value is added as a product is developed, from the extraction of the raw materials, through to the manufacture (the focus of the research here) and further into construction and retail. However, in the traditional linear economy "after a relatively short lifecycle the product's value is destroyed quickly and value rapidly goes downhill" (p. 4) at the recovery stage. While the first three RQs in this study focus on exploring the textile manufacturer and recovery actors within the value chain (Fig. 7.1), this final RQ explores how the flow can be reversed, and value can be retained (Fig. 7.9). To retain this value, circular strategies need to be employed. Fontell and Heikkilä (2017) highlight the four value cycles within a CE which this discussion will centre around:

1. Maintain and repair,
2. Re-use as product,
3. Remanufacturing and reuse as material, and
4. Recycle.

For this research, the focus will be on value cycles 1, 3 and 4 (discussed in the subsections below). Value cycle 2 (reuse as product) has been excluded, as the research specifically focused on the manufacture of textile materials and excluded solely the construction of these textiles into products. Furthermore, the focus of the research was not on consumption/use, and therefore the circular business models that deal with extending products' lives in their constructed state are not within the research scope.

To maximise these cycles, the Ellen MacArthur Foundation (EMF, 2013) describe the "power of circling longer" (p. 30) in which, a textile (in this case) should cycle around for as long as possible and through as many cycles as possible to extend its lifetime. To enable this, Fontell and Heikkilä (2017, p. 50) explain that circular

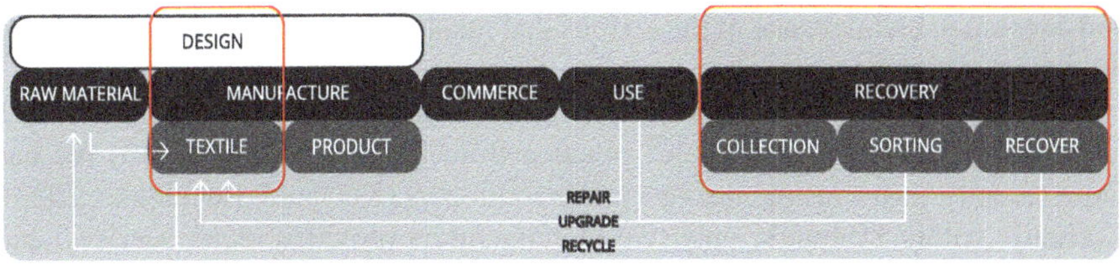

Fig. 7.9 Three circular business models, relevant to this study, within the textile value chain that involve textile manufacturing (based on Stamm et al., 2019 and Arponen et al., 2018)

business models "play a crucial role in a transition towards the circular economy of textiles".

In a discussion of their "circular business ecosystem of textiles", Fontell and Heikkilä (2017, p. 22) acknowledge the importance of the design of a product. They argue that it is first and foremost the design of a textile product that ensures that it can be used in its original format for as long as possible. In the research, while some manufacturers did allude to the design via "quality" and "durability" of the products produced, this was not a focus area for most of them. Manufacturers mainly create the textiles either at the specification of designers or create textiles for designers to turn into a product, which could explain the lack of emphasis on the design. However, conversations around extended use by design should be encouraged by manufacturers. This will provide clients with a view of the bigger picture to continue making informed choices beyond just the choice of a textile for a design.

Moreover, Fontell and Heikkilä (2017) suggest that users (those that purchase and/or wear textiles products) play a key role. The users, they rationalise, are at the centre of the ecosystem to maintain, reuse and share a textile product to extend its life. While the user themselves falls outside of the scope of the research, the engagement of the user is often conducted through businesses (including manufacture and recovery) that provide services to enable all four of the value cycles to function, such as repair, upgrade and recycling services that a user can access. Three circular business models that are enabled by textile manufacturing, relevant to this study (see Fig. 7.9), will be expanded in the discussion that follows.

7.4.1 Maintain and Repair

Fontell and Heikkilä's (2017) circular business ecosystem omits a consideration of maintenance, which would involve the care of a product/textile. Within a Danish context, there were a few examples, mainly carpet manufacturers, that provide care information. Notably, while the construction of products was excluded from this study, it was the manufacturers, permitted within this study, that created a product in which this occurred. Nevertheless, this points to an opportunity to communicate and encourage maintenance of the textiles that manufacturers produce, regardless of the products they become.

Repair forms a bigger opportunity within a Danish context. Repair of textiles has been described as a "humble" practice from the past (König, 2013) but has more recently seen a resurgence by craft practitioners (Kucher & Ravnløkke, 2023). Currently within Denmark, only two manufacturers provided repair services, and these provide bespoke hand embroidery services by a single practitioner. This type of repair (by hand) is a more typical method to mend textiles. However, of the 33 embroidery companies, 30 of them provided digital embroidery services. In the context of digitalisation, an opportunity exists for a different kind of embroidery repair service.

Other repair services exist in Denmark, such as dry-cleaning companies, who offer construction related repair and alterations, using a sewing machine (which was not included in the scope of this study). In addition to repair with a sewing machine and hand embroidering repairs, digital embroidery could be used. The network of digital embroidery machines found in this research is spread evenly across the country (Hall & Hasling, 2024). This is a circular opportunity, not because these embroidery manufacturers evidence a desire or skill to provide repair services, but rather that they have the know-how to deliver micro-embroidery manufacturing, designed by others, that could be implemented for repair. The fact that these companies are located sporadically across Denmark and are largely micro-sized enables this network to deliver decentralised, on-demand services, which if harnessed for a CE, has vast potential. Central to this circular opportunity is the design of digital embroidery repair. Once developed, this would enable a new circular business model, through which—by harnessing the manufacturing network—users could access digital embroidery repair services to extend the life of their textiles.

7.4.2 Remanufacturing and Re-use as a Material

Reuse as a material is described by Fontell and Heikkilä (2017) as

> Discarded textiles which are, for example, damaged so that they can no longer be easily repaired for re-use in their original shape and form, but still have good-quality fabric, could be used for this purpose (p. 24).

However, the methods of material reuse, they explain, are to cut a product into smaller materials and reconstruct them. This can be conducted by both individuals performing "handicrafts" or more professional "upcyclers" (which would fall outside of the scope of this study as a construction-based activity). However, what is omitted from Fontell and Heikkilä's (2017) model is a discussion around other forms of remanufacture, that falls between the direct reuse of a product (with no changes) and reuse of the materials into a new product. One such additional life-extension business model is described by Arponen et al. (2018) as "upgrading". They define upgrading as a method to "improve product performance by upgrading existing components with newer ones" (Arponen et al., 2018, p. 26). Yet for textiles, this definition does not always fit, as the textile within a product cannot be easily split into its component parts. The methods of upgrading, relevant to this research, are the application of finishing techniques, such as print, dye and embroidery to update a textile, making it more desirable and thus extending its usable life.

Across Denmark, 89 print, dye and embroidery manufacturers are currently in business. This represents 77% of all textile manufacturers in the study. Once again, as with the conversation regarding repair (Sect. 7.4.1), there is potential for decentralised and/or on-demand upgrading services. Currently, there is one example of a company (Rodina Generation) that is utilising print to create micro-manufacturing on demand. However, their services did not extend to upgrading to serve a CE. There was only one manufacturer that demonstrated a form of upgrading. Handdyed by Charlotte Spagner has used hand dying textile techniques on textile products (sweatshirts) to create a renewed aesthetic. The same principle can be applied to printing and embroidering either all over, or through specific placement, to upgrade a textile within a product. This also provides an example of how slower hand manufacturing techniques (such as hand-dying) can play a role in a CE. Overall, Denmark has a proven range of hand, digital and industrial manufacturing that, when combined, offers great potential for a thriving CE to function.

Finally, there are two ways these upgrading services could work. In the first, much like a repair, the user could take their tired and unloved reusable garment to a manufacturer to be upgraded. The second method requires discarded textiles to be first sorted for their upgrade potential, and a new business model created to upgrade this fraction of textile waste, and sell them to new users. In this second scenario, the collector and sorting network within Denmark would play a vital role, in which new sorting skills will be required to enable the CE to function.

7.4.3 Recycle

> Only when the quality of the fabric is such that it is no longer suitable for re-use, should the clothes and textiles be recycled to be used as raw materials for new products (Fontell & Heikkilä, 2017, p. 51).

In this final loop, worn-out textiles need to be separated from the reusable ones so they can flow into the most appropriate recovery routes, and this means that sorting plays a pivotal role. Within Denmark, the collection system is set up so that reusable and recyclable textiles are collected separately by a variety of actors. However, as Fontell and Heikkilä (2017) point out, consumers find it very difficult to accurately distinguish between reusable and non-reusable textiles. This is evident within the Danish collection system in which both types of textiles leak into the opposing streams. This demonstrates the importance of the small but vital role of both simple and detailed sorting practices in Denmark.

Automated sorting for reuse still faces many challenges, mainly due to the complexity of textiles (Pehrsson et al., 2023), and therefore it is still relevant to nurture hand sorting services for the CE. This is particularly true if the role of remanufacture increases. In this scenario garments that could be upgraded would need to be sorted out (as discussed in Sect. 7.4.2) and sorting this way is not yet achievable by machine. Automated sorting for recycling, however, has the potential for quicker and cost-effective sorting for worn-out textile waste. While the Danish sorting sector is small, the range of simple to detailed sorting which contains a mix of hand and automated technologies, provides a broad range of sorting businesses to service a CE.

This research has highlighted that most manufacturers are not using digital tools. However, one company promoted their implementation of digital software in preparation for the upcoming legislation on Digital Product Passports (DPPs). The introduction of DPPs is part of the EU's Eco-design directive that was approved on 27 May 2024, and will be enforced within 24 months after its upcoming publication (Council of the European Union, 2024). One of the benefits of a DPP is the information it will contain regarding a textile's material composition. This means, amongst other things, a textile with a DPP can be easily sorted for recycling.

While DPPs may not have been discussed by many manufacturers, traceability was highlighted as a key area. This was mainly due to the location of production within Denmark making the supplier chain less complicated than when it stretches across the globe. The focus on traceability is only set to increase as the European Commission (2023) is proposing to introduce mandatory Extended Producer Responsibility (EPR) schemes for textiles. They explain "EPR schemes require producers to take responsibility for the entire lifecycle of their products, in particular at the end of the product's life" (para. 23). The introduction of both EPR and DPP will certainly involve a digital element and serve sorters and recyclers to ensure the right textiles reach the correct end-of-life stream.

While sorting of post-consumer textiles waste is vital in a CE, pre-consumer waste is still a prevalent concern. Some efforts in Denmark are being made to sell on useable waste materials, such as yarn, but the route for other recyclable waste is less clear. Within the confines of the Danish borders, the types of fibres used in manufacturing are overwhelmingly oil-based, such as polyester or nylon. While it appeared that polyester was significantly more prevalent than nylon, this was because the data was not based on volumes. To illustrate, the embroidery manufacturers used polyester sewing threads and if waste embroidery threads from the Danish manufacturers' production were collected, the volumes would not match waste from other textile industries, such as salvage edges of weave manufacture, or offcuts from carpet manufacture that are bigger and bulker. Therefore, if fibre type data was collected by volumes alone, oil-based fibres might form less of a percentage of the waste. Nonetheless, the data captured tells us that polyester is estimated to form the largest share of the fibres used, whereas cellulose fibres and protein fibres (such as wool) formed a smaller share.

When this overview of fibre groups used in manufacture is matched to the recycling capabilities in Denmark, a clearer picture of textile end-of-life emerges. The local recycling actors, Convert and Haack Recycling, provide lower value recycling possibilities, transforming waste into non-woven materials (open loop). There is an opportunity for the pre-consumer waste created in Denmark to be used by these companies It should be noted that the types of products that Convert and Haack Recycling are creating are higher value (through design) than many of the global recycling companies that process mix fibres to produce non-wovens, such as insultation materials. However, while Convert and Haack Recycling can use mixed fibres, there are examples of Convert utilising mono-material waste—such as cotton or wool in the creation of furniture boards for the company "Kvadrat Really". Overall, the recycling options within the confines of Denmark are not many, yet the range between open and closed loop does provide a choice for end-of-life textiles.

The local recycling capacity is supplemented by companies working externally, such as NewRetex, Textile Pioneers, Hviid.nu and Retextil Global. While Textile Pioneers and Hviid.nu are mainly focused on pre-consumer waste produced abroad, the focus of Retextil Global and NewRetex is on post-consumer waste. While the ReTextil Global process transforms textiles into outdoor furniture (open loop), NewRetex focuses on recycling cotton and poly/cotton into yarns. Retextil Global can also recycle pre-consumer as their product does not require any specific sorting, but there is also potential for companies such as NewRetex, to process pre-consumer waste from Danish manufacturers in the future, if the quantities are large enough.

The main challenge is that Denmark does not have the necessary capacity to recycle all its waste textiles (pre and post-consumer) through its current recycling companies (Nordic Innovation, 2020). In addition, channels for pure waste streams such as polyester and nylon fibres need to be found (rather than mix them in with other fibres), and a question is raised whether Danish wool waste should be channelled into local mixed fibre recycling, or Denmark should utilise other networks, such as in the wool recycling industry in Prato (Hall, 2018) to create a CE. Furthermore, the results of the SATIN project suggest that Denmark might also have to rely on recycling capacity in Sweden and Norway in order to recover all its volumes of textile waste (Nordic Innovation, 2020).

Relevant here is the European Commission (2023) amendments to the waste framework directive that would see companies required to make financial contributions as part of the mandatory EPR schemes. These contributions would be used to ensure textiles are entering relevant recovery streams locally or abroad, which currently is under financed and not able to drive a demand for recycled materials. This type of legalisation would help to cement the Danish recovery routes internally or in collaboration with partners abroad. While these rulings are yet to be decided, the current circular landscape in Denmark does provide some opportunities for circular loops (closed and open), but a wider net would need to be cast for recycling to fully transition to a CE.

7.5 Conclusions

This chapter has explored the circular opportunities for the Danish textile manufacturing and recovery sectors. Firstly, the study investigated the nature of both areas. It found that Denmark's textile manufacturing companies comprise mainly micro-sized firms, chiefly providing textile finishing (embroidery, print and dye) across hand and industrial levels. It was estimated, with the information available, that the most predominate fibres used were oil based. The recovery sector was far smaller than manufacturing, but offers Denmark a variety of sorting levels (simple and detailed) and recycling services (non-woven and yarn).

Secondly, the research investigated how textile manufacturers were engaging with sustainability and circularity strategies. Over half of all Danish textile manufacturers in the study mention some sort of sustainable practices. These mentions on their website (the best method available) covered a range of activities within five themes: materials, use, production, recovery and social activities. Production related strategies were unsurprisingly most common, as the production of textiles is the main function of these companies. This was followed by materials, use, social and—least mentioned—recovery.

Thirdly, the research explored digitisation across textile manufacture and recovery. Here the study found very limited numbers of mentions of digital tools used for sustainable purposes. However, the limitations of the study, only exploring the company websites, also suggest that more digital technologies may be present than were mentioned. The mentions that did occur facilitated further discussions regarding the prior digitisation of the industry in which digital machinery, such as digital embroidery, is now commonplace. Furthermore, upcoming legislation was also mentioned by one company who was using digital tools to prepare for the implementation of DPPs.

The fourth and final point of discussion in the chapter drew together all the findings to pin-point Danish opportunities for a circular textile economy. Framed around three of Fontell and Heikkilä's (2017) four value cycles, the discussion covered repair, upgrading and recycling. The research found that Denmark has a decentralised digital embroidery network, which, if harnessed for a CE, has the potential to provide digital embroidery repair services for textile products. Furthermore, if this embroidery network was to be extended with print and dyers, it could in the same way provide decentralised upgrading services in a CE. This would also create an opportunity for sorters to diversify and sort textiles for upgrade to enter a new type of business model.

Finally, for recycling of textile waste, the Danish recovery sector, while small, is proven to offer a number of closed and open loop routes for textiles. However, to fully transition to a CE, Denmark, with its current solutions and infrastructure, would be required to utilise recovery routes outside of its boarders. Specifically, of particular relevance here, is the upcoming EU eco-design regulation which includes the introduction of DPPs, that aims to include important product-related data throughout the value chain that would hold vital information for the recovery sector. Alongside this, once the EU's proposed amendments to the waste framework directive are approved, EPR will be introduced for textiles which could finance and support the development of the European recovery network. Thus, the role of Denmark in this overall picture becomes more significant as it provides, albeit a small, contribution to this network.

While the research has made some interesting findings, there were limitations to the study. The main limitations were using websites, where information could have been missed due to the vast array of website constructions, companies' strategic choices may have been modestly expressed as regards to their sustainability actions for fear of greenwashing, or the sustainability initiatives may still be in the innovation and experimentation stage. In addition, searching for companies solely through DB07 codes (which are not wholly representative of company activities meaning some relevant businesses would not appear within the search) was a further limitation.

To overcome these shortcomings, further research is required to establish the companies missed in this study, as well as conducting interviews/questionnaires to obtain more accurate data regarding sustainability initiatives and digitalisation. Furthermore, the study could be expanded to investigate the Danish manufacturing network's readiness to adapt to the forthcoming ESPR regulation and examine what bottlenecks and opportunities they face to comply with the future legislative requirements. To further expand on the findings in this study, research could be conducted to explore the opportunities presented in this chapter. For example, exploring how the decentralised manufacturing network (such as embroidery) could be harnessed to progress new circular business models, and what types of products and design ideas could be applied to such a manufacturing network for upgrade and repair services. Finally, the role of sorting practices for new CE services could be profitably investigated. Overall, pursuit of these research initiatives would help to establish how Denmark can best transition to a circular textile economy.

Appendix

See Table 7.5.

Table 7.5 115 Manufacturers included in this study

Company name	Website	Type
Danspin A/S	https://www.danspin.com/	Yarn
H.Hjelholts Uldspinderi ApS	https://www.hjelholt.dk/	Yarn
Det Vilde Spinderi A/S	https://www.detvildespinderi.dk/	Yarn
Skive Garn ApS	http://www.henrichsens-uldspinderi.dk/7578/	Yarn
Nornac I/S	https://www.nornac.dk/	Yarn
Herning Special Væveri A/S	http://www.hsv.dk/dk.aspx	Weave
Eva Press—Vævestuen	https://evapressdesign.com/	Weave
Korneliusen	https://korneliusen.dk/	Weave
Danish Art Weaving A/S	https://daw.dk	Weave
KE Fibertec Væveri A/S	https://www.ke-fibertec.com/en/	Weave
Møns Vævehus	https://moensvaevehus.dk/	Weave
Kjellerup-Vaeveri	https://kjellerup-vaeveri.dk/	Weave
Gabriel A/S	https://www.gabrielfabrics.com/	Weave, knit
Fred B. Balsløw A/S	https://www.balslow.com/	Weave, print
Thygesen & Birk Fabrics A/S	https://tbfabrics.com/	Knit
Bedstestrik	https://bedstestrik.dk/	Knit
JV Knitwear ApS	https://jvknitwear.dk/	Knit
Bienbien	https://bienbien.dk/	Knit
Nadia Wire ApS	https://nadiawire.com/	Knit
Skovhuus Strik A/S	https://www.skovhuus-strik.com/	Knit
AL-X	http://al-x.dk/	Knit
H.H.S. & CO. ApS	https://dk.sns-herning.com/	Knit
Deima	https://deimaknitwear.com/	Knit
Læsø Garnspinderi	https://laesoe.com/	Knit, yarn
Henrik Ellerbæk A/S	https://www.ellerbaek.dk/	Knit, weave
Sontex DK ApS	https://sontexdk.dk/	Knit, dying
Herning Fabrics A/S	https://www.herningfabrics.dk/	Knit, dying
LC Textiltryk	https://tex-tryk.dk/	Print
Scan HTP A/S	https://scanhtp.com/en/	Print
Wendt Design	https://wendtdesign.dk/	Print
Miks Modetryk	https://miks-modetryk.dk/	Print
Personlige Bånd	https://personligebaand.dk/shop/frontpage.html	Print
Texprintcph ApS	https://texprint.dk/	Print
Tryk på tøj	https://xn--trykptj-ixa6o.dk/	Print
JP Textiltryk.Dk ApS	https://jp-textiltryk.dk/	Print
Omega Print	https://omegaprint.dk/	Print

(continued)

Table 7.5 (continued)

Company name	Website	Type
Unicorn Screen Printing ApS	https://unicornsp.dk/	Print
Textil Print A/S	https://textilprint.dk/	Print
Rodina	https://www.rodiniageneration.io/	Print
Logonålen	https://logonaalen.dk/	Print, embroidery
Trykfabrikken ApS	https://www.trykfabrikken.dk/	Print, embroidery
House of Shirts	https://houseoftshirts.dk/	Print, embroidery
KB Grindsted B ApS	http://peajack.dk/	Print, embroidery
Niki Textiltryk	https://nikishop.dk/	Print, embroidery
Sportstrykkeriet ApS	https://www.sportstrykkeriet.com/	Print, embroidery
Rønde Ruby ApS	https://www.rrtryk.dk/	Print, embroidery
Convert A/S	https://convert.as/	Non-woven
JP Air Tech ApS	https://www.jpairtech.com/	Non-woven
Ellepot A/S	https://www.ellepot.com/	Non-woven
Fibertex Nonwovens A/S	https://www.fibertex.com/	Non-woven
Fibertex Personal Care A/S	https://fibertexpersonalcare.com/	Non-woven
Foamtex ApS	https://foamtex.dk/	Non-woven, dying
BY Kiilerich	https://bykiilerich.com/	Dying
Plantefarvet Garn	https://plantefarvetgarn.com/	Dying
Yarnunik	https://yarnunik.dk/	Dying
Colorful by Buch	https://colorfulbybuch.dk/	Dying
Bossen Yarn	https://www.bossenyarn.dk/	Dying
Vildspire	https://vildspire.dk/pages/	Dying
Yarn by Sanders	https://www.yarnbysanders.dk/	Dying
Farvefantasier	https://www.farvefantasier.dk/	Dying
THYBO Hand Dyed	https://www.thybohanddyed.com/	Dying
Siestoe Fibres	https://siestoefibres.dk/	Dying
Westwooly	https://www.westwooly.dk/	Dying
CEPRA	https://ceprashop.dk/	Dying
Fru Krogh	https://frukrogh.dk/	Dying
Ladybug Yarn	https://ladybugyarn.company.site/	Dying
Garnfarveriet	https://www.garnfarveriet.dk/	Dying
Garngalleriet Mikrofarveri	https://garngalleriet.dk/	Dying
Galaxy Glow Yarn	https://galaxyglow.dk/	Dying
Garnlab	https://garnlab.dk/	Dying
Sysleriget ApS	https://sysleriget.dk/	Dying
Weeds & Wool	https://weedsandwool.dk/	Dying
Unika Garn	https://www.unikagarn.dk/	Dying
Københavns Plantefarveri ApS	https://kbhplantefarveri.dk/	Dying

(continued)

Table 7.5 (continued)

Company name	Website	Type
Painter's Yarn	https://paintersyarn.dk/	Dying
Handdyed by Charlotte Spagner	https://handdyed-denmark.com/	Dying
By Mehlsen	https://www.bymehlsen.com/	Dying
Nibe Garn	https://nibegarn.dk/pages/	Dying
Stampager	https://www.stampager.dk/	Dying
ALonelySockLadyYarn	https://www.alonelysockladyyarn.dk/	Dying
Jensen Broderidesign	https://jensenbroderidesign.dk/	Embroidery
MadeByTF	https://thenameshop.dk/	Embroidery
Ah Broderi ApS	https://www.ahbroderi.dk/	Embroidery
Reklame Brodering	https://reklamebrodering.dk/	Embroidery
JKL Design	https://www.jkldesign.dk/	Embroidery
ABC-Broderier	https://www.abc-broderier.dk/	Embroidery
UMM ApS	https://umm.dk/	Embroidery
SL Maskinbroderi	https://www.sl-maskinbroderi.dk/	Embroidery
O.N. Profil Broderi	https://profilbroderi.dk/	Embroidery
Ab Broderi ApS	https://abbroderi.dk/	Embroidery
Nat & Vic	https://www.natogvic.dk/	Embroidery
Micro Print	https://www.microprint.dk/	Embroidery
Berthe Bramsen Broderi	http://www.berthebramsenbroderi.dk/en	Embroidery
Lille Klode	https://lilleklode.dk/	Embroidery
Camee Broderi A/S	https://www.broderi-butik.dk/	Embroidery
Babybroderi.dk	https://babybroderi.dk/	Embroidery
Broderigavlen	https://broderigavlen.dk/	Embroidery
Nørklebixen	https://xn--nrklebixen-0cb.dk/	Embroidery
Needle	https://www.needle.dk/	Embroidery
DaisyKwiltsfotyou	https://daisyquiltsforyou.com/	Embroidery
Vision Broderi ApS	http://visionbroderi.dk/	Embroidery, print
Dane Broderi Design	https://danebroderi.dk/	Embroidery, print
Gultentorp Tryk	https://gultentorp-tryk.dk/	Embroidery, print
Svane Broderi	https://www.svane-broderi.dk/	Embroidery, print
Niller's Textiltryk	https://www.tryk-og-broderi.dk/	Embroidery, print
Protra A/S	https://protra-group.com/	Embroidery
Dansk Wilton A/S	https://danskwilton.com/	Carpet
Ege Carpets A/S	https://www.egecarpets.dk/	Carpet
Unica Carpets	https://www.unica-carpets.dk/	Carpet
Fraster ApS	https://fraster.dk/	Carpet
Bentzon Carpets ApS	https://bentzon.dk/	Carpet

(continued)

Table 7.5 (continued)

Company name	Website	Type
Danfloor A/S	https://www.danfloor.dk/	Carpet
Fletco Carpets A/S	https://www.fletcocarpets.dk/	Carpet
Dynamica Ropes ApS	https://dynamica-ropes.com/	Rope
Roblon Aktieselskab	https://www.roblon.com/	Rope

References

Achterberg, E., Hinfelaar, J., & Bocken, N. (2016). *Master circular business with the value hill* (White Paper). Circle Economy. https://hetgroenebrein.nl/wp-content/uploads/2017/08/finance-white-paper-20160923.pdf

Arponen, J., Juvonen, L., & Vanne, P. (2018). *Circular economy business models for the manufacturing industry—Circular economy playbook for Finnish SMEs*. Sitra, technology industries of Finland and Accenture strategy. https://www.sitra.fi/en/publications/circular-economy-business-models-manufacturing-industry/

Berthe Bramsen Broderi. (n.d.). *Workshop*. www.Berthebramsenbroderi.dk. http://www.berthebramsenbroderi.dk/embroidery-workplace

Brinkmann, S. (2014). Unstructured and semi-structured interviewing. In P. Leavy (Ed.), *The Oxford handbook of qualitative research*. Oxford University Press.

By Mehlsen. (n.d.). *About me*. By Mehlsen. https://www.bymehlsen.com/learn

Cepra. (2024). *2. Sortering*. Ceprashop. https://ceprashop.dk/collections/2-sortering

Council of the European Union. (2024). *Green transition: Council gives its final approval to the ecodesign regulation*. https://www.consilium.europa.eu/en/press/press-releases/2024/05/27/green-transition-council-gives-its-final-approval-to-the-ecodesign-regulation/

Danfloor. (n.d.). *Responsibility*. Danfloor. https://danfloor.com/responsibility/

Danmarks Statistik. (2014). *Dansk Branchekode DB07, v3:2014*. https://www.dst.dk/da/Statistik/dokumentation/nomenklaturer/db07

Dansk Wilton. (n.d.a). *RE:SHAPE material made from recycled carpets*. Dansk Wilton. https://danskwilton.com/reshape/

Dansk Wilton. (n.d.b). *Whistleblowing*. Dansk Wilton. https://danskwilton.com/about/whistleblowing/

Danspin. (n.d.). *Sustainability*. Danspin. https://www.danspin.com/sustainability/

Det Vilde Spinderi. (n.d.). *Forside [English: Home]*. Det Vilde Spinderi | Danmarks Første Minispinderi. https://www.detvildespinderi.dk

Ege Carpets. (n.d.a). *CircleBack*. Ege Carpets. https://www.egecarpets.dk/baeredygtighed/vores-baeredygtige-ambitioner/recycling-og-take-back/circleback

Ege Carpets. (n.d.b). *Instruction: How to maintain your carpet*. Ege Carpets. https://www.egecarpets.com/technical-info/how-to-maintain-your-carpet

EMF. (2013). *Towards the circular economy: Economic and business rationale for an accelerated transition* (1; Rethink the Future, pp. 23–44). Ellen Macarthur Foundation. https://www.ellenmacarthurfoundation.org/assets/downloads/publications/Ellen-MacArthur-Foundation-Towards-the-Circular-Economy-vol.1.pdf

EMF. (2017). *A new textiles economy: Redesigning fashion's future*. Ellen Macarthur Foundation and Circular Fibres Initiative. https://www.ellenmacarthurfoundation.org/assets/downloads/publications/A-New-Textiles-Economy_Full-Report.pdf

EU Commission. (2022). *Communication from the commission to the European parliament, the council, the European economic and social committee and the committee of the regions on making sustainable products the norm.* https://eur-lex.europa.eu/legal-content/EN/TXT/?uri=CELEX%3A52022DC0140&qid=1649112555090

European Commission. (2003). Commission Recommendation of 6 May 2003 concerning the definition of micro, small and medium-sized enterprises (Text with EEA relevance) (notified under document number C(2003) 1422). In *OJ L* (Vol. 124). http://data.europa.eu/eli/reco/2003/361/oj/eng

European Commission. (2022a). *Digital economy and society index (DESI) 2022.* https://digital-strategy.ec.europa.eu/en/policies/desi

European Commission. (2022b). *EU strategy for sustainable and circular textiles* [Communication from the commission to the European parliament, the council, the European economic and social committee and the committee of the regions]. EU Commission. https://ec.europa.eu/environment/system/files/2022-03/COM_2022_141_1_EN_ACT_part1_v8.pdf

European Commission. (2023). *Waste framework directive—European Commission.* https://environment.ec.europa.eu/topics/waste-and-recycling/waste-framework-directive_en

European Commission. (2024). *SME definition.* https://single-market-economy.ec.europa.eu/smes/sme-definition_en

Fibertex Nonwovens. (2022). *Fibertex Nonwovens ESG report 2022.* Fibertex Nonwovens. https://www.fibertex.com/sites/fibertex.com/files/2023-06/Fibertex%20Nonwovens%20ESG%20Report%202022.pdf

Fletco Carpets. (n.d.a). *100% Danish wind energy.* https://www.fletcocarpets.com/en/sustainability/100-danish-wind-energy

Fletco Carpets. (n.d.b). *Mindful materials.* Fletco Carpets. https://www.fletcocarpets.com/en/sustainability/mindful-materials

Foamtex. (n.d.). *Eco-friendly backing.* Foamtex. https://foamtex.dk/en/backing/eco-friendly-backing/

Fontell, P., & Heikkilä, P. (2017). *Model of circular business ecosystem for textiles* (VTT Technology 313; p. 62). VTT. https://publications.vtt.fi/pdf/technology/2017/T313.pdf

Fru Krogh. (2024). *Om Fru Krogh [English: About Fru Krogh].* Fru Krogh. https://frukrogh.dk/om-fru-krogh/

Gabriel. (n.d.a). *Loop.* Gabriel Fabrics. https://www.gabrielfabrics.com/inspiration/loop/

Gabriel. (n.d.b). *Sustainability.* Gabriel Fabrics. https://www.gabrielfabrics.com/sustainability/

Galaxy Glow Yarn. (2024). *Galaxy Glow Yarn.* Galaxy Glow Yarn. https://galaxyglow.dk/

Hall, C. (2018). Mixing it up in Prato: Identifying innovation hotspots within mechanical textile recycling. *Global Fashion Conference.* What's going on? A discourse on fashion, design and sustainability, London. https://docs.wixstatic.com/ugd/eb22a2_4a7b43538fbf45a7bc753a8a8bfb6aad.pdf

Hall, C. A. (2021). *Design for recycling knitwear: A framework for sorting, blending and cascading in the mechanical textile recycling industry* [Ph.D. thesis, University of the Arts London]. https://tinyurl.com/DesignForRecyclingKnitwear

Hall, C. A., & Hasling, K. M. (2024). *Textile manufacturers in Denmark 2024.* Google My Maps. https://www.google.com/maps/d/viewer?mid=10L-c0XHSoSeHedXFrgGYCrHsIEfL5eU

Hall, C. A., & Larsen, A. T. (2023). Recycled colour: An exploration of colour sorting strategies for the imperfect design of waste textiles. In *5th PLATE 2023 Conference Proceedings.* PLATE.

Hand dyed by Charlotte Spagner. (n.d.). *About Handdyed.* Handdyed.Dk. https://handdyed-denmark.com/about-handdyed/

Harmsen, P., Scheffer, M., & Bos, H. (2021). Textiles for circular fashion: The logic behind recycling options. *Sustainability, 13*(17), 9714. https://doi.org/10.3390/su13179714

Henrichsens Uldspinderi. (n.d.). *Henrichsens Uldspinderi—Fabriksudsalg.* http://www.henrichsens-uldspinderi.dk/7604

Herning Fabrics A/S. (2023). *Miljø politik.* Herning Fabrics. https://www.herningfabrics.dk/kopi-af-metervarer

H.Hjelholts Uldspinderi. (2021a). *Om os [English: About us].* Hjelholt.Dk. https://www.hjelholt.dk/om-os

H.Hjelholts Uldspinderi. (2021b). *Sådan laver vi garn [English: This is how we make yarn]*. Hjelholt.Dk. https://www.hjelholt.dk/s%C3%A5dan-arbejder-vi

JP Air Tech. (n.d.a). *About JP Air Tech*. JP Air Tech. https://www.jpairtech.com/about

JP Air Tech. (n.d.b). *Energy management* [JP Air Tech]. https://www.jpairtech.com/energy-management

JP Air Tech. (n.d.c). *Waste management*. JP Air Tech. https://www.jpairtech.com/waste-management

JV Knitwear ApS. (n.d.). *Home*. JV Knitwear. https://jvknitwear.dk/

König, A. (2013). A stitch in time: Changing cultural constructions of craft and mending. *Culture Unbound: Journal of Current Cultural Research, 5*, 569–585. https://doi.org/10.3384/cu.2000.1525.135569

Kucher, I., & Ravnløkke, L. (2023). Designing engagements with mending practices. In *The design school in sustainable transitions*. Design School Kolding. https://usercontent.one/wp/www.designforplanet.dk/wp-content/uploads/2024/04/Kucher-Ravnlokke-2023-Designing-engagements-with-mending-practices.pdf?media=1660828327

Lille Klode. (2024). *Om [English: About]*. Lille Klode. https://lilleklode.dk/pages/om

LogoNålen. (2023). *Om LogoNålen [English: About LogoNålen]*. LogoNålen. https://logonaalen.dk/om-logonålen

Marschinski, R., & Martinez Turegano, D. (2019). *Reassessing the decline of EU manufacturing: A global value chain analysis*. Publications Office of the European Union. https://data.europa.eu/doi/10.2760/30611

Miljøministeriet. (2024). *Sorteringskriterier / borgere—Textilaffald* (In English: Sorting criteria/citizens—Textile waste). https://edit.mst.dk/media/zjxfplkg/1_sorteringskriterier_tekstilaffald_a4.pdf

Naderifar, M., Goli, H., & Ghaljaie, F. (2017). Snowball sampling: A purposeful method of sampling in qualitative research. *Strides in Development of Medical Education, 14*(3). https://doi.org/10.5812/sdme.67670

Nadia Wire. (2024). *Knitwear production—Nadia Wire: Responsibly made in Denmark*. NadiaWire. https://nadiawire.com/madeindenmark/

NewRetex. (n.d.a). *Sortering af tekstilaffald*. NewRetex. https://newretex.dk/tekstil-sortering/

NewRetex. (n.d.b). *Teknologien bag tekstilsortering*. NewRetex. https://newretex.dk/teknologi/

Nordic Innovation. (2020). *Towards a sustainable textile circular system in the Nordic region* (SATIN Project). Nordic Innovation. https://www.nordicinnovation.org/programs/satin-towards-sustainable-textile-circular-system-nordic-region

OEKO-TEX. (2024). *OEKO-TEX® STANDARD 100*. https://www.oeko-tex.com/en/our-standards/oeko-tex-standard-100

Pehrsson, A., Köppe, G., Geldhäuser, S., Krichel, A., Pohlmeyer, F., Müller, K., & Breton, J. (2023). *Technology assessment—Redefining textile waste Sorting: Impulses and findings for the future of next-gen sorting facilities*. Texaid, International Centre for Sustainable Textiles, CETIA.

Rasmussen, K. (2019). *KORTLÆGNING AF DEN DANSKE TEKSTILBRANCHE* (In English: Mapping of the Danish Textile Industry). Dakofa. https://d1pdf7a38rpjk8.cloudfront.net/fileadmin/user_upload/documents/Footer/Publikationer/Udgivelser/KORTLAEGNING_AF_DEN_DANKSE_TEKSTILBRANCHE.pdf

Recover. (2024). *Frequently asked questions*. Recover™. https://recoverfiber.com/faqs

Rodinia Generation. (2023). *Home*. Rodinia Generation. https://www.rodiniageneration.io

Runnel, A., Raihan, K., Castle, N., Oja, D., & Bhuiya, H. (2017). *The undiscovered business potential of production leftovers within global fashion supply chains: Creating a digitally enhanced circular economy—Insight from research among fabric and garment factories of China and Bangladesh*. Reverse Resources. https://reverseresources.net/about/white-paper

Sandin, G., Roos, S., Spak, B., Zamani, B., & Peters, G. (2019). *Environmental assessment of Swedish clothing consumption—Six garments, sustainable futures*. MISTRA. https://doi.org/10.13140/RG.2.2.30502.27205

Scan HTP. (2024). *Textile printing: Digital printing versus conventional printing techniques*. https://Scanhtp.Com/En. https://scanhtp.com/en/digital-vs-conventional/

Stamm, A., Altenburg, T., Müngersdorff, M., Stoffel, T., & Vrolijk, K. (2019). *Soziale und ökologische Herausforderungen der globalen Textilwirtschaft: Lösungsbeiträge der deutschen Entwicklungszusammenarbeit* [In English: Social and ecological challenges of the global textile industry]. German Institute of Development and Sustainability (IDOS). https://www.idos-research.de/publikationen/mitarbeiter-sonstige/article/soziale-und-oekologische-herausforderungen-der-globalen-textilwirtschaft-loesungsbeitraege-der-deutschen-entwicklungszusammenarbeit/

Sunde, C., Hovda, E. V., Sheikhi, S., & Vestengen, S. (2023). *Method for picking analyses of textiles—REdu wasted textiles project*. https://s3-eu-west-1.amazonaws.com/avfall-norge-no/dokumenter/WT-REdu-final-report-Aug11-2023-revised-Aug17.pdf

Textil Print. (n.d.). *Om os [English: About us]*. Textil Print. https://textilprint.dk/om-os/

The Agency for Culture and Palaces. (2023, March 20). *Textile industry in Herning*. https://slks.dk/omraader/kulturarv/bevaringsvaerdige-bygninger-og-miljoeer/bevaringstemaer/industrikultur/industrihistoriens-danmarkskort/tekstilindustri-i-herning

The Danish Business Authority. (2024). *CSRD – Lovkrav til virksomheders bæredygtighedsrapportering | Virksomhedsguiden*. In English: CSRD—Legal requirements for corporate sustainability reporting | The business guide. https://virksomhedsguiden.dk/content/temaer/baeredygtig-omstilling/ydelser/csrd-lovkrav-til-virksomheders-baeredygtighedsrapportering/62c314cd-4e5f-422c-9f55-6f43873fd25e/?pk_campaign=Baeredygtighed&pk_source=google_ads&pk_medium=cpc&pk_keyword=CSRD&gad_source=1

Thygesen & Birk Fabrics. (n.d.). *Our CSR Policy*. tbfabrics. https://tbfabrics.com/csr/our-csr-policy/

UMM. (n.d.). *Home*. UMM. https://umm.dk/

Unicorn Screen Print ApS. (n.d.). *Om os* [English: About us]. Unicorn Screen Print ApS. https://unicornsp.dk/om-os/

Vildspire. (2024). *Om Vildspire* [English: About Vildspire]. Vildspire. https://vildspire.dk/pages/om-vildspire

Weeds and Wool. (n.d.). *Om* [English: About]. Weeds and Wool. https://weedsandwool.dk/om/

Wennberg, M. V., & Östlund, Å. (2019). *The Outlook report 2011–2019—Mistra Future Fashion final program report*. Mistra Future Fashion. http://mistrafuturefashion.com/wp-content/uploads/2019/10/the-Outlook-Report_Mistra-Future-Fashion-Final-Program-Report_31-okt-2019.pdf

WRAP. (n.d.). *WRAP textiles sorting and recycling database*. Airtable. https://airtable.com/appHr39OzGJywew0j/shr8Ngc3JEAGtGSDy/tbl7pBGFWDXP01FmS/viw5GB3B8Q1fqkDZQ

Cathryn Anneka Hall (cah@dskd.dk) is a postdoctoral researcher at Design School Kolding, Denmark, within the Lab for Sustainability and Design. Cathryn holds a Ph.D. in Circular Textile Design (University of the Arts London). Her research explores design for textile recycling systems (mechanical and chemical), design of textiles using recycled fibres as well as creating design tools for circularity and sustainability. Cathryn's research is conducted at the intersection of academia and industry, and she is currently working on projects exploring design for sorting/recycling textiles and their application in collaboration with the Danish textile industry.

Karen Marie Hasling (kmh@dskd.dk) is Associate Professor and Co-Head of Lab for Sustainability and Design at Design School Kolding. She is also Head of Program for the school's MA Design for Planet. With a background in textiles technology and design engineering, and building upon her Ph.D. on "Learning through Materials", her research centres on intersections between materials, design, sustainability and education. Recent research and development projects have investigated and developed perspectives within e.g., sustainable design methods, fashion design education, the textile technology sector and circular textile design on national as well as international scale.

Kerli Kant Hvass holds an industrial Ph.D. in circular business model innovation within fashion, and has worked with circularity in fashion, for over 15 years. She has extensive experience from practice in developing circular solutions and business models for start-ups, SMEs, corporations and the non-profit sector in the Nordic and Baltic countries, EU and USA. As a cofounder and consultant at an independent advisory firm REVALUATE, assistant professor in circular economy at AAU in Denmark and guest lecturer at several universities in the Baltic-Nordic region, she actively works for systemic changes in the textile industry to counteract overproduction and overconsumption. She is a member of the Estonian Climate Law Advisory Board, as well as a member of the Wasted Textiles consortium at Oslomet.

Nikola Kiørboe has been working with textiles in a circular economy context for more than thirteen years, and is an expert in textile collection, sorting and recycling. She has worked on a wide range of projects for both public and private actors who need expertise and advice within the field. Nikola's work includes facilitation, disseminating knowledge and promoting circularity within all parts of the value chain, with the aim of pushing a circular future.

Chapter 8
Portugal

Madalena Pereira, Teresa Raquel Barata, Marta Bicho, Assunção Mesquita, Helena Vilaça, Carla Joana Silva, Nuno Belino, Ana Santiago, Sandra Ferreira, and Rui Miguel

Abstract Textiles and clothing are an integral part of the global economy. We cannot persevere without the existence of textiles in their various forms, namely clothing, home textiles, personal protective equipment, transportation, construction and more. In Portugal, in 2022, exports from the textile and clothing industry amounted to €6.1 billion, a value never reached before, representing a 13% increase compared to

M. Pereira (✉) · T. R. Barata · M. Bicho · N. Belino · A. Santiago · S. Ferreira · R. Miguel
Fiber Materials and Environmental Technologies (FibEnTech-UBI), Universidade da Beira Interior, R. Marquês de D'Ávila e Bolama, 6201-001 Covilhã, Portugal
e-mail: madalena.pereira@ubi.pt

T. R. Barata
e-mail: traquel@ubi.pt

M. Bicho
e-mail: marta.bicho.fernandes@ubi.pt

N. Belino
e-mail: belino@ubi.pt

A. Santiago
e-mail: ana.santiago@ubi.pt

S. Ferreira
e-mail: sandra.ferreira@ubi.pt

R. Miguel
e-mail: rmiguel@ubi.pt

R. Miguel
CIAUD—University of Lisbon, Lisbon, Portugal

M. Pereira
UNIDCOM—IADE/UE, Lisbon, Portugal

A. Mesquita · H. Vilaça · C. J. Silva
CITEVE—Technological Centre for the Textile and Clothing Industry, 4760-034 Vila Nova de Famalicão, Portugal
e-mail: hvilaca@citeve.pt

C. J. Silva
e-mail: cjsilva@citeve.pt

M. Wynn and T. Wiegand (eds.), *Sustainability, the Circular Economy and Digitalisation in the European Textile and Clothing Industry*, SDGs and Textiles,
https://doi.org/10.1007/978-981-97-9116-3_8

2021. The national textile and clothing sector generated over €8.6 billion in revenue in 2022, reflecting an 8% growth compared to 2021 and involved around 129,000 workers across more than 11,000 companies. Research into the textile and clothing (T&C) industry, as found in various studies, is of growing significance to accelerate the implementation of sustainability and circularity in the industry—from products and processes, to human resources—and by optimising the use of new technologies in the digital era. Such initiatives will play a key role in meeting the United Nation's Sustainable Development Goals and those set by the European Union (EU) for 2030 and 2050. The objective of this chapter is to analyse sustainability strategies and circularity practices, and their relationship with the use of digital technologies in the Portuguese T&C sector. The findings identified a significant adoption of sustainability and circular practices which are currently evolving and often linked to Portuguese R&D projects. However, companies still face difficulties in establishing productive upstream and downstream partnerships to advance sustainability objectives. Digital technologies are seen as a means of increasing efficiencies and saving resources during the transition to sustainability and a circular economy. The T&C industry in Portugal is developing significant efforts in sustainable practices and in digitalisation for the transition to a circular economy, and to support alignment with the United Nations Sustainable Development Goals.

Keywords Sustainability · Circularity · Digital technologies · Textile and clothing industry · Portugal

8.1 Introduction and Background

In 2022, the Textile and Clothing (T&C) industry in Portugal achieved a business volume exceeding €8.6 billion, signifying an 8% increase compared to 2021. This sector employs nearly 129,000 workers across more than 11,000 companies. Within the manufacturing industry in Portugal, the T&C industry accounts for 8% of total production and 19% of employment. Exports from this sector surpassed €6.1 billion, a 12% rise from 2021, making up 9% of the country's total exports (Table 8.1). Meanwhile, imports amounted to around €5.4 billion, reflecting a 26% increase over the previous year (ATP, n.d.).

This growth in the sector, in terms of value, has been observed over the last 10 years, but became more significant after the decline caused by the pandemic, with current figures now surpassing pre-Covid levels. Most exports are destined for EU countries (73%), particularly Spain, France and Germany, while the United States is the largest market outside Europe. Clothing articles were the most exported products, followed by textiles and footwear. Some researchers analyse textiles and clothing together; however, in terms of production, they differ substantially. Textiles are capital-intensive, while clothing is labour-intensive (Pereira, 2008).

In recent years, Portugal's T&C sector has gained global recognition for its quality, responsiveness, resilience, creativity and innovation (CITEVE, 2023; Cunha et al.,

Table 8.1 Portuguese textile and clothing industry figures 2018–2022

	2023	2022	2021	2020	2019	2018
Production (millions €)	…	7,100.00	7,785.00	6,696.00	7,467.00	7,638.00
Turnover (millions €)	…	8,640.00	8,009.00	6,941.00	7,683.00	7,800.00
Exports (millions €)	5,753.00	6,106.00	5,413.00	4,653.00	5,215.00	5,313.00
Imports (millions €)	…	5,419.00	4,307.00	3,781.00	4,443.00	4,338.00
Employment	…	128,600	126,940	128,373	136,101	138,989

Source ATP (n.d.), Babo (2024)

2023; Sá et al., 2023). The Portuguese T&C industry encompasses the entire production chain in a close geographical proximity, from spinning to dyeing, finishing and manufacturing, with most of its facilities located in the northern and central regions of the country. It is a strongly innovative sector, with advanced sustainable practices, benefitting from financial support from the EU in terms of internationalisation and R&D projects, such as the ongoing GIATEX (Giatex, n.d.), be@t (be@t, n.d.), TEXP@CT (TEXP@CT, n.d.) and RDC@ITV (RDC@ITV, n.d.) projects. Additionally, company experience in exporting to demanding markets and customers has allowed for the evolution and sharing of technical and commercial knowledge. This, together with the work of business associations and technology centres, has contributed to a continuous improvement in the positioning of the Portuguese T&C industry in international markets over the last 30 years. The high-quality human resources developed in Academia in recent years have also contributed to this positioning.

The Portuguese T&C industry has an active and, in many cases, vertical production sector, which is not prevalent in other EU countries. The fact that the production system of this sector in Portugal encompasses all phases has also allowed for the development and production of a diverse and technically advanced range of products, in clothing, home textiles, personal protective equipment, and transportation and construction related materials. This has enabled the creation of products with greater added value for the market. Continued participation in top international European fairs, such as Techtextil, ISPO and Heimtextil in Germany, and Première Vision in France, has engendered an evolution in international markets and a growth in exports. At the same time, the growth and participation in industrial research and development (R&D) projects, with technological centres such as the Technological Centre for the Textile and Clothing Industry of Portugal (CITEVE), and with academia, and experience in export practices, has allowed a consolidated evolution in innovation and development of new products, business models, and the digitalisation of processes, strategies, marketing and communication.

More recently, the T&C industry in Portugal has seen a considerable investment in R&D projects supported by the Recovery and Resilience Plan (RRP), within the scope of the EU Recovery and Resilience Mechanism, framed in the Next Generation EU, for the period 2021–2026, in thematic areas such as the digitalisation in parallel with sustainability and circularity (Recuperar Portugal, n.d.). In this context, this chapter examines how Portuguese SMEs in the T&C industry are approaching digitalisation, sustainability and circular economy (CE) practices, and addresses the following research questions (RQs):

RQ1 How has the use of sustainability practices evolved in the Portuguese T&C industry in recent years and what practices are currently pursued?
RQ2 To what extent are T&C companies in Portugal adopting CE practices?
RQ3 Is the T&C sector in Portugal prepared for the era of digitalisation and industry 5.0, and how can this contribute to increasing sustainability and circularity practices?

This chapter comprises five main sections. First, in this Introduction, key background themes are explored, drawing upon the extant literature. The research method is then outlined, being centred around an analysis of relevant sources. In Sect. 8.3 (Findings), the RQs are directly addressed, and emergent themes are reflected upon in the ensuing Discussion Sect. 8.4. Finally, Sect. 8.5 provides a conclusion to the chapter and points up possible future research areas in this field of study.

8.1.1 Sustainability and Circular Economy in the Portuguese T&C Sector

In the T&C industry, the reduction of pre- and post-consumer waste, along with awareness of the reuse, recovery and recycling of materials, are crucial pillars for an effective CE strategy. The entire value chain in this sector must be aware of the origin of the fibres, materials and all the parts and components of products, as well as how they are obtained and processed, according to Bicho et al. (2024). The textile industry still uses a wide variety of fabrics, ranging from natural fibres like cotton, wool, silk and hemp to synthetic ones such as polyester, acrylic, nylon and others. All these materials come with an environmental impact, and future decisions regarding the use of natural versus synthetic fibres, for example, and their production methods, must be carefully considered (Baugh, 2011). The CE prioritises waste reduction throughout the entire production process, from limiting the use of raw materials to minimising waste at various stages of consumption and use. It promotes redesigning products to facilitate easier recycling or repurposing. By encouraging more reuse and recycling, the CE ultimately decreases the need for new production (Chen et al., 2021).

As the number of garments produced continues to rise, the concept of a CE is becoming increasingly relevant to the fashion industry to explore and implement. To successfully address this challenge, innovative approaches are required to manage limited resources and reduce waste. Indeed, the increasing demand for clothing compels all stakeholders, including designers, to adopt a sustainable mindset. Designers must be aware of the components that make up the textile products, as well as taking a broader industry perspective, encompassing the use of chemicals, water and energy consumption, as well as the waste generated from discarded

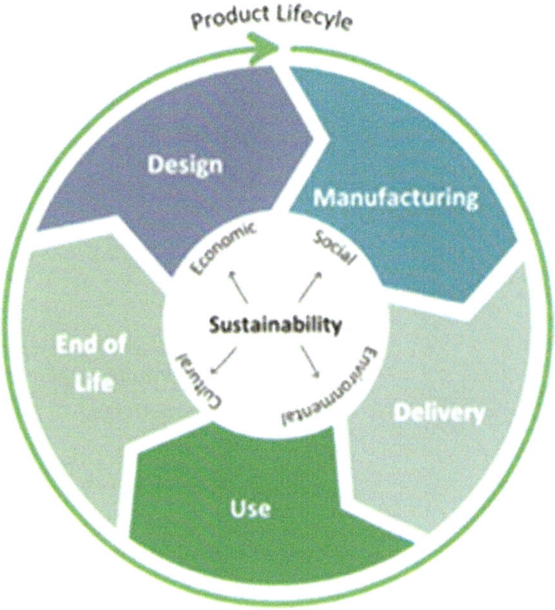

Fig. 8.1 A view of the circular economy in textile production. *Source* Adapted from Cunha et al. (2023)

textile products. This is where designers play a critical role in the fashion industry, by selecting more efficient fibres and fabrics that minimise their environmental impact (Bhatnagar & Niinimaki, 2024). The linear "take-make-dispose" model poses significant risks to the planet, and is both costly and highly polluting. The use of harmful chemicals in textile production not only impacts the environment but also threatens the health of farmers, factory workers and surrounding communities.

Traditional consumption and production methods are nearing their limits, and our planet is facing a depletion of natural resources. The CE offers clear environmental and macroeconomic benefits. It is characterised by a business model that aims to maintain the value of products and materials for as long as possible (Fig. 8.1). Key benefits and strategies in this model include stabilising raw material prices and mitigating supply risks, fostering new relationships with consumers through take-back programmes and new business models, enhancing economic competitiveness, and contributing to nature preservation by reducing emissions, minimising waste and combating climate change (Anicet et al., 2011; Gueye, 2021).

8.1.2 Waste in the T&C Industry

Each year, the T&C Industry in Europe disposes of 5 million tonnes of material, generating 20 to 35 jobs for every 1,000 tonnes of textiles collected for reuse. However, only 1% of the material from clothing is recycled into new garments. A major issue with

Table 8.2 Materials comparison in the Higg index

	Production (%)	Higg index
World production of fibres in 2022: 106.84 million tonnes (91.1% of total fibre production)		
Polyester	54.0	44
Cotton	27.0	98
Viscose	5.8	62
Polyamide	5.0	60
Wool	4.0	82
others	4.2	
World production of recycled fibres in 2022: 9.16 million tonnes (7.9% of total fibre production)		
Recycled polyester	14.0	35
Recycled wool	7.0	49
Recycled polyamide	2.0	36
Recycled cotton	1.0	39
Recycled viscose	0.5	43

Source Hansson Reuter and Zetterlund (2018), Textile Exchange (2023)

post-consumer waste is the lack of standardisation for recycled materials and their quality, along with the absence of established end-of-waste criteria, which creates significant bottlenecks (EURATEX, 2022; European Commission, 2022). Textile waste has a substantial environmental impact.

According to a recent T&C industry study (EURATEX, 2022; European Commission, 2022), the EU-27 currently generate 7–7.5 million tonnes of textile waste annually, of which 2–2.5 million tonnes (30–35%) are collected. Of this, 1 million tonnes are available for recycling, with the potential to increase this figure by 60%. To achieve this, it is crucial that countries invest in recycling infrastructure and the recycling industry. To this end, the EU has recommended improvements in mechanical recycling technologies and other strategies. In 2020, Portugal produced 72,861 tonnes of textile waste, according to Eurostat (2024), although some reports suggest the actual figure may be higher.

The concepts of the CE and recycling have gained increasing importance as the fashion industry has actively sought ways to improve its ecological footprint. One of the key steps in the CE is the recycling of textile fibres, where the finished products are repurposed. Nonetheless the production of mechanically adequate recycled fibres is still of low importance globally. Recently, several methodologies to classify fibres according to their impact have emerged. Among them, the Higg index classification allows a quantitative analysis of the level of sustainability of different materials, in which the higher the index is, so the less sustainable the material is, in terms of water and energy consumption, chemical substances, CO_2 emissions and land use (Table 8.2).

Textile recycling involves breaking down and recovering used materials to design new ones (Rathinamoorthy, 2018), and is recognised as a viable strategy for transition to the CE. Waste is generated at various stages in the life cycle of textiles and can be classified into different categories: (a) Post-industrial waste: waste produced during the manufacturing process of textile products and their precursor materials; (b) Pre-consumer waste: waste resulting from retail, such as unsold textiles; and (c) Post-consumer waste: textiles disposed of after use by consumers or commercial and industrial organisations. Textile production waste is classified as post-industrial waste, but it can also be classified as a by-product, if it is re-used in the textile sector or another industrial sector.

Waste from post-consumer products can be grouped into three main categories: (1) Clothing: this includes clothing items such as t-shirts, sweatshirts, jackets, trousers and accessories; (2) Home textiles: includes items such as curtains, bed linen and carpets and (3) Professional textiles: this category includes textiles used in industries such as the automotive sector, agro-textiles and others (Huygens et al., 2023).

Utilising recycled materials instead of virgin raw materials provides a significant opportunity to reduce the use of non-renewable resources and lessen the negative impacts of the industry. Although textile recycling has been practised for over 250 years, there are still considerable limitations. The textile waste recycling process can be divided into chemical, biochemical, thermal and mechanical methods, with mechanical recycling currently making up most of the processes—around 43% (Damayanti et al., 2021; Dissanayake et al., 2020). In mechanical recycling, materials are shredded into fibres, which typically results in fibres of lower quality compared to virgin fibres (Hawley, 2015). In this form of recycling, it is crucial that each garment is properly labelled and that materials are correctly identified to ensure accurate sorting for recycling. The main stages of mechanical recycling involve sorting batches of textile waste by colour and material type; the materials are then crushed and shredded, which can include single or mixed fibres, such as cotton and polyester; once the fibres are broken down, they are carded, mixed with virgin fibres for mechanical improvement and transformed into yarns (Hawley, 2015).

While the benefits of textile recycling are well-documented, there is a need for greater transparency regarding the properties and specifications of recycled materials to better align supply and demand. Open communication between recyclers, textile mills and brands could help harmonise the main properties required for the materials in various applications. For companies that recycle fibres, the quality of the recycled fibres may not match those of virgin fibres, but they can still meet the requirements for most clothing applications (typically in mixtures, with virgin fibres constituting up to 70%). Figure 8.2 provides an overview of the current waste scenario within the Portuguese T&C industry, and future goals.

Fig. 8.2 T&C waste generation, collection and future development goals in Portugal in 2022. *Source* European Union (2022a), Eurostat (2024a), APA (n.d.), APA (2022)

8.1.3 Regulations and Directives for Sustainability and the CE

The EU is at the forefront of promoting sustainability and CE practices within the textile sector. Through a comprehensive framework of regulations, directives and certifications, the EU is focused on minimising environmental impacts, improving resource efficiency and encouraging sustainable production and consumption. These initiatives are a crucial part of the EU's Circular Economy Action Plan, which aims to ensure that by 2030, all textiles sold in the EU are durable, repairable, recyclable, made with maximum recycled content, free from hazardous substances and produced with respect for social rights (European Commission, 2019, 2022). To support this transformation, the EU is implementing mandatory criteria for textiles, including eco-design standards, Extended Producer Responsibility (EPR) schemes, a stronger emphasis on sustainable production and efforts to combat fast fashion. Additionally, the development of a Digital Product Passport (DPP) is underway, which will provide essential information on the sustainability and circularity of textile products (Legardeur & Hospital, 2024).

Table 8.3 provides a summary of the current regulations applicable to the T&C sector in terms of sustainability and circularity.

According to European Commission reports (Council of the European Union, 2023; European Commission, 2019) on the EU strategy for sustainability and circularity in the textile industry, textile production and consumption are anticipated to rise significantly. By 2030, the consumption of textiles and footwear is projected to increase by 63%. In response, the EU has implemented several strategies to promote the sustainability and circularity of textile materials. Regulation plays a crucial role in this sector, particularly concerning chemicals like those regulated under REACH (European Parliament & Council of the European Union, 2024), which aims to protect both the consumer and the environment.

Table 8.3 Regulations and directives

Regulations and directives—circular economy policy dossiers	Regulations and directives—chemicals policy dossiers
Eco-design Requirements for Sustainable Products (ESPR) Waste Framework Directive (WFD) Extended Producer Responsibility (EPR) End-of-waste criteria Waste shipment Microplastics Green claims	REACH regulation with restrictions or limitations (or proposals of) to several substances commonly used in textiles, namely in the following groups: CMRs Skin sensitisers PFASs (e.g. PFHxS, PFOA) Alkylphenol ethoxylates (APEO) (e.g. NPEO) Bisphenols Flame retardants Azocolourants, azo dyes, carcinogenic amines Phthalates Chlorinated paraffins (SCCPs and MCCPs) Polycyclic aromatic hydrocarbons (PAHs) Metals Chemical strategy for sustainability Biocides Regulation (BPR) Waste Framework Directive (WFD)

Source Adapted from EURATEX (2023a), EURATEX (2023b)

The EU seeks to foster economic growth through sustainable practices that minimise resource usage while ensuring no individual or region is socially or environmentally marginalised. These efforts are supported by regulations with clearly defined targets (Council of the European Union, 2023; Economic Commission for Europe, 2022; European Commission, 2019, 2022; United Nations, 2015). Table 8.4 lists key legislation applicable to the T&C industry, focusing on eco-engineering, eco-design and digitalisation.

Although the sustainability concept has existed for more than 50 years, research in this area has only recently increased, partly justified by political pressure with the development and definition of a set of directives and regulations recently established by the EU. These require companies to reformulate their short, medium and long-term strategies to ensure that resources for future generations are preserved and the EU targets for 2030 and 2050 are achieved (Fig. 8.2).

8.1.4 Certifications for Sustainability and Circularity

Standards, certifications and eco-labels, although not mandatory, play a critical role in the T&C industry, offering consumers vital insights into the origins, quality and environmental impacts of products. These certifications are designed to raise awareness of environmental and social concerns, ethical practices and animal welfare.

Table 8.4 Regulations and directives for sustainability and the CE for textiles and clothing at European and national levels

Regulations and directives for sustainability and circular economy of textiles		
Regulation	Directive	References/Source
Extended Producer Responsibility (EPR) 11/12/2017	Decreto-Lei n.º 152-D/2017	Ambiente (2017)
End-of-Waste Criteria 30/05/2018	Directive (EU) 2018/851 of the European of the Parliament and the Council	European Union (2018)
Waste Framework Directive (WFD) 07/06/2019	Commission Implementing Decision (EU) 2019/1004	European Union (2019)
Sustainability and Circularity 11/12/2019	The European Green Deal	European Commission (2019)
Sustainable and Circular Textiles 30/03/2022	EU Strategy for Sustainable and Circular Textiles	European Commission (2022)
Corporate Sustainability Reporting Directive (CSRD) 14/12/2022	"Directive (EU) 2022/2464 of the European Parliament and of the Council"	European Union (2022b)
National Strategy for Green Public Purchasing 2030—ECO360 10/02/2023	Resolução do Conselho de Ministros n.º 13/2023	Presidência do Conselho de Ministros (2023)
Right to Repair 25/05/2024	Proposal for a Directive of the European Parliament and the of the Council, common rules promoting the repair of goods and amending Regulation (EU) 2017/2394, Directives (EU) 2019/771 and (EU) 2020/1828"	Council of the European Union (2024)
Net-Zero Industry Act 27/05/2024	Industrial policy of the Council	European Commission (2024)
Ecodesign Requirements for Sustainable Products (ESPR) 13/06/2024	Regulation (EU) 2024/1781 of the European Parliament and of the Council	European Union (2024a)
End-of-Waste Criteria 13/06/2024	Directive (EU) 2024/1799 of the European Parliament and of the Council	European Union (2024b)
Digital Product Passport (DPP) 28/06/2024	Digital product passport for the textile sector—study	Legardeur and Hospital (2024)
Textiles Labelling Regulation 06/2024	Revision of the Textile Labelling Regulation (planned for the 4th quarter 2024)	European Parliament (2024)

Source Chapter authors

By focusing on key areas such as the use of organic materials, sustainable production methods, ethical labour practices and responsible resource management, they serve as important tools for promoting more conscious consumer choices and driving industry-wide improvements in sustainability (Diekel et al., 2021; Harsanto et al., 2023) (Table 8.5).

In recent years, companies have faced increasing pressure to secure green certifications on an annual basis. This demand is driven not only by government-supported environmental labelling initiatives that encourage green procurement policies, but also by international brands that use green certification as a standard for assessing their suppliers. These brands set environmental goals and require suppliers to obtain specific certifications to meet these objectives. As a result, these regulations, certifications and initiatives are guiding the European textile industry towards a more sustainable and circular economy, where products are designed with their entire life-cycle in mind—from production to disposal and beyond (Plakantonaki et al., 2023; Ranasinghe & Jayasooriya, 2021).

Certifications inherently rely on traceability, which is increasingly enabled by digitalisation and advanced information systems. A key element in this process of transparency is ensuring that consumers have access to essential product information. The DPP, which has warranted and gained significant attention on the EU's political agenda and within the T&C industry, exemplifies this effort. The DPP aims to provide detailed information on a product's entire life cycle—covering its origin, composition, repair options and end-of-life management (Council of the European Union, 2023). When combined with eco-design, the DPP can mark a pivotal shift towards more sustainable production systems worldwide. The digitalisation of the entire supply chain, from raw materials to the final consumer, promises to strengthen the relationships between brands, manufacturers and consumers, fostering greater trust and encouraging responsible practices throughout the industry (Adisorn et al., 2021).

8.1.5 Digitalisation in the T&C Industry

Digitalisation involves implementing digital technologies to transform resources and business models, creating new revenue streams and value-added opportunities. It is a key component of Industry 4.0, also called smart factory and originally the "Connected Enterprise" concept created in 2011 in Hanover in Germany. Industry 4.0 encompasses the application of nine critical digital technologies in industrial settings, essential for revolutionising industrial production: simulation, cloud computing, additive manufacturing, augmented reality, cybersecurity, the industrial Internet of Things (IoT), autonomous robots, vertical and horizontal system integration, and big data and analytics (Buhr, 2015; Gartner, n.d.; Morrar et al., 2017). These contribute to industry's increased flexibility, efficiency and productivity and can increase innovation and improve sustainability (Müller, 2023; Müller et al., 2018; Pereira et al., 2022; Pozzi et al., 2021; Schöggl et al., 2023; Strandhagen et al., 2020; Taddei et al.,

Table 8.5 Examples of certifications applied in the T&C industry

Certifications	Raw material	Process	Environment	Social impact	Chemical requirements	Traceability	Source
ISO 9001:2015		x				x	ISO 9001:2015 (2015)
ISO 14001:2015			x				ISO 14001:2015 (n.d.)
GOTS	x	x	x	x	x	x	GOTS—Global Organic Textile Standard (n.d.)
OEKO TEX 100	x				x	x	OEKO-TEX® Service GmbH (n.d.-a)
OEKO TEX STEP	x	x	x	x	x	x	OEKO-TEX® Service GmbH (n.d.-b)
Better Cotton	x		x	x		x	Better Cotton (2023)
OCS	x					x	Textile Exchange (2024a)
GRS	x	x	x	x	x	x	Textile Exchange (2024b)
RCS	x					x	Textile Exchange (2024b)
bluesign	x	x	x		x	x	bluesign (2024)
B Corp Certified			x	x		x	Bcorporation (n.d.)
Cradle to Cradle Certified			x	x			Cradle to Cradle Certified (n.d.)
Ethical Trade Initiative				x			Ethical Trading Initiative (n.d.)
Amfori Business Social Compliance Initiative				x			Amfori (n.d.)

(continued)

Table 8.5 (continued)

Certifications	Raw material	Process	Environment	Social impact	Chemical requirements	Traceability	Source
Fairtrade				x			Fairtrade International (n.d.)
Leather Working Group	x			x			Leather Working Group (n.d.)
Worldwide Responsible Accredited Production				x			Worldwide Responsible Accredited Production (2024)
Woolmark	x					x	The Woolmark Company (n.d.)

Source Chapter authors with information from the source material

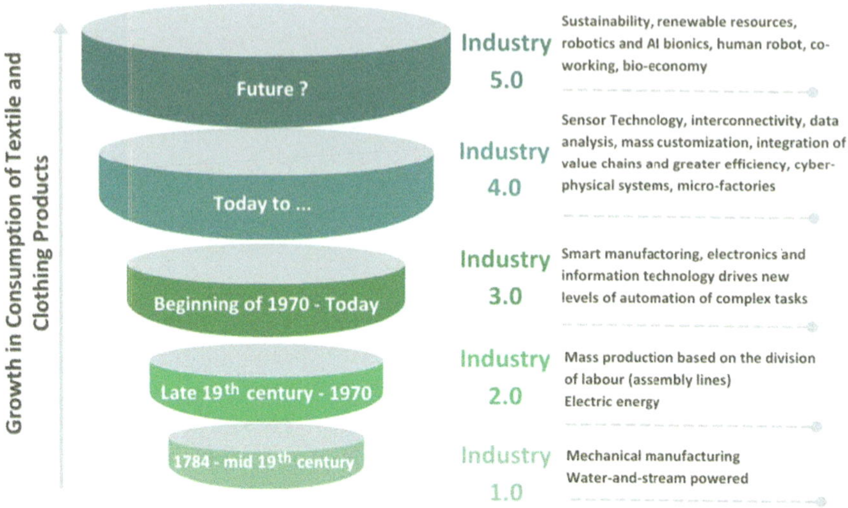

Fig. 8.3 Industrial revolutions from 1784 to the present day. *Source* Adapted from Fraunhofer and Ingenics (2014), Davis (2016), Demir et al. (2019)

2022; Wiegand & Wynn, 2023). Today, Industry 4.0 is being superseded by 5.0 (Fig. 8.3).

European Commission (2024) statistics from Eurostat identify indicators for the digital economy and society, for digital or e-business companies, analysing the: (1) integration of internal processes; (2) integration with customers/suppliers and supply chain management; (3) use of cloud computing services; (4) big data analysis; (5) 3D printing and robotics; (6) internet of things and (7) artificial intelligence (AI). There are few studies that analyse the impact of digitalisation on the implementation of a CE in companies. In some literature, both themes are analysed separately, but their relationship requires more research, data and information.

Schöggl et al. (2023) analysed the importance of four digital technologies (DTs) in implementing the CE in industry. The sample included companies in the textile sector also. They concluded:

> Of the four DTs, Internet of Things (IoT) technology is most widely implemented, followed by big data analytics, artificial intelligence (AI), and blockchain technology. However, their use in sustainability management is still very limited and is currently mainly in pilot phases. (…) IoT technology is most frequently used for collecting data from production processes, AI for predictive maintenance, big data analytics for demand forecasting, and blockchain technology for tracking product origins (p. 401).

In Table 8.6, based on the findings of Schöggl et al. (2023), examples of applications of digital technologies in sustainability management for a CE are shown.

One point here, though, is that for digital transformation to be effective and lasting, it requires human resources with appropriate digital skills. The T&C industry in Portugal has more than 50% of its workers above 50 years old, and the digital skills level is relatively low.

Table 8.6 Digital technologies and examples of their applications in sustainability management in a CE

Digital technology	Examples of applications in sustainability management in a CE
Internet of things (IoT)	"Track the life cycle data of products to improve their maintenance; Improve recycling efficiency using additional data; Address a company's energy management and sustainable production practices; Product status monitoring; Collect use-phase data to inform product design; Life cycle data collection for data-driven recycling"
Big data analytics	"Green supply chain management and environmental impact assessment; Preventive and predictive maintenance activities so as to facilitate the lifetime extension of products; Increase equipment energy efficiency; Analyse consumer behaviour; Enhance data quality and completion of data in sustainability assessment; Inventory management, optimise transportation, demand analysis; Energy and operations management; Collect social sustainability data; Knowledge discovery in databases to enhance the efficiency of the design, production and service processes"
Artificial intelligence (AI)	"Help manufacturers to select the best possible sustainable supplier and to enhance the manufacturers' sustainability; Help in environmental impact assessment; Establishing traceability along a product's life cycle; DPP; Energy and environmental impact predictions"

Source Schöggl et al. (2023)

8.2 Research Method

The data used in this chapter was obtained from relevant academic research in the areas of sustainability, circularity and digitalisation, and from the first sustainability report for the Portuguese T&C industry, developed by CITEVE (the Technological Centre for the Textile & Clothing Industry), in collaboration with several companies in the sector, within the framework of the be@t project (be@t, n.d.), in which the Department of Textile Science and Technology at the University of Beira Interior is a project partner. This report aims to present a quantitative assessment of the sector in terms of general sustainability (focusing on Environmental, Social and Governance data), as well as highlighting initiatives implemented to promote sustainability practices in the textile sector in Portugal. The information disclosed in the report refers to the performance consolidated from a sample of 77 companies in the sector, which agreed to participate in this study. The data reported here was collected over the four-year period from 2019 to 2022. It includes publicly available data for some companies, evidenced in corporate sustainability reports for example, but here sustainability indicators are used to analyse this data in a consistent manner. Complementary data sources included third-party databases, reports, statistics from Eurostat and surveys of companies in Portugal.

8.2.1 Sample of Portuguese T&C Industry

In Portugal, the T&C industry comprises 11,000 companies and 129,000 workers. The sample considered in this research for analysing sustainable and circularity practices and digitalisation is 77 companies, employing almost 12,000 workers, more than 9% of the total employed in the industry in Portugal. The authors maintain this size of sample is adequately representative of the sector as a whole. With regards to processes, the sample is made up of companies that cover the different subsectors of the industry value chain, there being six main process typologies: (1) spinning, weaving, dyeing and finishing; (2) knitting and/or weaving; (3) finalisation, including dyeing, finishing, printing, laundry; (4) cutting and confection processes; (5) companies with several processes (which do not fit into the previous groups), including vertical companies (textile production processes from spinning to confection) and (6) other processes, which include specialised companies, not covered by the previous typologies, such as rolling and coiling, among others.

8.2.2 Data Collection and Analysis

The analysis methodology used in this study consisted of data collection followed by its statistical treatment and interpretation. First, however, an identification and definition of key themes relating to the RQs was conducted by the CITEVE team through a benchmarking process and consultation with several relevant sustainability references. This included carrying out a comparative analysis of several T&C companies, at national and international levels, to understand which are the most relevant sustainability themes globally, as well as a comparative analysis of various standards and benchmarks, to identify the challenges commonly faced by the sector. Then, companies were invited to provide data relating to these themes, including Key Performance Indicators (KPIs) developed for the study, based on the previous analysis, through a specific online platform that was developed for this research project. This platform was able to collect the data for each company according to the defined KPIs and a range of key themes. This chapter includes qualitative and quantitative data from the 77 companies that accepted the challenge of participating in this study and duly submitted their KPIs, although not all companies in the sample responded to all performance indicators, as is reflected in the analysis below.

8.3 Findings

The three RQs are addressed in this section. The first aim at this point is to answer the initial questions raised by analysing the results obtained from different sources (Eurostat, and data from Portuguese companies collected by CITEVE).

8.3.1 RQ1. How Has the Use of Sustainability Practices Evolved in the Portuguese T&C Industry in Recent years and What Practices Are Currently Pursued?

Data collected by CITEVE on the use of different sustainability techniques (including eco-design, eco-engineering, digitalisation, best available techniques and others) used in companies from 2019 to 2022 was analysed to address RQ1. Production, manufacturing and management methods that considered the preservation of resources, the reduction of waste and the mitigation of negative environmental impacts were assessed. These processes encompassed choosing more sustainable materials or efficiency in the use of resources, such as water and energy. The development of circular business models, reduction in plastic use and process certifications were aspects addressed to evaluate and quantify sustainability techniques in the companies' processes. The findings demonstrate a positive evolution from 2019–2022 in the application of sustainability techniques in manufacturing processes (Fig. 8.4).

The percentage of companies that did not report any of the sustainability techniques decreased over the four years of the collected data (from 14% in 2019 to 10% in 2022). On the other hand, during this period, the use of eco-design (17%), eco-engineering (13%) and digitalisation (23%) was increasingly adopted by companies in 2022, albeit this remained at a fairly low level. Additionally, in the period reported in the study, the most adopted sustainability technique was clearly the Best Available Techniques (BAT)—this being the available techniques which are the best for preventing or minimising emissions and impacts on the environment (European Commission, n.d.). However, its applicability has decreased over the years (43% in 2019 and 36% in 2022).

The "cutting and confection" (35%) and "knitting and weaving" (55%) process companies showed the greatest application of digitalisation technologies in 2022, while eco-design and eco-engineering were prominent in "several processes" (42% and 29%, respectively), in "spinning, weaving, dyeing and finishing" (25% for both techniques) and in "knitting and weaving" (18% and 36%, respectively). This is displayed graphically in Fig. 8.5 (note that companies are normally using more than one of the techniques and/or pursuing digitalisation simultaneously).

Another important aspect of sustainable practices is the certifications adopted by companies, which ensure compliance with the practices established by each standard or certification from a legal perspective. For example, regarding certifications and

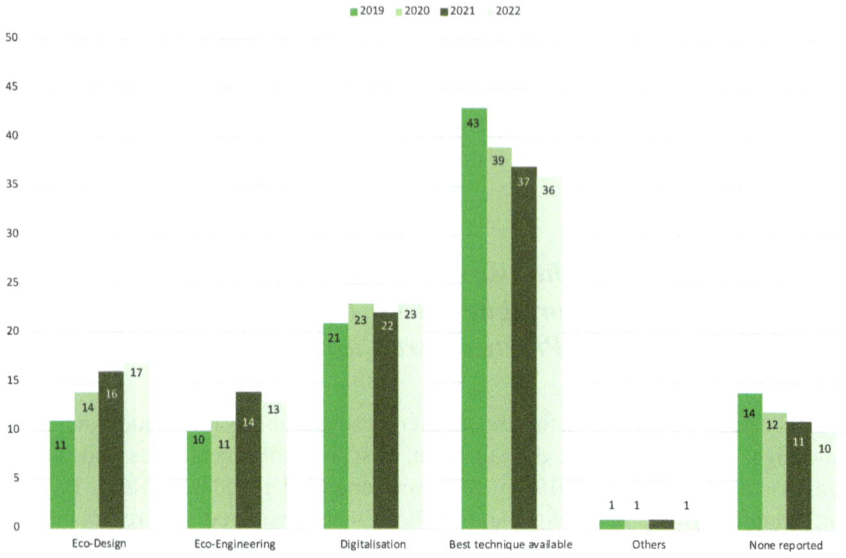

Fig. 8.4 Application of sustainability techniques and digital technologies? in manufacturing processes 2019–2022 (by %). *Source* CITEVE (2023)

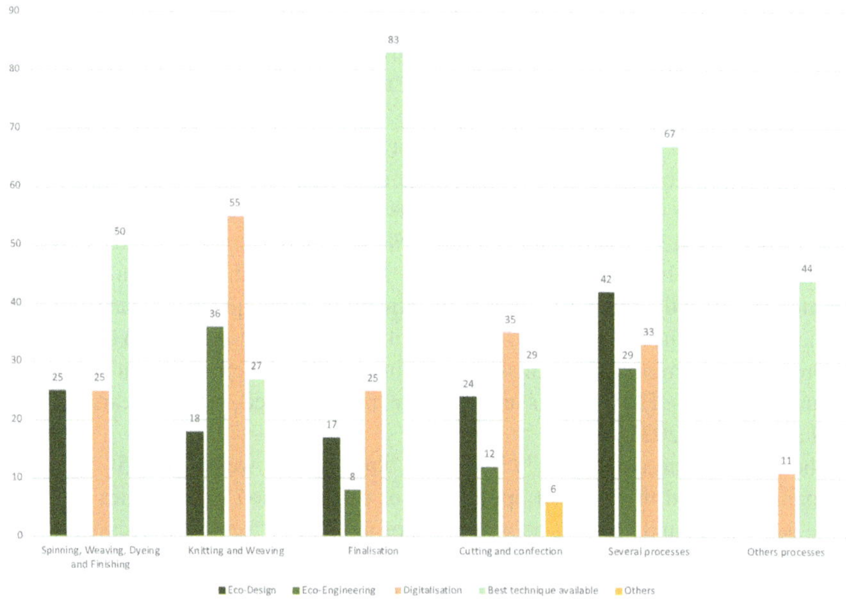

Fig. 8.5 Application of sustainability techniques and digitalisation in manufacturing processes in 2022 (by %). *Source* CITEVE (2023)

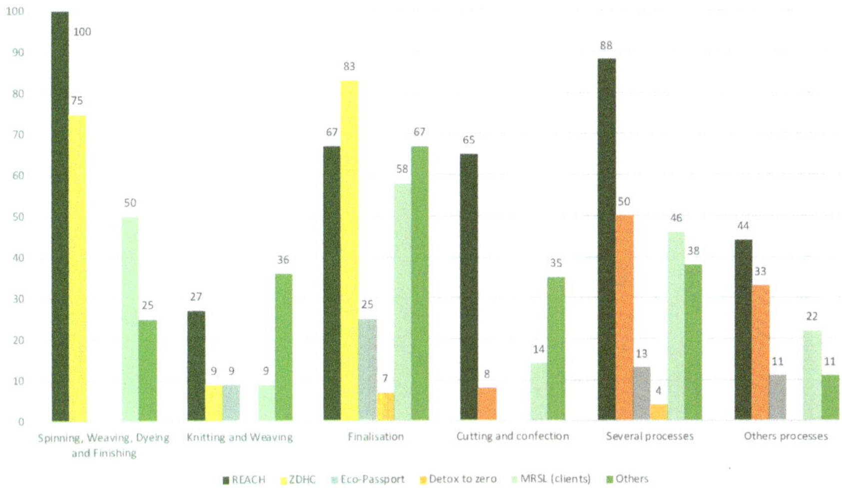

Fig. 8.6 Compliance with certifications associated with good practices in the use of chemical products in the different T&C company processes (by %). *Source* CITEVE (2023)

regulations associated with good sustainability practices in terms of chemical products used in the company processes, there is a high percentage in all areas of use in accordance with the REACH regulation followed by ZDHC and to a lesser extent by Eco-passport, Detox to zero and Manufacturing Restricted Substances List (MRSL) defined by the company clients (Fig. 8.6).

Among the certifications identified in Table 8.5, ISO 14001 stands out, with 29% of the companies surveyed complying with environmental management practices in a systematic way (Fig. 8.7). Additionally, 31% of the companies were certified, in 2022, under OEKO-TEX® STeP, which guarantees several pillars of sustainability—management of chemical products, environmental performance, environmental management, social responsibility, quality management, and occupational health and safety. However, the most common certification in the Portuguese T&C industry is ISO 9001, related with quality management systems.

These findings support the view that there has been a positive evolution since 2019 in the use of sustainable practices in eco-design and eco-engineering in the T&C industry in Portugal. The results from this online survey also suggest progress towards digitalisation has been consistent over the four-year period. This is examined in greater detail in Sect. 8.3.3.

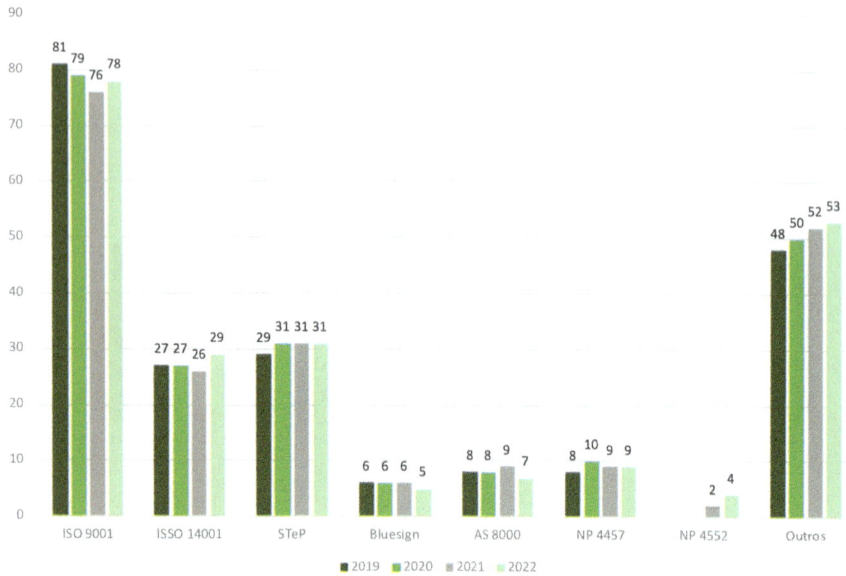

Fig. 8.7 Compliance by Portuguese T&C companies with certifications associated with good practices 2019–2022 (by %). *Source* CITEVE (2023)

8.3.2 RQ2. To What Extent Are T&C Companies in Portugal Adopting CE Practices?

Currently, the Portuguese T&C industry export market is increasingly demanding ecological labels. This is predominantly characterised by the private label model with added value services for clients (ATP, 2021), which requires companies to adhere to and obtain certifications associated with sustainable practices in both processes and products. At international fairs such as Première Vision, Techtextil and Milano Unica among others, companies are requested and identified on platforms by the types of ecological labels they possess. According to EU regulations, brands' communication of sustainable practices will be prohibited in 2026 on products unless they are associated with a certification or eco-label, to prevent greenwashing. The Portuguese T&C industry includes 1,826 companies that use the most requested labels, such as GOTS, OCS, GRS, RCS, BCI and OEKO-TEX® standard 100, with the latter being the most used, followed by GOTS, GRS and BCI (Fig. 8.8 and Table 8.5).

The use of more environmentally friendly materials, such as organic, recycled or bio-based, is another important aspect that characterises the national T&C industry, with an increase in 2022 of around 17% in recycled materials and 9% in organic/biologic materials in production compared to 2019. Based on the findings from the be@t project (be@t, n.d.), there is an increase in the use of these types of material, but in relation to the high volume of the total production in Portugal, which in 2022 was around 60 million tonnes, this growth is insufficient. One of the measures of the

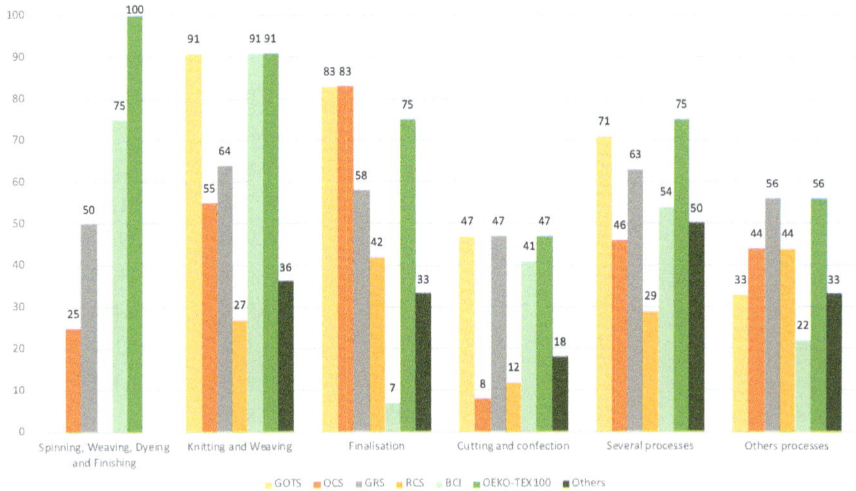

Fig. 8.8 Ecological labels used in Portuguese T&C industry processes (by %). *Source* CITEVE (2023)

be@t project is to promote and develop a set of more sustainable materials for the T&C sector in Portugal, for which there is a consortium of 58 partners, most of them being companies, with an investment of €128 million (be@t, n.d.).

Another important aspect of circular practices in the T&C sector is the management of waste. The results demonstrate greater concern in the management and reduction of different types of waste. Although there was a reported growth of 9% between 2019 and 2022 in report information, there was in fact from 2019 to 2022 decrease by 13% of generated waste. Also, in terms of water and effluent management, the T&C industry demonstrates increased concern for its monitoring, and the implementation of treatment technologies to reuse water, reduce waste and treat effluents. A practical example is the GIATEX project (GIATEX, n.d.).

From Eurostat indicators from 2014 to 2020, the total waste in textiles indicates a mix of increases and decreases across this period. 2020 was something of an atypical year, due to the pandemic (Table 8.7).

In summary, the research on circularity was limited to a few main aspects. Results indicate an increase in recycling, and a decrease in waste generation across the 4-year period 2019–2022. In addition, there is widespread use of ecological labels across all main T&C industry processes, supporting the view that companies in the last 4 years have increased their investment in circularity practices.

Table 8.7 Tonnes of textile waste in European countries 2014–2020

GEO (labels)	2014	2016	2018	2020
EU—27	2,010,000	1,980,000	2,170,000	1,950,000
EU—28	2,220,000	2,190,000	2,370,000	*
Denmark	2,932	18,134	26,854	18,276
Germany	343,757	391,752	338,342	281,041
France	175,000	209,999	238,998	239,086
Italy	439,752	465,925	519,214	424,268
Portugal	**70,144**	**81,715**	**103,954**	**76,412**
Romania	25,697	30,360	41,150	31,120
Sweden	5,743	8,646	10,696	12,307
United Kingdom	211,805	209,545	199,813	*
Türkiye	214,324	206,441	231,836	273,667

Source Eurostat (2024a)
* Not available

8.3.3 RQ3. Is the T&C Sector in Portugal Prepared for the Era of Digitalisation and Industry 5.0, and How Can This Contribute to Increasing Sustainability and Circularity Practices?

One of the indicators that allows the digitalisation of companies to be assessed is the Eurostat digital intensity index (DII), which is based on a scale of 0–12 points, a point being achieved whenever one of the following conditions is met (Eurostat, 2024b): (1)—"Enterprises where more than 50% of the persons employed have access to the internet for business purposes"; (2)—"The maximum contracted download speed of the fastest fixed internet connection is at least 30 Mb/s"; (3)—"Enterprises with e-commerce sales of at least 1% turnover"; (4)—"Enterprises where web sales are more than 1% of the total turnover and business-to-consumer (B2C) web sales more than 10% of the web sales"; (5)—"Enterprises have document(s) on measures, practices or procedures on information and communication technologies (ICT) security"; (6)—"Enterprises make employed persons aware of their obligations in ICT security related issues"; (7)—"Use at least 3 ICT security measures"; (8)—"Enterprise provided training to their personnel to develop their ICT skills"; (9)—"Employ ICT specialists"; (10)—"Use any AI technology"; (11)—"Provide to the persons employed remote access to the enterprise's e-mail system, documents or applications" and (12)—Enterprises, which conducted remote meetings via the internet. The sum of the verified conditions indicates the digital intensity of companies in four groups: (a) very low digital intensity (0–3 verified conditions); (b) low digital intensity (4–6 verified conditions); (c) high digital intensity (7–9 verified conditions) and (d) very high digital intensity (10–12 verified conditions).

This index provides a means of assessing the level of digitalisation of Portugal's T&C industry against other EU countries (Table 8.8). This indicates that in 2023, Sweden and Denmark were the countries in the EU which reported having 9.8% and 22% of the T&C companies with very low digital intensity, whereas France have 28.1% of the companies had at very low digital intensity. Portugal was at the other end of the scale with 65.2% of T&C companies having very low digital intensity, 29.5% with low digital intensity, 5.2% high, although 0.1% of companies were adjudged to have a high level of digital intensity.

According to Eurostat (2024a), Portugal was, in 2023, the country in the EU where the T&C industry most applied AI technologies for production processes (Fig. 8.9) but just the 18th in the use of cloud technologies (Fig. 8.10), below EU's average. The countries leading in the use of cloud computing were Sweden, Finland, Norway, the Netherlands, Denmark and Italy.

The fact that Portugal was ranked first in the use of AI technologies in the T&C industry but fairly low down in the use of cloud computing appears to be something of an anomaly. Further, the integration with customers/supplier processes (Fig. 8.11) indicates modest performance in this aspect with only 11.2% of companies qualifying for this measure in 2017, and much the same in 2023. Some researchers have suggested this reflects the increase in cyberattacks in recent years and thus the reluctance of T&C companies—particularly SMEs—to expose their data and systems to the perceived vulnerabilities of cloud computing and business to business e-commerce (Lamine et al., 2020). Some of these risks can be addressed via codes of conduct and establishing responsible and ethical relationships between all interested parties, as well as ensuring data transparency. Data protection policies are essential to allow digitalisation practices to contribute to the growth of sustainable practices

Table 8.8 Digital intensity of T&C companies in EU countries 2023

GEO (labels)	Very low digital intensity	Low digital intensity	High digital intensity	Very high intensity
EU	60.6	25.5	11.4	2.5
Denmark	22.0	32.7	43.8	1.4
Germany	40.6	30.4	25.4	3.6
France	28.1	*	*	*
Italy	66.6	23.8	7.4	2.2
Portugal	65.2	29.5	5.2	0.1
Romania	85.0	12.5	2.5	0
Sweden	9.8	36.6	43.6	10.0
UK	*	*	*	*
Türkiye	*	*	*	*

Source Eurostat (2024a). Companies are those manufacturing textiles, clothing and leather (NACE, Rev 2)
* Data not available

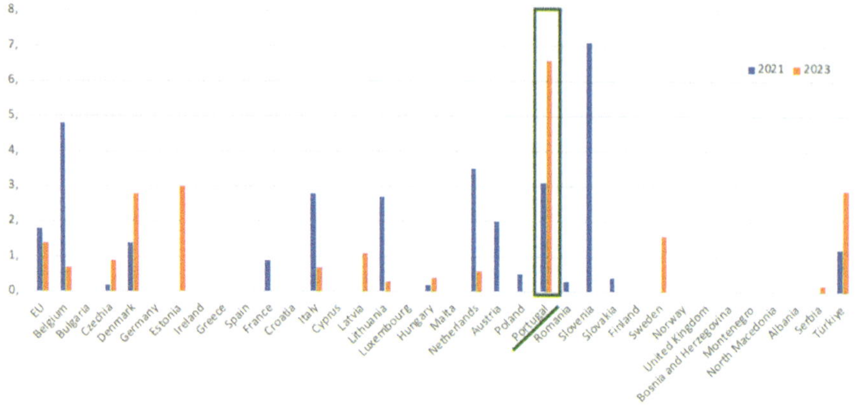

Fig. 8.9 Percentage of T&C companies in European countries using AI technologies for production processes in 2021 and 2023. (Data not available for some countries). *Source* Eurostat (2024a). Companies are those manufacturing textiles, clothing and leather (NACE, Rev 2)

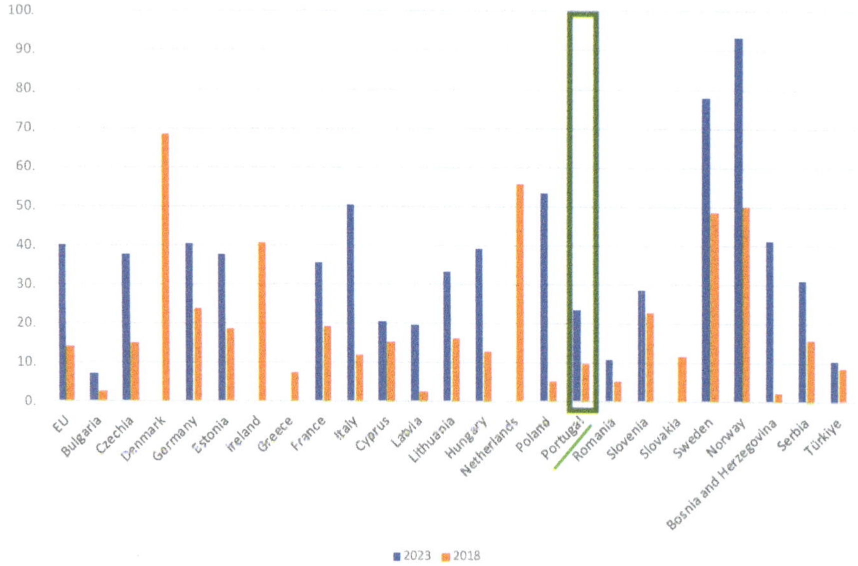

Fig. 8.10 Percentage of T&C companies in EU countries using cloud computing services in 2018 and 2023. *Source* Eurostat (2024a). Companies are those manufacturing textiles, clothing and leather (NACE, Rev 2)

and sourcing with integrity (Alam et al., 2023; CITEVE, 2023; Jones & Comfort, 2021).

Other indicators of digitalisation—use of 3D-printing, service robotics, IoT and data analytics (Figs. 8.12, 8.13, 8.14 and 8.15)—show Portugal close to the EU

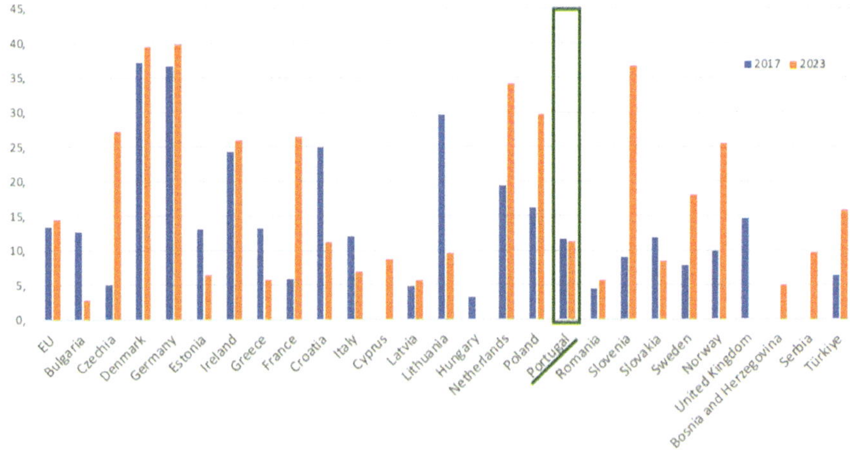

Fig. 8.11 Percentage of T&C industry companies in EU countries whose business processes are automatically linked to those of their suppliers and/or customers in 2017 and 2023. *Source* Eurostat (2024a)

average overall. These indicators, along with the projects that are taking place in partnership between technology centres, industry and academia in Portugal, suggest that the necessary conditions exist in the country to comply with the EU's strategic plan for textiles and clothing.

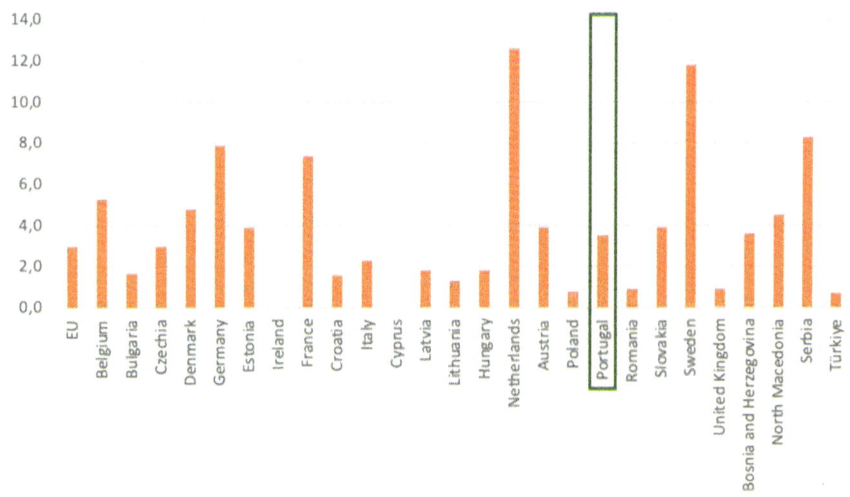

Fig. 8.12 Percentage of companies in EU countries using 3D Printing in 2020. *Source* Eurostat (2024a). Companies (10 persons employed or more) manufacture textiles, clothing, leather and related products (NACE, Rev 2)

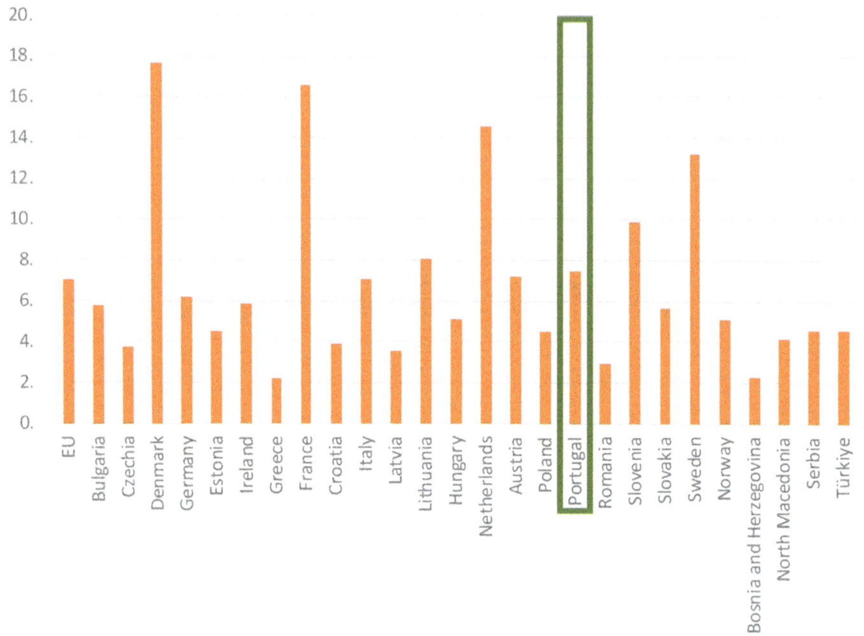

Fig. 8.13 Percentage of companies in EU countries using industrial or service robots in 2022 per country. *Source* Eurostat (2024a). Companies (10 persons employed or more) manufacture textiles, clothing, leather and related products (NACE, Rev 2)

8.4 Discussion

8.4.1 Sustainability in the Textile Industry: Challenges and Opportunities

Sustainability in the textile industry has been a central topic of debate in recent decades, especially with the increasing awareness of the negative environmental impacts associated with traditional textile production. Studies like that of Niinimäki (2018) argue that sustainable fashion must integrate all sectors of the value chain, from material selection to the final disposal of products, promoting the adoption of a CE as an alternative model to the traditional linear production and consumption model. In the context of the Portuguese T&C industry, the results obtained in this study indicate that, although there is an increase in the adoption of sustainable practices, there is still a long way to go before these practices are universally adopted and effectively integrated. Eco-Design and Eco-Engineering practices, for example, have shown positive evolution but still represent a fraction of the techniques applied in the sector. This finding aligns with the existing literature, which suggests that the transition to more sustainable practices is often limited by factors such as high costs, resistance to change and lack of adequate infrastructure (Bressanelli et al., 2022).

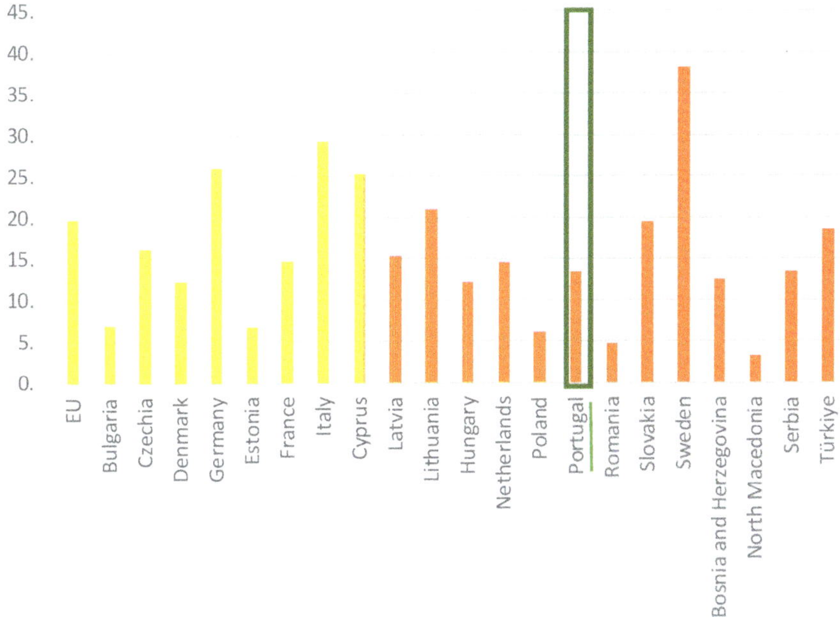

Fig. 8.14 Percentage of companies in EU countries using IoT devices as of 2021. *Source* Eurostat (2024a). Companies (10 persons employed or more) manufacture textiles, clothing, leather and related products (NACE, Rev 2)

However, given the number of RRP projects underway in this area, the results should be monitored annually, and a review of this issue should be conducted in 2027.

8.4.2 Circular Economy: Implementation and Structural Challenges

The CE has been widely recognised as a crucial component of long-term sustainability. However, the implementation of the CE in the textile industry faces significant challenges, such as the need for greater collaboration among stakeholders, the development of new recycling technologies and the creation of value-added markets for recycled materials. It was observed at the recent international fairs in July 2024 that most exporting companies already presented various solutions on this matter.

The results of the present study, which indicate that circularity is still in its early stages of development in the Portuguese T&C industry, corroborate the observations of Kirchherr et al. (2023), who identified the lack of integration between different levels of implementation (micro, meso and macro) as a critical barrier to the adoption of the CE. Furthermore, the difficulty in establishing effective partnerships upstream and downstream in the textile value chain is a problem identified in several studies,

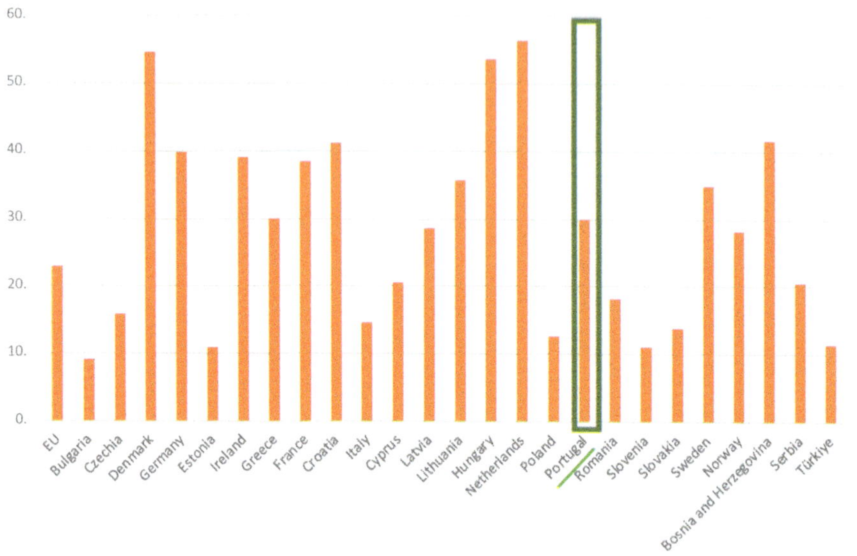

Fig. 8.15 Percentage of companies in EU countries using Data analytics for the enterprise is performed by the enterprise's own employees or by an external provider as of 2023. *Source* Eurostat (2024a). Companies (10 persons employed or more) manufacture textiles, clothing, leather and related products (NACE, Rev 2)

which point to the need for public policies that encourage cooperation between companies and promote the creation of support infrastructures (Chen et al., 2021).

8.4.3 Impact of Environmental Policies and Certifications

Public policies and environmental certifications play a critical role in promoting sustainability and circularity in the textile industry. Studies such as that by Bressanelli et al. (2022) highlight that ecological certifications, such as GOTS, OEKO-TEX® standard 100, OEKO-TEX® standard 100 STeP and ISO 14001, not only enhance consumer confidence but also drive companies to adopt more sustainable practices. However, complying with these certifications is often challenging, particularly for small and medium-sized enterprises, due to the associated costs and the complexity of the legal requirements.

The present study revealed that adherence to environmental certifications is growing in the Portuguese T&C sector, reflecting a positive trend towards sustainability, but also with a social component, with GOTS certification in the knitting and weaving industry in front (Table 8.5). Nevertheless, significant barriers remain, particularly concerning awareness and the accessibility of certifications, given the slow growth during the period analysed, which may have been influenced by the pandemic. This point is supported by literature suggesting that, without adequate support, many companies may struggle to meet the required standards, potentially limiting the overall effectiveness of the environmental policies (Guedes & Buest, 2018).

8.4.4 Digitalisation as a Facilitator of Sustainability and Circularity

Digitalisation in industry, particularly within the scope of Industry 4.0 and the transition to Industry 5.0, has been widely discussed in the literature as a catalyst for sustainability and the CE (Buhr, 2015; Wiegand & Wynn, 2023). The application of digital technologies, such as AI, the IoT and cloud computing, enables more efficient resource management, waste reduction and the optimisation of production processes.

In the case of the Portuguese T&C industry, the adoption of these technologies has been gradual, with promising results compared to other countries, but still falling short of their maximum potential, as indicated by Eurostat data on the use of digital technologies in the sector. This scenario is consistent with the findings of Pozzi et al. (2021), who point out that digitalisation, though essential, faces obstacles such as high implementation costs and the lack of digital skills in some companies. On the other hand, the study by Wiegand and Wynn (2023) on the German textile industry suggests that digitalisation not only facilitates sustainability but also offers new opportunities for the development of innovative business models, which should be explored in the Portuguese context. Considering the investments Portuguese companies have been doing in digitalisation, it is expected to be observed an increase in the results of its application in the next few years.

Finally, the most important aspect of this discussion is the interconnection of the impact of digitalisation as a facilitator of sustainability and circularity, which in most studies is analysed separately. This issue is crucial as digitalisation not only supports the transition to more sustainable practices but can also revolutionise the textile industry by creating new opportunities for innovation, efficiency and competitiveness. Furthermore, digitalisation is directly linked to the emerging concepts of Industry 4.0 and 5.0, which are fundamental to the future of industrial production in general.

8.5 Conclusions

The challenge of this work was to investigate the concept of sustainability and how it is implemented within companies in conjunction with the adoption of CE practices. Given the increasing questions surrounding this topic and the rising interest it has sparked in society, there was an opportunity to gather insights and explore the subject more deeply. This also provided a chance for theoretical enrichment and reflection on our responsibilities as citizens towards the environment.

The literature review facilitated a clearer understanding of the themes and concepts. The research reveals several gaps that need to be addressed regarding circularity and digitalisation practices in the industry in Portugal. The CE emerges as a novel business model that, despite requiring additional investment from companies, offers significant advantages, such as drastically reducing environmental impact and fostering economic growth. For the T&C industry to be truly sustainable, it must ensure cross-company process and technology integration, while also encouraging new concepts and behaviours among all stakeholders and providing the necessary training and upskilling for the workforce.

There is an urgent need to find solutions for more sustainable materials, increase recycling rates and develop value-added products and services—with only 7.9% of waste currently being recycled (Textile Exchange, 2023). While some sustainable practices in Portuguese T&C companies are already implemented, and CE initiatives are already being adopted by some companies, there is a need to devise and promote new strategies to enable small and medium-sized businesses, in particular, to embrace this business model. Companies and society also have a responsibility to educate consumers about these new practices, including creating projects and pathways that allow consumers to participate and contribute to reduce environmental impacts, such as the project be@t is doing. The T&C industry in Portugal is developing significant initiatives in sustainable and digitalisation practices, and undertaking related R&D projects for the transition to the CE, and consequently to support compliance with the SDG objectives. These conclusions are in line with the results obtained by the CITEVE (2023) study, within the scope of the be@t project, regarding SDG objectives related to environmental (7, 9, 12, 13, 14 and 15), governance (8, 9, 12 and 17) and social (8 and 10) issues.

In conclusion, there are clearly opportunities in Portugal to increase the efficiency in sustainability practices and circularity through more effective deployment of digital technologies in the industry. This can underpin the development and optimisation of new processes for design and production, thereby reducing resource consumption. Finally, it is important for future strategy and work that all system actors—policymakers, industrialists, technology centres, academia, consumers and society in general—are aligned with the objectives defined by the EU, to support the transitioning of T&C manufacturing companies, as regards the internal and external application of regulations in Tables 8.3 and 8.4. Only in this way will companies be successful in transitioning to a new paradigm and the Sustainable Development Goals for 2030 and 2050, whilst also remaining financially viable within their new business models. It is also considered important as a strategy in the future to monitor the results of practices in sustainability, circularity and digitalisation after the conclusion of ongoing R&D projects, and to undertake a comparative analysis between the T&C sector companies, with and without ongoing R&D projects.

Acknowledgements The authors acknowledge the financial support from integrated project be@t—Textile Bioeconomy (TC-C12-i01, Sustainable Bioeconomy No. 02/C12-i01.01/2022), promoted by the Recovery and Resilience Plan (RRP), Next Generation EU, for the period 2021–2026.

The authors are very grateful for the support granted by the Research Unit of Fiber Materials and Environmental Technologies (FibEnTech-UBI), through the Project reference UIDB/00195/2020, funded by the Fundação para a Ciência e a Tecnologia, IP/MCTES through national funds (PIDDAC), and DOI: 10.54499/UIDB/00195/2020 (https://doi.org/10.54499/UIDB/00195/2020).

References

Adisorn, T., Tholen, L., & Götz, T. (2021). Towards a digital product passport fit for contributing to a circular economy. *Energies, 14*(8), 2289. https://doi.org/10.3390/en14082289

Alam, Md., Hosen, I., Mridha, J., Chowdhury, S., & Rahman, Md. (2023). Assessing the barriers of integrating technological innovations in textiles sector: Implications towards sustainable production. *Green Technologies and Sustainability, 1,* 100039. https://doi.org/10.1016/j.grets.2023.100039

Ambiente. (2017). "Decreto-Lei n.º 152-D/2017" Diário da República n.º 236/2017, 2º Suplemento, Série I de 2017–12–11, pp. 88–135. https://diariodarepublica.pt/dr/detalhe/decreto-lei/152-d-2017-114337042

Anicet, A., Bessa, P., & Broega, A. C. (2011). Recycling of textile wastes through textile collage. In *CIPED VI International Congress on Design Research.* https://repositorium.sdum.uminho.pt/bitstream/1822/14644/2/15041852_ENG.pdf

APA. (n.d.). *Dados sobre resíduos urbanos.* Agência Portuguesa do Ambiente. https://apambiente.pt/residuos/dados-sobre-residuos-urbanos

APA. (2022). *Relatório dos resíduos urbanos 2022.* Agência Portuguesa do Ambiente. https://apambiente.pt/sites/default/files/_Residuos/Producao_Gest%C3%A3o_Residuos/Dados%20RU/RARU_2022_V1.2.pdf

ATP. (n.d.). *Estatísticas [Statistics].* Associação Têxtil e Vestuário de Portugal. https://atp.pt/pt-pt/estatisticas/2022-2/

ATP. (2021). *Visão Prospetiva e Estratégias ITV 2030 Contributo para um Plano Estratégico para o Setor Têxtil e Vestuário Português até 2030.* Associação Têxtil e Vestuário de Portugal. https://atp.pt/wp-content/uploads/2021/07/Estudo-2030-digital.pdf

Babo, L. (2024, February 12). *Após recorde em 2022, as exportações de Têxteis e Vestuário sofrem quebra em 2023* [After a record in 2022, textile and clothing exports suffer a drop in 2023]. ATP—Associação Têxtil e Vestuário de Portugal. https://atp.pt/wp-content/uploads/2024/02/PressRelease-Exportacoes-da-ITV-2023.pdf

Baugh, G. (2011). *The fashion designer's textile directory* (1st ed.). Paul Carslake. Barrons Educational Series. ISBN 0764146289.

Bhatnagar, A., & Niinimäki, K. (2024). Advancing integration of CSR and social life cycle assessment in the textile and fashion industry. In S. S. Muthu (Ed.), *Corporate social responsibility in textiles and fashion. Sustainable textiles: Production, processing, manufacturing & chemistry.* Springer. https://doi.org/10.1007/978-3-031-61099-8_5

Bicho, M., Pereira, M., Barata, T. R., Belino, N., & Miguel, R. (2024). Sustainable fashion product development: The importance of skills in new materials and processes in the academy and industry context. In J. Cunha, A. C. Broega, H. Carvalho, & B. Providência (Eds.), *Advances in fashion and design research II* (pp. 287–301). CIMODE 2023. Springer. https://doi.org/10.1007/978-3-031-43937-7_26

Bressanelli, G., Visintin, F., & Saccani, N. (2022). Circular economy and the evolution of industrial districts: A supply chain perspective. *International Journal of Production Economics, 243,* 108348. ISSN 0925-5273. https://doi.org/10.1016/j.ijpe.2021.108348

Buhr, D. (2015). *Social innovation policy for Industry 4.0*. Friedrich-Ebert-Stiftung. https://library.fes.de/pdf-files/wiso/11479.pdf

Chen, X., Memon, H. A., Wang, Y., Marriam, I., & Tebyetekerwa, M. (2021). Circular economy and sustainability of the clothing and textile industry. *Materials Circular Economy, 3*(12). https://doi.org/10.1007/s42824-021-00026-2

CITEVE. (2023). *Relatório de Sustentabilidade do Setor Têxtil e Vestuário 2023_be@t* [Textile and clothing sector sustainability report 2023_be@t]. CITEVE. https://static.citeve.pt/share/2024-03/31eef65c-a86f-4a52-ac20-325df1d0b2d9/e878fbf3-b07c-4163-a63a-ab697a7a5251/en_gb__file/286ac85d-b50b-486a-a68a-9507fc174de1/20240327142740/Relatorio_Sustentabilidade_STV_be@t_vEN_.pdf

Council of the European Union. (2023). *Ecodesign regulation: Council adopts position, Council of the EU, Press release*, 22 May 2023. https://www.consilium.europa.eu/en/press/press-releases/2023/05/22/ecodesign-regulation-council-adopts-position/

Council of the European Union. (2024). Proposal for a directive of the European Parliament and of the Council on common rules promoting the repair of goods and amending Regulation (EU) 2017/2394, Directives (EU) 2019/771 and (EU) 2020/1828. https://eur-lex.europa.eu/legal-content/EN/TXT/PDF/?uri=CONSIL:ST_9219_2024_INIT

Cunha, A., et al. (2023). Next-generation fashion ecosystem: A STVgoDigital approach. In C. F. da Silva Portela (Ed.), *Sustainable, innovative and intelligent societies and cities* (pp. 355–388). EAI/Springer Innovations in Communication and Computing. Springer. https://doi.org/10.1007/978-3-031-30514-6_15

Damayanti, D., Supriyadi, D., Amelia, D., Saputri, D. R., Devi, Y. L. L., Auriyani, W. A., & Wu, H. S. (2021). Conversion of lignocellulose for bioethanol production, applied in bio-polyethylene terephthalate. *Polymers, 13*, 2886. https://doi.org/10.3390/polym13172886

Davis, N. (2016, January 19). *What is the fourth industrial revolution*. World Economic Forum. https://www.weforum.org/agenda/2016/01/what-is-the-fourth-industrial-revolution/

Demir, K. A., Döven, G., & Sezen, B. (2019). Industry 5.0 and human-robot co-working. *Procedia Computer Science, 158*, 688–695. https://doi.org/10.1016/j.procs.2019.09.104

Diekel, F., Mikosch, N., Bach, V., & Finkbeiner, M. (2021). Life cycle based comparison of textile ecolabels. *Sustainability, 13*(4), 1751. https://doi.org/10.3390/su13041751

Dissanayake, D., Kuruppu, S., Qian, W., & Tilt, C. (2020). Barriers for sustainability reporting: Evidence from Indo-Pacific region. *Meditari Accountancy Research, 29*(2), 264–293. https://doi.org/10.1108/MEDAR-01-2020-0703

Economic Commission for Europe. (2022). *Enabling sustainability and circularity in the garment and footwear sector: Policy developments and industry perspectives on traceability and transparency (ECE/TRADE/C/CEFACT/2022/10)*. United Nations. https://unece.org/sites/default/files/2023-07/ECE_TRADE_C_CEFACT_2022_10E_0.pdf

EURATEX. (2022). *Facts & key figures 2022 of the European textile and clothing industry*. EURATEX, Economic and Statistics. https://euratex.eu/wp-content/uploads/EURATEX_FactsKey_Figures_2022rev-1.pdf

EURATEX. (2023a, December 1). *Circular economy—EURATEX*. https://euratex.eu/sustainable-businesses/circular-economy/

EURATEX. (2023b, December 20). *Chemicals policy: Providing the best product and ensuring safety for human health and environment—EURATEX*. https://euratex.eu/sustainable-businesses/chemicals-policy/

European Commission. (2019). The European Green Deal. Communication from the commission to the European parliament, the European council, the council, the European economic and social committee and the committee of the regions, Brussels. https://eur-lex.europa.eu/resource.html?uri=cellar:b828d165-1c22-11ea-8c1f-01aa75ed71a1.0002.02/DOC_1&format=PDF

European Commission. (2022). *EU strategy for sustainable and circular textiles*. COM(2022). https://eur-lex.europa.eu/resource.html?uri=cellar:9d2e47d1-b0f3-11ec-83e1-01aa75ed71a1.0001.02/DOC_1&format=PDF

European Commission. (2024). *The Net-Zero Industry Act: Accelerating the transition to climate neutrality*. Internal Market, Industry, Entrepreneurship and SMEs. https://single-market-economy.ec.europa.eu/industry/sustainability/net-zero-industry-act_en

European Commission. (n.d.). *BAT reference documents*. https://eippcb.jrc.ec.europa.eu/reference/

European Parliament. (2024). *Legislative train: A European green deal revision of the textile labelling regulation—Q4 2023*. https://www.europarl.europa.eu/legislative-train/carriage/textile-labeling-regulation/report?sid=8201#/stage/6301/theme/100/file/45901/

European Parliament & Council of the European Union. (2024). Regulation (EC) No. 1907/2006 of the European Parliament and of the Council of 18 December 2006, amended version from 6 June 2024. https://eur-lex.europa.eu/legal-content/EN/TXT/PDF/?uri=CELEX:02006R1907-20240606

European Union. (2018). The Council Directive (EU) 2018/851 of the European Parliament and of the Council amending Directive 2008/98/EC on waste. *Official Journal of the European Union*, L 150, 14 June 2018, 109–140. https://faolex.fao.org/docs/pdf/eur178599.pdf

European Union. (2019). Commission Implementing Decision (EU) 2019/1004 of 7 June 2019 laying down rules for the calculation, verification and reporting of data on waste in accordance with Directive 2008/98/EC of the European Parliament and of the Council and repealing Commission Implementing Decision C (2012) 2384. *Official Journal of the European Union*, L 163, 20 June 2019, 66–100. https://eur-lex.europa.eu/legal-content/EN/TXT/PDF/?uri=CELEX:32019D1004

European Union. (2022a) Strategy for sustainable and circular textiles. In *Energy, climate change, environment—Textiles strategy*. European Commission. https://environment.ec.europa.eu/publications/textiles-strategy_en

European Union. (2022b). Directive (EU) 2022/2464 of the European Parliament and of the Council of 14 December 2022 amending Regulation (EU) No 537/2014, Directive 2004/109/EC, Directive 2006/43/EC and Directive 2013/34/EU, as regards corporate sustainability reporting. *Official Journal of the European Union*, L 320, 16 December 2022, 15–80. https://eur-lex.europa.eu/legal-content/EN/TXT/PDF/?uri=CELEX:32022L2464

European Union. (2024a). Regulation (EU) 2024/1781 of the European parliament and of the council of 13 June 2024 establishing a framework for the setting of ecodesign requirements for sustainable products, amending Directive (EU) 2020/1828 and Regulation (EU) 2023/1542 and repealing Directive 2009/125/EC. *Official Journal of the European Union*, L series, 28 June 2024. https://eur-lex.europa.eu/legal-content/EN/TXT/PDF/?uri=OJ:L_202401781

European Union. (2024b). Directive (EU) 2024/1799 of the European parliament and of the council of 13 June 2024 on common rules promoting the repair of goods and amending Regulation (EU) 2017/2394 and Directives (EU) 2019/771 and (EU) 2020/1828. *Official Journal of the European Union*, L series, 10 July 2024. https://eur-lex.europa.eu/legal-content/EN/TXT/PDF/?uri=OJ:L_202401799

EUROSTAT. (2024a). *Digital economy and society statistics—Households and individuals*. https://ec.europa.eu/eurostat/statistics-explained/index.php?title=Digital_economy_and_society_statistics_-_households_and_individuals

EUROSTAT. (2024b). *Digitalisation in Europe—2024 edition*. https://ec.europa.eu/eurostat/web/interactive-publications/digitalisation-2024 Fonseca, A., Ramalho, E., Gouveia, A, Henriques, R., Figueiredo, F., & Nunes, J. (2023). Systematic insights into a textile industry: Reviewing life cycle assessment and eco-design. *Sustainability, 15*(21), 15267. https://doi.org/10.3390/su152115267

Fraunhofer, I. A. O., & Ingenics, A. G. (2014). *Industrie 4.0—Eine Revolution der Arbeitsgestaltung*. Wie Automatisierung und Digitalisierung Unsere Produkte Verändern Warden.

Gartner. (n.d.). *Information technology—Gartner glossary*. Gartner. https://www.gartner.com/en/information-technology/glossary

Guedes, M. G., & Buest, A. (2018). Fashion designer, fashion consumer, fashion learner. *IOP Conference Series: Materials Science and Engineering, 459*, 012043. https://doi.org/10.1088/1757-899X/459/1/012043

Gueye, S. (2021, June 22). *Circular economy in the fashion industry*. Ellen MacArthur Foundation. https://ellenmacarthurfoundation.org/articles/the-trends-and-trailblazers-creating-a-circular-economy-for-fashion

Hansson Reuter, M., & Zetterlund, L. (2018). *Managing for sustainable fashion—A comparative review of the Higg Index 2.0 brand and retail module and the UN sustainable development goals*. Master thesis, Faculty of Engineering at Lund University. Lund University Libraries. http://lup.lub.lu.se/student-papers/record/8949900

Harsanto, B., Primiana, I., Sarasi, V., & Satyakti, Y. (2023). Sustainability innovation in the textile industry: A systematic review. *Sustainability, 15*(2), 1549. https://doi.org/10.3390/su15021549

Hawley, J. (2015). Economic impact of textile and clothing recycling. In J. Hethorn & C. Ulasewicz (Ed.), *Sustainable fashion what's next?: A conversation about issues, practices and possibilities* (pp. 204–230). Fairchild Books. https://doi.org/10.5040/9781501312250.ch-009

Huygens, D., Foschi, J., Caro, D., Patinha, C. C., Faraca, G., Foster, G., Solis, M., Marschinski, R., Napolano, L., Fruergaard, A. T., & Tonini, D. (2023). *Techno-scientific assessment of the management options for used and waste textiles in the European Union*. JRC Publications Repository. https://doi.org/10.2760/6292

Jones, P., & Comfort, D. (2021). The sustainable development goals and leading European retailers. *Athens Journal of Business & Economics, 7*(1), 105–122. https://www.athensjournals.gr/business/2021-7-1-5-Jones.pdf

Kirchherr, J., Yang, N. N., Schulze-Spüntrup, F., Heerink, M. J., & Hartley, K. (2023). Conceptualizing the circular economy (revisited): An analysis of 221 definitions. *Resources Conservation and Recycling, 194*, 107001. https://doi.org/10.1016/j.resconrec.2023.107001

Lamine, E., Thabet, R., Sienou, A., Bork, D., Fontanili, F., & Pingaud, H. (2020). BPRIM: An integrated framework for business process management and risk management. *Computers in Industry, 117*, 103199. https://doi.org/10.1016/j.compind.2020.103199

Legardeur, J., & Hospital, P. (2024). *Digital product passport for the textile sector*. Panel for the Future of Science and Technology (STOA)—European Parliamentary Research Service. European Union. https://doi.org/10.2861/947638. https://www.europarl.europa.eu/RegData/etudes/STUD/2024/757808/EPRS_STU(2024)757808_EN.pdf

Morrar, R., Arman, H., & Mousa, S. (2017). The fourth industrial revolution (Industry 4.0): A social innovation perspective. *Technology Innovation Management Review, 7*(11), 12–20. https://doi.org/10.22215/timreview/1117

Müller, H. (2023). Creative solutions in a saturated market. How individual players develop successful digital solutions in the textile and apparel industry. In R.C. Geibel & S. Machavariani (Eds.), *Chances and challenges of digital management. ISPC 2022. Springer proceedings in business and economics*. Springer. https://doi.org/10.1007/978-3-031-45601-5_19.

Müller, M., Jürgens, J., Redaèlli, M., Klingberg, K., Hautz, W., & Stock, S. (2018). Impact of the communication and patient hand-off tool SBAR on patient safety: A systematic review. *British Medical Journal Open, 8*, e022202. https://doi.org/10.1136/bmjopen-2018-022202

Niinimäki, K. (2018). Sustainable fashion in a circular economy. In K. Niinimäki (Ed.), *Sustainable fashion in a circular economy* (pp. 12–42). Aalto ARTS Books. https://core.ac.uk/download/pdf/301138773.pdf

Pereira, F., Carvalho, V., Vasconcelos, R., & Soares, F. (2022). A review in the use of artificial intelligence in textile industry. In J. Machado, F. Soares, J. Trojanowska, & S. Yildirim (Eds.), *Innovations in mechatronics engineering. ICIENG 2021. Lecture notes in mechanical engineering*. Springer. https://doi.org/10.1007/978-3-030-79168-1_34

Pereira, M. M. R. (2008). *Análise de Estratégias e Práticas de Gestão da Produção para um Melhor Desempenho na Cadeia Têxtil Portuguesa*. Doctoral thesis, Universidade da Beira Interior.

Plakantonaki, S., Kiskira, K., Zacharopoulos, N., Chronis, I., Coelho, F., Togiani, A., Kalkanis, K., & Priniotakis, G. (2023). A review of sustainability standards and ecolabeling in the textile industry. *Sustainability, 15*(15), 11589. https://doi.org/10.3390/su151511589

Pozzi, R., Rossi, T., & Secchi, R. (2021). Industry 4.0 technologies: Critical success factors for implementation and improvements in manufacturing companies. *Production Planning & Control, 34*(2), 139–158. https://doi.org/10.1080/09537287.2021.1891481

Presidência do Conselho de Ministros. (2023). "Resolução do Conselho de Ministros n.º 13/2023". Diário da República n.º 30/2023, 1.ª série, 10 de fevereiro de 2023:193–224. https://files.diariodarepublica.pt/1s/2023/02/03000/0019300224.pdf

Ranasinghe, L., & Jayasooriya, V. M. (2021). Ecolabelling in textile industry: A review. *Resources, Environment and Sustainability, 6*(2), 100037. https://doi.org/10.1016/j.resenv.2021.100037

Rathinamoorthy, R. (2018). Sustainable apparel production from recycled fabric waste. In S. S. Muthu (Ed.), *Sustainable innovations in recycled textiles* (pp. 19–52). Textile science and clothing technology. Springer. https://doi.org/10.1007/978-981-10-8515-4_2

Recuperar Portugal. (n.d.). *Recovery and resilience plan* (PRR). https://recuperarportugal.gov.pt/

Sá, M., Guise, C., Costa, F., Rodrigues, P., Oliveira, J., Barros, A., Silva, R., Alves, T., Santos, M., Da Cruz, A. M. R., Alves, M., & Silva, C. J. (2023). STVGoDigital: A digital product passport solution. In T. Guarda, F. Portela, & J. M. Diaz-Nafria (Eds.), *Advanced research in technologies, information, innovation and sustainability. ARTIIS 2023. Communications in computer and information science*, 1937. Springer. https://doi.org/10.1007/978-3-031-48930-3_28

Schöggl, J., Rusch, M., Stumpf, L., & Baumgartner, R. J. (2023). Implementation of digital technologies for a circular economy and sustainability management in the manufacturing sector. *Sustainable Production and Consumption, 35*, 401–420. https://doi.org/10.1016/j.spc.2022.11.012

Strandhagen, J. W., Buer, S., Semini, M., Alfnes, E., & Strandhagen, J. O. (2020). Sustainability challenges and how Industry 4.0 technologies can address them: A case study of a shipbuilding supply chain. *Production Planning & Control, 33*(9–10), 995–1010. https://doi.org/10.1080/09537287.2020.1837940

Taddei, E., Sassanelli, C., Rosa, P., & Terzi, S. (2022). Circular supply chains in the era of industry 4.0: A systematic literature review. *Computers & Industrial Engineering*, 170, 108268. https://doi.org/10.1016/j.cie.2022.108268

Textile Exchange. (2023, December 1). *Materials market report 2023*. https://textileexchange.org/app/uploads/2023/11/Materials-Market-Report-2023.pdf

United Nations. (2015) *Transforming our world: The 2030 Agenda for Sustainable Development (A/RES/70/1)*. Resolution adopted by the General Assembly on 25 September 2015. https://documents.un.org/doc/undoc/gen/n15/291/89/pdf/n1529189.pdf

Wiegand, T., & Wynn, M. (2023). Sustainability, the circular economy and digitalisation in the German textile and clothing industry. *Sustainability, 15*, 9111. https://doi.org/10.3390/su15119111

Webpages

Amfori. (n.d.). *amfori BSCI*. https://www.amfori.org/en/solutions/social/about-bsci

Bcorporation. (n.d.). *B Corp Certification demonstrates a company's entire social and environmental impact*. https://www.bcorporation.net/en-us/certification/

be@t. (n.d.). *be@t—bioeconomy at textiles*. https://www.bioeconomy-at-textiles.com

Better Cotton. (2023, September 28). *The world doesn't just need cotton, it needs Better Cotton.* https://bettercotton.org/

bluesign. (2024, July 3). *bluesign®—Solutions and services for a sustainable textile industry—bluesign*. Bluesign. https://www.bluesign.com/en/

Cradle to Cradle Certified. (n.d.). *Cradle to Cradle Certified®—Cradle to Cradle Products Innovation Institute*. https://c2ccertified.org/the-standard

Ethical Trading Initiative. (n.d.). *ETI*. https://www.ethicaltrade.org/

Fairtrade International. (n.d.). *Textile standard*. Fairtrade International. https://www.fairtrade.net/standard/textile
GIATEX. (n.d.). *Giatex—Reducing water consumption in the T&C sector through intelligent management*. https://www.giatex.pt
GOTS—Global Organic Textile Standard. (n.d.). https://global-standard.org/
ISO 14001:2015. (n.d.). https://www.iso.org/standard/60857.html
ISO 9001:2015. (2015, September 1). ISO. https://www.iso.org/standard/62085.html
Leather Working Group. (n.d.). *Responsible leather—Leather Working Group*. https://www.leatherworkinggroup.com/
OEKO-TEX® Service GmbH. (n.d.-a). *OEKO-TEX® STANDARD 100*. https://www.oeko-tex.com/en/our-standards/oeko-tex-standard-100
OEKO-TEX® Service GmbH. (n.d.-b). *OEKO-TEX® sustainable textile and leather production (STEP)*. https://www.oeko-tex.com/en/our-standards/oeko-tex-step
RDC@ITV. (n.d.). *RDC@ITV—On the path to decarbonization of the textile and clothing industry*. https://www.rdc-itv.pt/
TEXP@CT. (n.d.). *TEXP@CT—Improving innovation in Portugal*. https://www.texpact.pt
Textile Exchange. (2024a, May 7). *The organic content standard aims to increase organic agricultural production*. https://textileexchange.org/organic-content-standard/
Textile Exchange. (2024b, May 8). *The RCS and GRS are designed to boost the use of recycled materials*. https://textileexchange.org/recycled-claim-global-recycled-standard/
The Woolmark Company. (n.d.). *The Woolmark company*. https://www.woolmark.com/
Worldwide Responsible Accredited Production. (2024, July 30). *What we do—WRAP Compliance*. WRAP Compliance. https://wrapcompliance.org/en/about/what-we-do/

Madalena Pereira (madalena.pereira@ubi.pt) started her professional activity in 1989 in the clothing and textile industry and spent over 10 years in product development and sourcing. Madalena has a Ph.D. in Textile Engineering from the University of Beira Interior (UBI), Portugal, and is the author and co-author of numerous publications on the textile and clothing industry, sustainability, and branding fashion design. She is currently Professor in Sustainability and Circular Economy, President of the Faculty Council (Engineering) and Vice President of the Department of Textile Science and Technology in UBI, and was previously Director of the Fashion Design course and Director of the Masters degree in Branding and Fashion Design. Her areas of expertise are the textile and clothing industry, fashion design, sourcing and sustainability in the textile and clothing industry, fashion accessories, technologies for the clothing industry and fashion trends. Madalena is a Member of the Portuguese Textile Cluster.

Teresa Raquel Barata (traquel@ubi.pt) is Visiting Assistant Professor in the Department of Textile Science and Technology at the University of Beira Interior (UBI), Portugal. Her research interests include nanofibers, digitalisation, 3D-CAD, sustainability, sourcing, and fashion design education for the Portuguese textile and clothing industry. She is currently participating in the Be@t—Bioeconomy project and GIATEX—Intelligent water management in the textile and clothing industries project. Teresa is also the coordinating auditor (for environmental, quality and safety management systems) at TÜV Rheinland, Portugal.

Marta Bicho (marta.bicho.fernandes@ubi.pt) is Visiting Assistant Professor in the Department of Textile Science and Technology at the University of Beira Interior (UBI), Portugal and a project coordinator in the Portuguese textile and clothing industry, having also held management positions in the clothing industry. Marta is currently participating in the Be@t—Bioconomy project led by CITEVE and involving around 56 partners from the Portuguese textile and clothing Industry and academia.

Assunção Mesquita has a degree in Industrial Engineering and Management, a post-graduate degree in Quality Engineering and an advanced degree in Innovation Management. As a senior consultant at CITEVE for over 25 years, she has supported companies in their process of organisational evolution, focusing on the satisfaction of the different stakeholders, intervening in the definition, improvement and auditing of processes with a view to their certification, whether in Quality Management, Environment, Sustainable Textile Production-STeP, Innovation Management, Social Responsibility, Management of Reconciliation between professional, family and personal life, Ecological Labels for Textile Products. She is a trainer at CITEVE in the areas of organisation, sustainability and innovation, and participates in national and international R&D projects. She is manager of Initiative 9—Society Pillar of the be@t project, responsible for managing the ESG sustainability report for the Portuguese Textile and Clothing sector.

Helena Vilaça (hvilaca@citeve.pt) has a degree in Chemistry (2008, University of Minho), a Master's in Medicinal Chemistry (2010, UM) and a Ph.D. in Sciences-Chemistry (2015, University of Minho). In CITEVE, she is a Laboratory and R&D Project Manager at the Department of Chemistry and Biotechnology. Her work is focused on textile functionalisation using bio-additives, sustainability, circular economy, healthcare textiles, legislation on chemicals and their safe use. She has 2 international patents, 12 peer-reviewed articles, 1 book chapter, and has supervised several students.

Carla Joana Silva (cjsilva@citeve.pt), has a degree in Biological Engineering (1996, UMinho), a Master's in Industrial Biotechnology (1999, USP, Brazil) and Ph.D. in Textile Chemistry (2005, UMinho). From 2007 to 2019 she worked as a senior researcher and later as Chief Technology Officer at CeNTI (Centre for Nanotechnology and Smart Materials), and from 2019 she has lead the Chemical and Biotechnological Department at CITEVE. With more than 20 years of experience in applied and industry-oriented research, 11 patents and more than 80 published papers in peer reviewed journals, she is/has supervised 5 Ph.D. theses, and more than 30 MSc. and 10 BSc. dissertations. She has been responsible for the management and coordination of several national and international R&D projects covering various research topics, ranging from surface engineering, biotechnology, (nano)coatings, processes optimization, sustainability, textile recycling, bio and circular economy and development of innovative nanomaterials, fibres and polymers.

Nuno Belino (belino@ubi.pt) is currently the head of the Department of Textile Science and Technology at the University of Beira Interior (UBI), Portugal, where he is also director of the Sustainable Fashion Technology and Product course. He holds a Ph.D. in Textile Engineering Science and a Master's degree in computer sciences. He was the production director at Nova Penteação e Fiação da Covilhã—a vertical textile company with line responsibility for the industrial automation, weaving and finishing departments. He was also involved in consulting and research activities related to industrial automation, computer science and new technologies applied in the textile industry at Fitecom, Comercialização e Industrialização Têxtil S.A., another vertical textile company.

Ana Santiago (ana.santiago@ubi.pt) is a Ph.D. candidate in Fashion Design, with research focused on new sustainable textile materials and processes, at the University of Beira Interior (UBI). She completed an internship in the textile and clothing industry, working in various areas within product development. After her internship she took part in two projects as a research fellow. She recently completed a short-term mobility internship at the Dikniga (development of innovative knitted textiles and garments) Laboratory, University of West Attica, Greece.

Sandra Ferreira (sandra.ferreira@ubi.pt) is a Ph.D. candidate in Industrial Engineering and Management at the University of Beira Interior. She holds a Bachelor and Master's degrees in Textile Engineering, more specifically in the clothing branch. For 24 years, she worked in a clothing factory and decided to apply to a research schollarship, being involved in a be@t project—Bioeconomy for Textiles and Clothing, led by CITEVE, and involving around 56 partners from the Portuguese textile and clothing Industry and academia. This project is dedicated to the sustainability and circularity of textiles and it supports the industry in training stakeholders. She is a collaborative member of the FibEnTech/UBI Research Unit.

Rui Miguel (rmiguel@ubi.pt) is an Associate Professor at the Faculty of Engineering of the University of Beira Interior (UBI). He holds an undergraduate degree and a Ph.D. in Textile Engineering from UBI. For 8 years, he worked as a textile engineer in the wool industry. In 1987 he started his academic career at UBI in the areas of Textile Engineering and Fashion Design, and since 2000, he has been primarily responsible for the creation and development of UBI's Fashion Design courses. He was President of the Department of Textile Science and Technology, a member of the General Council of UBI, Senate and Scientific Council of the Faculty of Engineering at UBI. He has also acted as a consultant to the Portuguese Air Force for the design and performance of fabrics. He was Director of the Ph.D. Courses in Textile Engineering and Fashion Design and the Master's Courses in Fashion Design and Branding and Fashion Design at UBI. He is a member of the Advisory Council of ATP—the Textile and Clothing Association of Portugal. He is the author and co-author of around 250 scientific papers including book chapters, journal articles and communications at international conferences in the fields of textile physics, smart textiles and clothing and fashion design. He has supervised several Ph.D. and Master's theses, notably in fashion design and textile engineering.

Chapter 9
Bulgaria

Radoslav Radev

Abstract The main purpose of this chapter is to generate a comprehensive understanding of the transition to sustainability and the circular economy (CE) in the Bulgarian textile and clothing (T&C) industry, and the role that digital technologies play in this transition. The methods used in this research centre on a descriptive-analytical method, an examination of pertinent literature, comparative analysis, field observation and questionnaire/survey. Empirical material is garnered from thematically selected scientific reports and articles, statistical data and the websites of Bulgarian companies. The 35 companies included in this research were randomly selected from online listings of companies in the field of the T&C industry in Bulgaria, and the websites of the mentioned companies were thoroughly studied. The study found that only twelve out of 35 surveyed companies develop sustainability-related actions and activities, with five of these companies pursuing all the eight listed sustainability-related activities. Eight out of 35 companies actively participate in one or more of the CE-related activities, and three companies perform all seven activities. The majority of companies are aware of the benefits of using digital technologies and appreciate the importance of digital technologies in the processes related to sustainability and CE. The findings indicate that a significant part of the studied Bulgarian companies lag behind in adopting policies, strategies and activities related to sustainability, the CE and digital technology deployment. Sustainability activities are better developed in a greater percentage of Bulgarian companies compared to activities related to the CE.

Keywords Textile industry · Clothing industry · Sustainability · Circular economy · Digital technologies

R. Radev (✉)
University of Economics, Varna, Bulgaria
e-mail: r.radev@ue-varna.bg

M. Wynn and T. Wiegand (eds.), *Sustainability, the Circular Economy and Digitalisation in the European Textile and Clothing Industry*, SDGs and Textiles,
https://doi.org/10.1007/978-981-97-9116-3_9

9.1 Introduction and Background

By the second quarter of the nineteenth century, the lands of Bulgaria were one of the most important industrial areas of the Ottoman Empire. The introduction of a regular army in 1826 and the policy of centralised supply of uniforms acted as the catalyst for the development of Bulgarian textile handicraft production. The exact amount of military supplies is unknown, but Ottoman Turkish documents from the early 1830s speak of about 240,000 m^2 of textile materials (Ivanov, 2021).

The creation of the textile industry in Bulgaria marks the beginning of factory production in the country. In 1834, in the town of Sliven, the industrialist Dobri Zhelyazkov founded the first cloth factory in the European part of the Ottoman Empire (Rusev, 1995, 1998). It served as the foundation for the first capitalist industrial textile company in Sliven, which produced cloth for Rumelia and Turkey (Patarchanov, 2016). By 1879 there were 20 industrial enterprises in Bulgaria, of which 4 were textile factories. According to quoted authors, publications and textbook editions, the number of factories in the Bulgarian lands before the Liberation (1878) varies between 11 and 40. However, there are several reasons for this variation, the main one being the different meaning that is attached to the term "factory" (Rusev, 2016).

In 1887, there were 92 factories in Bulgaria, in 1894–130, and in 1900–143 with 5848 workers. By 1894, 16 factories had been built in Sliven, and in Veliko Tarnovo, there were silk factories and a dye factory. In Tryavna, there were two factories for woolen cloth and one for cordage, in Svishtovsko—two knitting mills, in Karlovo—the most modern silk mill of its time, and in Kazanlak—a textile factory. From the Liberation in 1878 to the unification of the country in 1885, 12 industrial enterprises were built in Plovdiv, whilst in Gabrovo, a mechanical wool-textile spinning mill was constructed, and in Varna, the "First Bulgarian Anonymous Privileged Company for Cotton Yarns" appeared, using high-performance imported machines (Patarchanov, 2016).

In the first decade of the twentieth century, the textile industry in Bulgaria experienced dynamic growth. The significant changes that occurred in the textile industry were evident not only in the increase in the number of workers, but also in improving the technical base of Bulgarian enterprises. Until the outbreak of the Balkan Wars in 1912, all wool processing enterprises worked on mechanical looms. During this period, the other branches of the sector—carpet weaving, knitting, rope weaving, and the production of linen, silk and cotton fabrics—developed rapidly (Popova, 2019).

In 1911, there were 72 textile factories amongst Bulgaria enterprises with 4257 workers in total. Nevertheless, there remained large disparities between sub-sectors within the industry—for example in 1909, of the 61 enterprises in the textile industry, 46 were for processing wool, and only 15 for processing cotton, linen, jute and silk. The slower development of cotton textile production was connected with the need for more capital investment for building modern factories that would be competitive. For this reason, until the Balkan Wars, cotton textile factories were mainly created with foreign capital (Popova, 2019).

During the first 40–50 years of the twentieth century, the textile industry in Bulgaria became one of the country's leading industrial sectors, and continued to develop steadily in the second half of the century, notably in the 1970s and 80s, when the volume of textile and clothing (T&C) production nearly doubled, thanks to huge investments made in new and modern equipment in textile production (Patarchanov, 2016). After 1989 the transition to a market economy started, in which the Bulgarian T&C industry was one of the few industrial sectors that maintained good levels of competition on the international stage in the new economic conditions of the 1990s and early 2000s. The development of the T&C industry in Bulgaria was largely driven by the desire for European integration. The competitive environment changed after 2007, when Bulgaria became a member of the EU. The liberalisation of the European market led to a significant restructuring of the T&C sector, which led to the loss of many jobs. The negative trends in the industry deepened after the global financial crisis of 2008–2009, and the industry suffered further decline following the COVID-19 pandemic (Zhelev, 2021).

Figure 9.1 depicts the imports of T&C into Bulgaria in the period from 2007 to 2020, during which period the import of textiles was consistently dominant over clothing. The COVID-19 pandemic led to a sharp reduction in these imports, declining by 11% in 2020. Exports have also declined since the pandemic (Fig. 9.2), but to a lesser degree (9.2% in 2020), thus maintaining a positive trade balance in the T&C industry, which is characteristic of Bulgaria for the period under review (Texman, 2022).

In fact, textile exports increased by 9.7% in 2020 as some Bulgarian enterprises were among the first to switch to the production of protective masks, which were very important during the pandemic period. It should be noted, however, that Bulgaria largely specialises in the production of clothing. For this reason, the decline in consumer demand for clothing and the decline in its exports (by 15.5%) cannot

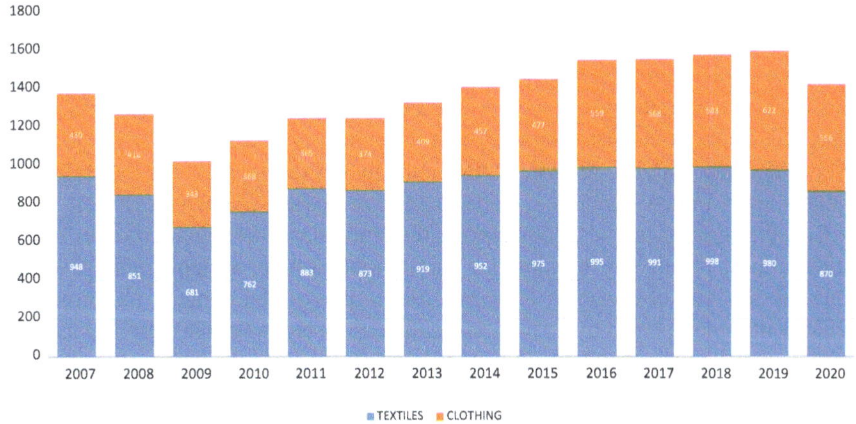

Fig. 9.1 Imports of textiles and clothing into Bulgaria (2007–2020, € million). *Source* Chapter author, based on Texman (2022)

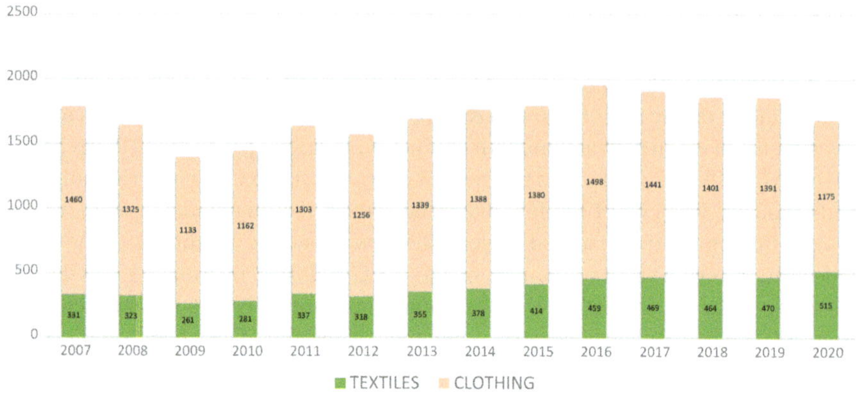

Fig. 9.2 Exports of textiles and clothing from Bulgaria (2007–2020, €million). *Source* Chapter author, based on Texman (2022)

be compensated for by the increased demand for personal protective equipment such as protective masks (Texman, 2022). In 2021, the top partner countries to which Bulgaria exports textiles and clothing (in US$) include Italy (404 million), Germany (394 million), France (230 million), Greece (210 million), Romania (97 million) and Turkey (85 million) (WITS, n.d.).

In recent years, the industry has faced a number of challenges related to delaying or stopping the supplies of materials to manufacturers; redundancies in textile companies due to reduced orders (the return of these workers is difficult); closing of shops due to reduced demand for clothes; increase in inventory levels because of reduced sales; increase in company debt and difficulties related to the COVID-19 crisis. Survey analysis (Angelova, 2022) indicates a reorientation in purchases from suits and dresses in the high-price segment to casual clothes in the low-price segment. The second reorientation is from the production of textiles mainly for clothing, to production of textile products, notably as face masks and protective clothes.

As the Bulgarian economy began to overcome the effects of the COVID-19 pandemic, new challenges emerged, particularly the war in Ukraine. Subsequent sanctions led to an increase in energy prices, and the acceleration of inflation negatively impacted both the economic development of Europe, and also the Bulgarian T&C industry in new dynamic conditions of development (Angelova, 2022).

Consumption of textile goods among the Bulgarian population tends to follow world trends. Similar to the situation throughout Europe, a significant growth in consumption was observed in Bulgaria, due in part to the demand for protective equipment, accompanied by an increase in the cost of clothing. This also had a major impact on the environmental footprint of the T&C industry in terms of production and waste generation (Bulgarian Association Circular Textile, 2022). Over the last 10 years in Bulgaria, sales of clothing have grown by 40%, with a 50% increase in the amount of T&C waste amounting to over 100,000 tonnes thrown away per year—equal to 5 times the weight of the National Palace of Culture building in Sofia! (Bulgarian Association Circular Textile, 2022).

Despite the gradual decrease in total household waste for the period 2012–2019, the amount of discarded textiles has increased significantly. Between 2012 and 2019, the amount of textile waste increased by as much as 54% (reaching 7 kg/inhabitant in 2019). As a share of total household waste, textile waste increased to 5% in 2019 (Bulgarian Association Circular Textile, 2022).

Bulgaria, as a member State of the European Union, is obliged to create a system for the separate collection of textiles and clothing by the beginning of 2025. At the moment, the mechanism for introducing a system for the separate collection of textiles in Bulgaria is not fully regulated, being carried out by private companies, on a competitive market basis. Most of these companies are members of the Bulgarian Association of Circular Textiles (BACT), which has developed some expertise in advising on the separate collection and processing of used textiles in Bulgaria. The activities carried out by BACT represent the basis on which a functioning and effective system can be built for separate collection of textile waste, thus fulfilling EU requirements (Bulgarian Association Circular Textile, 2022).

Although the value of imports and exports of T&C has remained fairly consistent over the past decade, there has over this period been a significant reduction in the percentage of GDP generated by the industry, declining from over 13% in 2013 to just over 9% in 2021 (Fig. 9.3). The reasons for this reduction in GDP are complex and can be found in a number of issues: the structure of enterprises in the industry; quantities produced; development or lack of development of markets; levels of competition; financing and investment opportunities; price levels, etc.

In 2020, the number of enterprises (Table 9.1) engaged in the clothing industry reduced by 437 companies (−10.38%) compared with 2019. This is the smallest number of enterprises operating in this industry during the observed period (Statista,

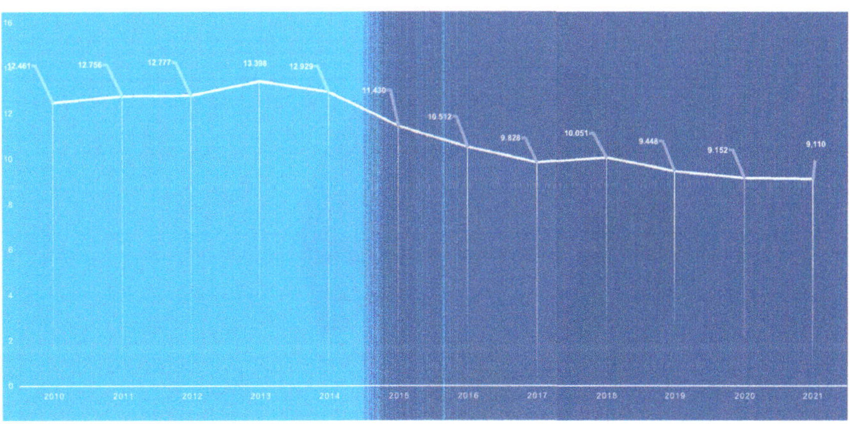

Fig. 9.3 % Contribution of T&C industry to GDP in Bulgaria 2010–2021. *Source* Chapter author, based on World Bank (2022)

2023a). At the same time, the number of enterprises in the textile industry in Bulgaria increased by 22 enterprises (+3.19%) in that year—with 711 enterprises, the number of enterprises in textiles reached its highest level for the observed period (Statista, 2023b). This is likely explained by the COVID crisis in 2020, which increased the need for protective textile masks and decreased the need to purchase clothing, as a large part of the population spent the majority of their time at home. This impacted the number of enterprises operating in the T&C industry in the period 2019–2020 (Fig. 9.4).

In the context of the EU, Bulgaria's employment in the T&C industry is 7% of the total EU workforce employed in this industry, behind Italy (22%), Romania (11%), Poland (11%), Portugal (9%) and Germany (9%), but ahead of Spain (6%),

Table 9.1 Number and percentage of enterprises in the T&C industry in Bulgaria 2011–2020

Year	Enterprises in the production of textiles	Enterprises in the production of clothing	Total enterprises in the T&C industry
2011	591 (11.89%)	4379 (88.11%)	4970
2012	583 (11.88%)	4324 (88.12%)	4907
2013	589 (11.79%)	4405 (88.21%)	4994
2014	589 (11.85%)	4381 (88.15%)	4970
2015	602 (12.10%)	4373 (87.90%)	4975
2016	653 (12.85%)	4428 (87.15%)	5081
2017	667 (13.11%)	4419 (86.89%)	5086
2018	692 (13.65%)	4379 (86.35%)	5071
2019	689 (14.06%)	4211 (85.94%)	4900
2020	711 (15.85%)	3774 (84.15%)	4485

Source Adapted from Statista (2023a, 2023b)

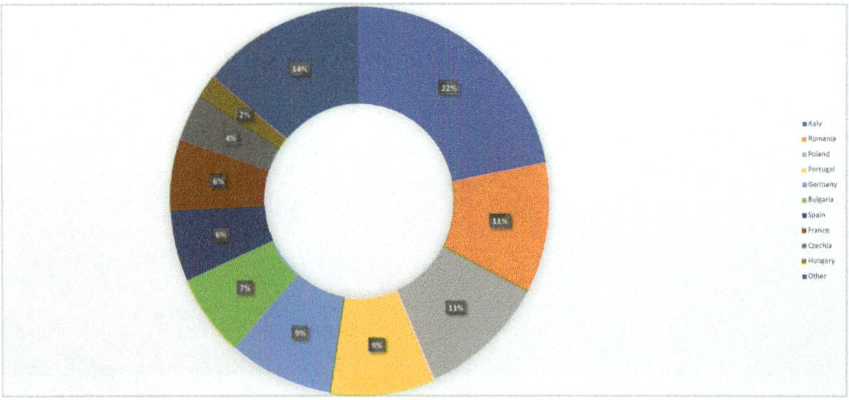

Fig. 9.4 Percentage of total workforce employed in the T&C industry in the EU, by country, 2021. *Source* Chapter author, based on Euratex (2022b)

France (6%), the Czech Republic (4%) and Hungary (2%). In other EU countries, employment in the industry accounts for 14% of the total. The total number of employees in the industry in Bulgaria is about 80,000 (including the leather and shoe industry), representing a decrease of 20% over the the last 4 years. The annual revenue from the T&C manufacturing industry in Bulgaria is estimated at €1.9 billion, which was ranked 8th in Europe in 2024 (out of a total of 26 EU countries). The average T&C manufacturing business in Bulgaria has 18.1 employees, an increase over the past 5 years (IBIS World, 2024).

The EU Strategy for Sustainable Textiles, focusing on environmental and social sustainability, maintains that all interested parties must unite and a level playing field must be maintained between European (including Bulgarian) and non-European companies (Euratex, 2022a). In conditions of tough global competition, significant new investment in technology and people will be needed to ensure that the Bulgarian T&C sector can become greener and more digital. Modernisation of the T&C industry would also help make the sector more attractive to Bulgarian workers, thereby reducing the need to hire workers from abroad. The T&C industry (including the leather and footwear industry) is one of the 14 industrial priority sectors in the EU (Euratex, 2022b).

Sustainable development (including for the T&C industry) is an important priority in Bulgaria's national policy. During the European Presidency of the Council of the EU in 2018, Bulgaria defined the sustainable development of the EU and Bulgaria as an important and integral element of the agenda. The adoption of the 2030 Program based on the United Nations SDGs and the next steps for a sustainable European future were discussed. Bulgaria's focus centred on progress towards specific SDGs, notably involving environmental action as regards air quality (SDGs 3 and 11), water management (SDG 6), marine protection (SDG 14) and nature and biodiversity protection (SDG 15). The country is also focusing on creating a green and energy efficient economy (SDG 7) (European Environment Agency, 2020).

At the end of 2020, the Council of Ministers adopted the National Development Program "BULGARIA 2030", which will address three main strategic objectives: accelerated economic development, demographic growth and reducing inequalities. The program outlined 13 national priorities in five areas of development. A "green and sustainable Bulgaria" will include development in the areas of a circular and low-carbon economy (SDGs 7, 8, 12), clean air and biodiversity (SDGs 3, 8, 11, 12, 13, 14, 15) and sustainable agriculture (SDGs 2, 14, 15). The program reflects a cross-government approach to sustainable development (including sustainability in the T&C industry) and examines the interrelationships between the various SDGs (European Environment Agency, 2020).

The National Development Program of Bulgaria 2030 states that one of the main goals related to the circular economy (CE) is the reduction of the resource intensity of the economy of the country, and increasing the efficiency of the materials used (including textile raw materials). The program mandates taking action to increase the productivity of resources during their entire life cycle and the rate of circular (secondary) use of materials (including textile materials) in the Bulgarian economy. In addition, the extension of the life of the product itself (including T&C goods), the reduction of waste (including textile waste) and the control of the need to extract new resources are specified. These measures are aimed at supporting enterprises (including companies in the T&C industry) in the introduction of waste-free technologies, and in reducing the amount of waste generated in the production process (National Development Program of Bulgaria 2030, 2020).

The transition to new business models assumes new significance, underpinning the remodelling of the processing and production of raw materials and products throughout the supply chain, as well as the development of design, reuse and recycling strategies to ensure longer product-life. A special focus is placed on the development of the scientific and technical potential and innovations related to the CE, and support for the development and introduction of green business models (National development program of Bulgaria 2030, 2020).

Digital technologies are constantly evolving and have hitherto been represented by the acronyms SMAC (Social Media, Mobile, Analytics/Big Data and Cloud) and BRAID (Blockchain, Robotics, Knowledge Work Automation/Artificial Intelligence, IoT and Digital Manufacturing). The scientific literature includes different perspectives and definitions of the terms "digitalisation" and "digital transformation", and they are sometimes used interchangeably. No universally accepted, robust conceptual framework has been found that can help businesses, practitioners and academics understand the constructs of digitalisation, digital business model transformation and digital innovation. However, digitalisation is seen as the implementation of digital technologies to support improvements in existing processes, while digital transformation is seen as constituting more of a significant transition to a new business model or at least a new way of working in a significant part of the business (Wynn & Felser, 2023).

Digital transformation holds enormous potential for economic growth. Bulgaria can build on its strengths in industries such as the T&C industry by using advanced digital technologies, including 5G networks, Internet of Things, Big Data, robotics and artificial intelligence, blockchain, 3D printing and others. The use of digital technologies will make it possible to maintain high levels of competition and be successful in emerging markets for the products of the future (Ministry of Transport and Communications, 2020). Currently, many companies in Bulgaria do not take full advantage of new digital technologies, which reduces the possibility of developing innovative business models. The difference in the level of digitalisation in Bulgarian industry is particularly marked between the high-tech and the traditional sectors (such as the T&C industry). In order to ensure higher levels of digitalisation in Bulgarian industry, enterprises must be able to make the right decisions related to the creation of new products based on digital technologies, including artificial intelligence (Ministry of Transport and Communications, 2020). Digital transformation will also enable higher levels of innovation and economic growth (Ministry of Transport and Communications, 2020).

Wiegand and Wynn (2023) put forward a conceptual framework for the transition to a sustainable and circular T&C industry based on digital technologies (See Fig. 1.3 in the Introduction chapter of this book). Important for this transformation is the creation and implementation of sustainable strategies and CE practices in accordance with European policies through the deployment of digital technologies. This change affects business organisations in terms of strategic alignment with new sustainability and CE policies and improvements in the company's business model, and digital transformation also normally involves changes in products and/or services. In this context, this chapter addresses the following research questions (RQs):

RQ1: How are Bulgarian T&C companies addressing sustainability in their corporate strategies and activities?
RQ2: What strategies and activities relating to the CE are being pursued in the Bulgarian T&C industry?
RQ3: What role are digital technologies playing in the transition to sustainability and the CE in the Bulgarian T&C industry?

Following this introduction, Sect. 9.2 focuses on the research methods used in the study, noting the randomly selected Bulgarian companies (C1–C35), as well as the product categories of the T&C goods covered by the study. In Sect. 9.3, the findings are presented and the RQs are addressed. The discussion in Sect. 9.4 includes an in-depth review of the main results of the study and identifies the key emergent themes. Finally, the conclusion Sect. 9.5 provides a brief summary of the study and suggestions for future possible research in this area.

9.2 Research Method

The main purpose of the study is to generate a comprehensive understanding of the transition to sustainability and the CE in the Bulgarian T&C industry, and the role that digital technologies play in this transition. The methods used in this research centre on a descriptive-analytical method, an examination of pertinent literature, comparative analysis, field observation and questionnaire/survey. Empirical material is garnered from thematically selected scientific reports and articles, statistical data and the websites of Bulgarian companies.

Source material was found in various scientific and publicly available databases and sources related to specific Bulgarian and European institutions in order to ensure a broad and representative coverage of the relevant research. The inclusion of retrieved publicly available information does not violate any ethical or legal requirements regarding the use of copyrighted material. The author has correctly, clearly and accurately cited the original sources of the literature used in accordance with academic conventions and intellectual property regulations.

The 35 companies included in this research were randomly selected from online listings of companies in the field of the T&C industry in Bulgaria, and the websites of the mentioned companies were thoroughly studied. For some of the companies, the sites had rather sparse content, which somewhat limited the analysis, but for most companies, the sites had the necessary information. Since all the identified information needed for the analysis is in the public domain, the author took the view that it is not necessary to contact the selected companies to obtain official permission before conducting the research. Company names are not provided, but companies are allocated codes C1–C35 (Table 9.2).

The research focused on these randomly selected companies from the T&C industry in Bulgaria, with the majority of companies located in Sofia, Plovdiv, Gabrovo and Kyustendil. The size of the companies shows that 18 medium-sized companies (51.43%) predominate (50–249 staff), followed by 12 small-sized companies (34.28%) (10–49 staff), 4 micro-sized companies (11.43%) (1–9 staff) and only 1 large-sized company (2.86%) (250+ staff).

The two main product categories in the studied companies were "work clothes" and "outer clothing" (Table 9.3). The "Other T&C goods" category (16.39%) is something of a "catch-all", and includes functional clothing, shoulder pads, sleeve wicks, textile bands, auxiliary materials for clothing, technical textiles—felts (non-woven textile materials), accessories, home textiles, industrial needs, hotels and restaurants, specialised clothing, stage costumes, textile labels and textile tapes, special and protective clothing.

In order to answer RQ1 and RQ2, the information from the websites of the selected Bulgarian companies was mainly used and their activities were analysed using content analysis (Fig. 9.5). The content analysis included identification of material relevant to the following themes: (a) information relating to sustainability, including integration of sustainability into corporate culture/structure; adoption of UN SDGs; cross-supply chain transparency; sustainability risk management; supplier selection and influence; emissions and chemicals reduction; sustainable products and materials; and adopting sustainable business models; and (b) information relating to the CE, including reduction of emissions and waste in operations; use of recycled materials; production of recyclable/biodegradable products; activities to extend product use phase; reverse logistics processes/take back schemes; reuse or recycling of materials or products and cooperation/collaboration with partners. The surveyed companies do not have Corporate Social Responsibility (CSR) or Environmental, Social, and Governance (ESG) reports.

In order to answer RQ3 on digitalisation, a questionnaire comprising seven questions was prepared and made available online. Questions concerned: whether Bulgarian companies are familiar with the need for the use of digital technologies in the T&C industry in Bulgaria; how do Bulgarian companies assess the importance of digital technologies as regards processes related to sustainability; how do Bulgarian companies assess the importance of digital technologies as regards processes related to the CE; do Bulgarian companies participate in projects related to the development of digital technologies; what is the role of digital technologies in the transition to sustainability in Bulgarian companies; what is the role of digital technologies in the transition to a CE in Bulgarian companies and which digital technologies are used by Bulgarian companies?

As regards RQ3, just six returns were received via the online questionnaire, providing information on the role of digital technologies in the transition to sustainability and CE in the Bulgarian T&C industry. The questionnaire responses reflected

Table 9.2 Bulgarian companies included in the research (C1–C35)

Company code	Company location	Company size	Product category
C1	Sofia	Micro	Work clothes
C2	Sofia	Medium	Men's and women's clothing, workwear, uniforms and functional clothing
C3	Kyustendil	Medium	Classic shirts, sports T-shirts, uniforms
C4	Pleven	Small	Shoulder pads, sleeve wicks, textile bands, auxiliary materials for clothing
C5	Veliko Tarnovo	Small	Technical textiles—felts (non-woven textile materials)
C6	Ruse	Medium	Knitted goods
C7	Ruse	Small	Work clothes and bed linen
C8	Plovdiv	Medium	Work clothes
C9	Haskovo	Micro	Textiles for the home, for industrial needs, hotels and restaurants, as well as wholesale sales of fabrics
C10	Kyustendil	Medium	Outer clothing
C11	Troyan	Medium	Wool and wool type fabrics
C12	Dupnitsa	Medium	Outerwear for leisure and products with goose down
C13	Zlatograd	Medium	Work clothes
C14	Gabrovo	Micro	Outer clothing
C15	Zlatograd	Medium	Cotton and cotton type fabrics
C16	Kazanlak	Medium	Wool fabrics in various compositions, mixtures with viscose, linen and cotton
C17	Sofia	Small	Bedroom sets
C18	Smolyan	Small	Women's and men's clothing, uniform clothing, accessories
C19	Gabrovo	Medium	Fabrics, military-police uniforms, work clothes, clothes for leisure and sports
C20	Kazichene	Medium	Medical and work clothing
C21	Asenovgrad	Small	Women's and men's outer clothing
C22	Troyan	Large	Home textiles
C23	Burgas	Medium	Work clothes
C24	Kyustendil	Small	Outer clothing (women's, men's, children's)
C25	Shumen	Medium	Outer clothing and uniform clothing
C26	Kardzhali	Medium	Outer clothing
C27	Stara Zagora	Small	Work, specialised and uniform clothing, women's and men's clothing, bed linen
C28	Plovdiv	Small	Bulgarian traditional costumes and stage costumes
C29	Gabrovo	Small	Textile labels and textile tapes

(continued)

Table 9.2 (continued)

Company code	Company location	Company size	Product category
C30	Varna	Medium	Women's, men's and children's clothing, bulgarian traditional costumes, uniform clothing
C31	Kazanlak	Medium	Fabrics from linen, hemp and other natural textile materials
C32	Sofia	Small	Women's clothing
C33	Plovdiv	Medium	Men's, women's and children's underwear and knitwear
C34	Sofia	Micro	Corporate work, special and protective clothing
C35	Haskovo	Small	Sports clothes

Source Chapter author

Table 9.3 Product categories in the studied T&C companies

Product categories of T&C goods	% of Total
Work clothes	18.03
Outer clothing	18.03
Uniform clothing	11.47
Fabrics	9.84
Bed linen	8.20
Outerwear	6.56
Sports clothes	3.28
Knitted goods	3.28
Bulgarian traditional costumes	3.28
Men's, women's and children's underwear	1.64
Other T&C goods	16.39

Source Chapter author

the view of just one individual in each company; thus, it cannot be seen as a fully comprehensive analysis but rather a subjective snapshot of the status quo and future possibilities of each company. The remaining 28 companies did not respond to the sent questionnaire. One company does not have an email address on its website and therefore the questionnaire was not sent to it.

9.3 Findings

Of the 35 companies surveyed in the T&C industry, ten companies (28.57%) provide sustainability information on separate pages on their websites, one company has a specific document available on its website related to sustainability, and one other

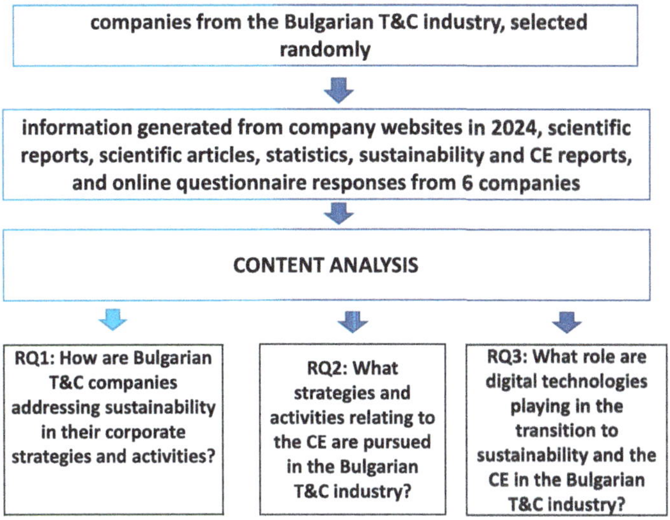

Fig. 9.5 Research process

company has a page on its website related to corporate social responsibility. There were no reports where sustainability and corporate social responsibility were integrally linked. The main reports are related to the financial status of the company concerned (i.e., the company annual reports).

9.3.1 RQ1. How Are Bulgarian T&C Companies Addressing Sustainability in Their Corporate Strategies and Activities?

In the analysis of information from web sources and reports, some main themes emerged that collectively provide an overview of how sustainability is considered and implemented in Bulgarian T&C companies.

9.3.1.1 Integration of Sustainability into Company Culture and Structure

The significance of company culture and structure in the management of business organisations in Bulgaria has recently been researched by Minkov (2009, 2023). In this research, it was found that only 20% of the surveyed companies (C2, C8, C18, C19, C20, C22 and C24) clearly managed to integrate sustainability into their company culture and structure. The integration of sustainability into their company culture has involved the identity of the business organisation, encompassing a number of aspects including: the individual personality and the atmosphere of these companies; working conditions; leadership style; the mission; values; the norms; traditions; customs; expectations; beliefs; communication and professional conduct. Other companies could follow their lead to improve their internal organisational structure and company culture in the context of sustainability.

9.3.1.2 Adoption of the UN SDGs

Only seven companies (14.29%) claim to be aligned with (some of) the UN SDGs. Company C2 supports the UN SDGs as part of the company's strategy, culture and day-to-day operations, and has participated in projects that promote the achievement of sustainable development goals within the framework of UN policies. C18 strives to be aware of the environmental footprint that the company leaves behind in its operations, and has introduced modern technology solutions to optimise the production process and achieve resource efficiency. C19 has embarked upon the "technological modernisation of the enterprise" project, which aims to increase the efficiency of production processes, achieve higher productivity, reduce production costs and optimise the supply chain. Investments within the project comply with the requirements of the principle of "not causing significant damage" to the environment and will contribute to the ecological transition of the country. C20 has a clear responsibility for, and policies aimed at, protecting the environment, and the company has made a commitment to invest in protecting the planet from excessive pollution from textiles. The company concurs that the textile industry is one of the main factors for environmental pollution. Therefore, as a socially responsible company, it will continue its "green" policies. C22 works for sustainable development through which it can have a positive impact on people, the environment and the ability of future generations to meet their needs using renewable energy. Five companies (C2, C18, C19, C20, C22) refer to SDGs 8, 12, 13 and 17 as the most relevant and these are accordingly supported through appropriate measures and policies. C22 also supports policies related to SDG7.

9.3.1.3 Cross-supply Chain Transparency

In only nine (25.71%) of the studied Bulgarian companies (C2, C8, C17, C18, C19, C20, C22, C24, C29) is there relevant information about the transparency of the supply chain. Much of the information presented on company websites regarding supply chain transparency is general and is difficult to analyse in depth. However, the importance of full transparency of the supply chain regarding the origin of basic raw materials and the textile products produced from them in the context of EU sustainable policies (European Commission, 2022) is evidenced in some websites. C19, for example, is a vertically integrated company that produces raw fabrics, owning production lines for finishing, dyeing and printing of fabrics, with a variety of finishing technologies, including hydro and oleophobic, lamination, antibacterial, non-flammability and anti-mosquito treatment. The cycle closes with the production of ready-made garments from the fabrics produced in this way, intended for military police uniforms, work clothes of all kinds and clothes for leisure and sports. The company can influence several stages of the supply chain, and work with a manageable number of suppliers based on long-term agreements. Transparency in the supply chain also allows for more effective tracking and optimisation of business operations.

9.3.1.4 Sustainability-Related Risk Management

Only 20% of the studied Bulgarian companies (C2, C8, C17, C18, C19, C20, C22) clearly and distinctly demonstrated that practices for risk management related to sustainability are being adopted. Bulgarian companies assess sustainability risk management by identifying, analysing and implementing certain activities, practices and policies regarding protection from potential negative environmental and social impacts from the use of certain types of textile materials and certain production technologies.

There are no risks associated with human rights violations through child labour, as observed in other studies (International Labour Organization, 2023; Unicef, 2020). However, there are some serious labour employment issues in the Bulgarian T&C industry: the minimum wage is sometimes only met through working overtime; there is systematic non-payment of the net minimum wage for an eight-hour working day, without bonuses; non-payment of insurance contributions; abuses of part-time contracts; non-payment of overtime; illegal levels of overtime; annual leave is not guaranteed, but the length of sick leave is illegally restricted; failure to pay workers' compensation benefits due after the closing or bankruptcy of enterprises; employers force employees to borrow from them. The T&C industry in Bulgaria is among the sectors with the lowest wages in the country. Employers institute bans on unionisation and refuse to participate in collective agreement negotiations. This leads to the low levels of unionisation in the sector—about 5%. Sometimes management uses intimidation, pressure and mental harassment (Medarov et al., 2019). These issues were reported in the publications referenced in this subsection, but no such data have been identified in the companies that are the subject of this study.

9.3.1.5 Supplier Selection and Influence

Manufacturers of textiles and clothing goods can influence the social and environmental impacts associated with the production of their products, and the supply and processing of the main raw materials. All the companies in the study have established business relationships with suppliers of key raw materials, and claim to adhere to a code of conduct aligned with internationally recognised principles, guidelines and practices in the context of sustainability. C2, through its production department, plans, directs and controls the supply of basic raw materials that are needed for the production of textile garments in the context of sustainable environmental protection policies. C8 has business relationships with suppliers of basic raw materials that enable the company to create workwear that is carefully designed with sustainable materials—biodegradable or compostable—or with ecological packaging that is in line with environmental sustainability policies. C17 works with suppliers of basic raw materials, which are necessary for the production of ecologically clean, safe and sustainable bedding sets. C18 strives to build long-lasting partnerships with its suppliers through high professionalism and trust, which is achieved only by sourcing, processing and manufacturing raw materials of the highest quality for sustainable women's and men's clothing, and uniform clothing.

Due to well-established long-term relationships with suppliers of basic raw materials, C19 produces fabrics, military-police uniforms, work clothes, clothes for leisure and sports in accordance with sustainable development policies. C20 has committed to look for and use every single opportunity to improve all processes and products (medical and work clothing) that the company offers, including activities related to suppliers and sustainability. C22 ensures that all processes are fully transparent and traceable, and that suppliers are reliable companies that share the vision of the company in terms of sustainable development of home textile production, when purchasing raw materials that are safe for people and nature. C24 works with verified and reliable suppliers of quality fabrics and materials from all over the world, observing the principles of creativity, responsibility and sustainability in its practices. C29 states that the quality and safety of textile products largely depends on the company's suppliers. For this, the company carries out strict incoming control of the delivered materials, which is also important in the context of sustainable policies. Many of the companies monitor compliance with social and environmental requirements at suppliers, relying on external support through audits and certification.

Nine of the companies studied (C2, C8, C17, C18, C19, C20, C22, C24, C29) rely on long-term relationships with suppliers, and support them in implementing social and environmental standards. These standards and measures support the achievement of SDG8 by these companies.

9.3.1.6 Emissions and Chemical Reduction

The reduction of harmful emissions and chemical components plays an important and fundamental role in protecting the environment and related climate change. C2 applies a set of internal rules, processes and policies related to environmental protection by preventing or reducing the adverse effects of harmful emissions and chemical components from economic activity. In addition, some attempts are made to reduce the risks related to harmful emissions and chemical pollution, through preventing the adverse impact on the environment throughout the life cycle of textile clothes. C8 produces eco-friendly personal protective equipment, reflecting the corporate commitment to sustainability, including limiting harmful emissions and chemical pollution that contaminates the environment. C16 tests all textile raw materials for harmful substances in the production of woolen fabrics in various compositions, mixtures with viscose, linen and cotton, which will help reduce chemical pollution and the release of regular emissions. C17 (producing bedroom sets) and C19 (fabrics, military-police uniforms, work clothes, clothes for leisure and sports) manufacture products which are environmentally friendly and harmless. These two companies use production processes and technologies guaranteeing the absence of toxic substances, which leads to the limitation of harmful emissions and chemical pollution.

C18 pursues a policy of establishing, analysing and reducing their ecological footprint (including limiting harmful emissions and chemical pollution). For this reason, the company introduced modern technological solutions to optimise their production process and achieve resource efficiencies. C20 has introduced policies related to responsibility for environmental protection, as well as investing in protecting the planet from excessive pollution with textiles (including limiting harmful emissions and chemical pollution) by continuing the company's "green" policy of creating fabrics from recycled raw materials. C22 claims to work hard for the sustainable development of its company, through which it can have a positive impact on the environment and the needs of future generations in accordance with Bulgarian and international legislation. In the creation of home textiles, the company promotes the use of recycled materials, the control and disposal of waste in accordance with national regulations so as not to harm the environment, as well as making continuous efforts to reduce the amount of resources used by reducing of their waste. C29 offers textile labels and textile tapes, constantly updating its technologies, optimising the production processes of its business and making it even more ecological and environmentally friendly (including limiting harmful emissions and chemical pollution), which has a positive impact on the environment. The activities and policies noted above (for companies C2, C8, C16, C17, C18, C19, C20, C22, C29) contribute to the SDGs, in particular SDGs 13, 14 and 15.

9.3.1.7 Sustainable Products and Materials

Eleven of the surveyed companies (31.43%) have focused their efforts on the use of sustainable products and/or materials, thereby reducing the negative impact of the production of conventional T&C goods. For example, C2 offers sustainable clothing products produced from organic textile materials; in creating technical textile products, C5 uses recycled textile materials; C8 offers a new line of green/eco-friendly personal protective equipment developed from sustainable materials, incorporating innovative designs, and being biodegradable or compostable and with eco-friendly packaging; C16 offers fabrics that comply with requirements related to the content of harmful substances, which is in line with the use of sustainable textile materials; C17 offers sustainable textile products with guaranteed absence of substances harmful to human health such as allergens, formaldehydes, pesticides, chlorophenols, unapproved azo dyes or heavy metals; C18 is implementing a project that introduces modern technological solutions to optimise the production process and to achieve resource efficiency, thereby ensuring the supply of sustainable clothing and thus reducing the ecological footprint.

High-tech equipment has been introduced, which uses zero-waste and automated technologies and specialised machines for the processing of textile waste and their recycling, which ensures a closed production cycle; C19 invests in increasing the efficiency of production processes in the production of sustainable fabrics and clothing to achieve higher productivity, reduce production costs, optimise the production supply chain and protect the environment; C20 uses recycled fabrics for the production of medical and work clothing, and C22 uses textile raw materials that are safe for consumers and the environment, as the company also promotes the use of recycled textile materials for the manufacture of sustainable textile products to help save energy worldwide; C26 produces some of its outer clothing from organic textile materials, which is proof of the business organisation's desire to produce sustainable textile products; C29 uses recycled polyester materials for the production of textile labels and textile tapes, which illustrates the strict policies that have been implemented. These measures align with SDG12 in particular.

9.3.1.8 Adopting Sustainable Business Models

Possibilities for the implementation of new or changed business models were observed in eight (22.86%) of the studied companies (C2, C8, C17, C18, C19, C20, C22 and C24). The business models of these companies are based on sustainability to some degree, notably in the integration of sustainability into company culture and structure, adoption of the UN SDGs, cross-supply chain transparency, sustainability-related risk management, supplier selection and influence, emissions and chemical reduction, and sustainable products and materials, which will increase their competitiveness in the Bulgarian and European markets. However, the majority of businesses studied have not yet faced up to the full implications of transitioning to more sustainable strategies and policies by 2030, and the impact this may have on the way companies are structured and organised.

9.3.2 RQ2. What Strategies and Activities Relating to the CE Are Being Pursued in the Bulgarian T&C Industry?

Eight companies (22.86%) have activities and policies related to the CE, which is a major component of their sustainability strategies. C2, C5, C8, C18, C19, C20, C22 and C29 adopt certain measures related to the CE, which supports the general movement towards sustainable practices. Seven types of activities related to the CE were identified. Whilst these activities are not all specifically concerned with circular practices, they all support circularity to some degree, even if indirectly. Reducing emissions and waste in the production process is listed as a current activity by six (17.14%) companies. C2 develops its activity in accordance with the regulated policies of the United Nations (United Nations Global Compact, n.d.), working to protect the environment by reducing harmful emissions and waste from the production process by maintaining their system of environmental management, according to the appropriate ISO standard (ISO 14001:2015; Stoyanova, 2020). C8 markets some of its products as being in accordance with having a positive impact on the world (including by reducing harmful emissions and waste). C18 also makes efforts relating to the reduction of harmful emissions from its activities.

In addition, C18 uses high-precision equipment using zero-waste and automated technologies, as well as specialised machines for the processing of textile waste, with which it has closed its production cycle, which shows responsibility in terms of the management and reduction of textile waste. C19 has invested in certain projects that comply with the requirements of the principle of "not causing significant harm" to the environment (including by reducing harmful emissions and waste) and will contribute to Bulgaria's ecological transition in the field of the T&C industry. Responsibility for the protection of the environment is deeply embedded in the policies of C20, which has accepted the need to invest in the protection of the planet from excessive pollution by textiles and harmful emissions. Waste on the site of C22 is controlled and disposed of in accordance with national regulations so that it does not harm the environment.

The use of recycled materials in the production of T&C products was noted by six (17.14%) of the studied companies. C5 produces technical textiles—felts (non-woven textiles)—some of which contain recycled materials. Recycled materials (polyester, nylon) are used in the production of green/environmentally friendly personal protection products by C8. C18 is implementing a project that introduces modern technological solutions to optimise the production process and achieve resource efficiency through the use of recycled textile materials. In the production of its products, C20 uses recycled materials to improve the quality of life on Earth with positive consequences regarding social and economic significance in the textile sector. C22 promotes the use of recycled materials and raw materials to help save energy worldwide, whilst in the production of some textile labels and textile tapes, C29 uses recycled textile materials such as polyester.

Six companies (17.14%) (C5, C8, C18, C20, C22 and C29) are involved in the production or sale of easily recyclable or biodegradable products. For example, C8 develops personal protective equipment from sustainable materials, with innovative design, being biodegradable or compostable, or with ecological packaging. The protective gloves offered by C8 are certified RCS (Recycled Claim Standard), which underpins the production of products containing recycled materials without compromising on comfort, performance or safety. Some of these gloves are compostable and biodegradable within 180 days in industrial composting facilities. Another type of glove produced by C8 is biodegradable up to 91% within 490 days. C20 calls on every user not to throw their unwanted clothes in the garbage, but to take them for recycling to the places designated for this, and also points out that textile recycling is one of the most environmentally friendly processes in the recycling of materials overall. C5, C22 and C29 are also involved in the production or sale of easily recyclable or biodegradable textile products, thus illustrating their commitment to environmental policies for the development of their business. Only 5 companies (14.29%) (C2, C8,

C18, C20, C22) pursue activities to extend the life of their products; C18, C20 and C22 had reverse logistics processes for product readmission; C18, C20, C22 and C29 reuse or recycle materials or products, whilst C2, C8, C18 and C22 cooperate with partners in the field of CE.

Although the above section provides some useful information to address RQ2, it is evident that this is an incomplete analysis due to the scarcity of relevant details available on company websites and associated documents. However, in summary, the studied companies appear to have relatively low levels of development of CE practices and policies. These findings are in line with those reported by the European Court of Auditors (European Court of Auditors, 2023), which further highlight the limitations and slow progress achieved by the Bulgarian economy in the transition to a CE.

9.3.3 RQ3. What Role Are Digital Technologies Playing in the Transition to Sustainability and the CE in the Bulgarian T&C Industry?

Digitalisation can be defined as the use of digital technologies to change a business model and provide new opportunities for value creation, transforming entire industries (including the textile and apparel industry) at different speeds and to varying extents (Marinova, 2022). The role of digital technologies in general and its importance in the transition to a sustainable and circular T&C industry is advocated in a number of studies. Pal and Jayarathne (2022) research the impact of digitalisation on the entire value chain of the T&C industry, including issues related to sustainable production and consumption and the potential to adopt CE principles. Akhtar et al. (2022) focus on evaluating Amazon's digital fashion initiatives through the use of artificial intelligence (AI) powered by Amazon's advanced web services (AWS), presenting new digital solutions for the T&C industry. Another study found that digitalisation in the T&C industry is helping to advance product design through the integration of CAD software in garment creation. Digital technologies also help streamline manufacturing processes and support online retail platforms and e-commerce websites (Kim, 2023).

The following figures and tables present the results (from six responding companies) of a questionnaire survey on the role of digital technologies in the transition to sustainability and CE in the Bulgarian T&C industry. The study indicates that the majority (83.30%) of the companies are aware of the benefits of using digital technologies in their activities in general. However, only three of the six companies saw the role of digital technologies in progressing sustainability as being very significant. Of the other three companies, one judged digital technologies as being of some significance, one thought digital technologies were of no significance, and one had no opinion, in the context of adopting sustainability practices (Fig. 9.6).

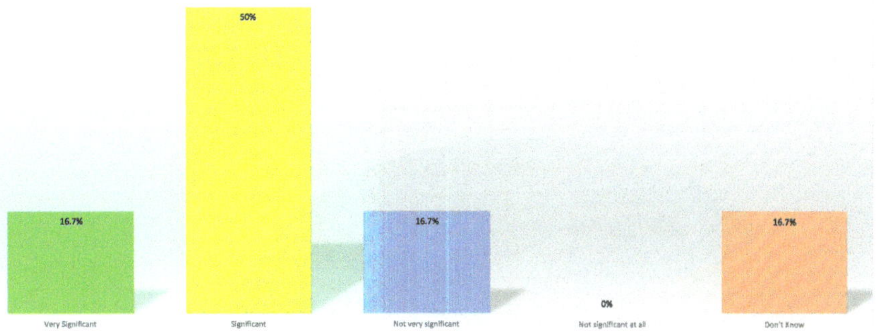

Fig. 9.6 How do you assess the importance of digital technologies in relation to sustainability in your company? (Question 2 in the online questionnaire)

Figure 9.7 indicates the assessment of the importance of digital technologies in the transition to the CE in the six companies that responded. This shows a similarity with Fig. 9.6. Four of the six companies answered very significant or significant, suggesting that the majority of companies appreciate the importance of digital technologies in the processes related to the CE in their companies.

Figure 9.8 shows 50% of the companies participate in projects related to the implementation of digital technologies, 16.67% of them do not participate in projects, and for another 16.67% of the companies, this is not important for their business organisation, and one company (16.67%) did not answer question 4. Table 9.4 summarises the results of questions 5 and 6 on the questionnaire regarding the role of digital

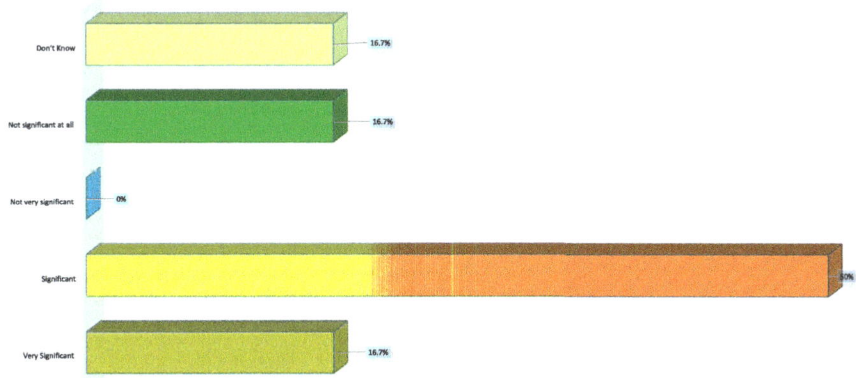

Fig. 9.7 How do you assess the importance of digital technologies in the transition to the CE of your company? (Question 3 in the online questionnaire)

technologies for the transition to sustainability and CE in each of the responding companies.

Four of the six respondents gave a positive perspective on the significance of digital technologies in the transition to sustainability in the studied companies. As regards

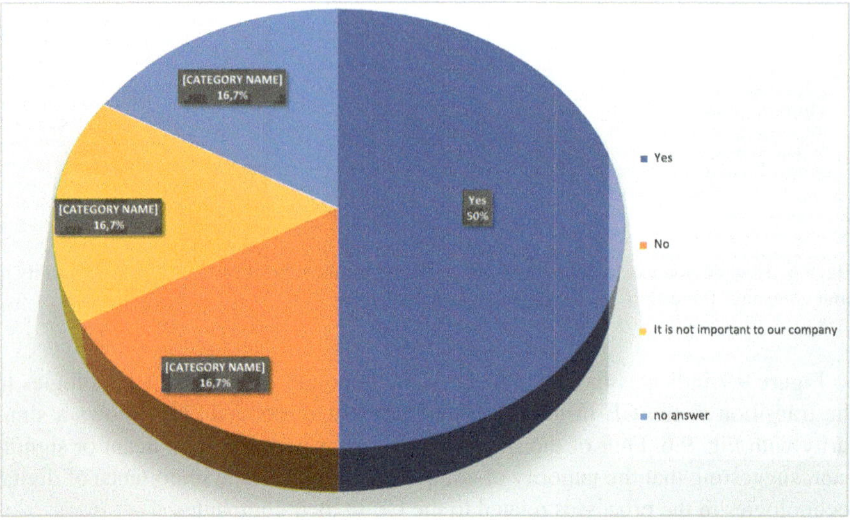

Fig. 9.8 Do you participate in projects related to the implementation of digital technologies in your company? (Question 4 in the online questionnaire)

Table 9.4 What is the role of digital technologies in the transition to sustainability and CE in your company? (Questions 5 and 6 in the online questionnaire)

Evaluation The role is none (0)/very significant (10)	Q5. What is the role of digital technologies in the transition to sustainability in your company?	Q6. What is the role of digital technologies in the transition to a CE in your company?
10		
9	**33.3% (2)**	**16.67% (1)**
8		
7	**33.3% (2)**	**50% (3)**
6		
5	**16.7% (1)**	
4		
3		
2	**16.7% (1)**	**16.67% (1)**
1		
0		16.67% (1) no answer*

Source Chapter author

Table 9.5 Which digital technologies do you use in your company? (Question 7 in the online questionnaire)

Types of digital technologies	The digital technology is currently being used	The digital technology will be used in the future	We are not familiar with this type of digital technology	No answer
Social media	**66.67% (4)**		**16.67% (1)**	16.67% (1)
Big data and analytics	**33.33% (2)**	**33.33% (2)**	**16.67% (1)**	16.67% (1)
Digital development and manufacturing	**16.67% (1)**	**50% (3)**	**16.67% (1)**	16.67% (1)
Artificial intelligence		**50% (3)**	**33.33% (2)**	16.67% (1)
Blockchain			**50% (3)**	50% (3)
Augmented reality		**16.67% (1)**	**50% (3)**	33.33% (2)
Mobile computing	**16.67% (1)**	**16.67% (1)**		66.67% (4)
Internet of things	**50% (3)**	**16.67% (1)**		33.33% (2)
Robotics	**16.67% (1)**	**33.33% (2)**	**16.67% (1)**	33.33% (2)

Source Chapter author

the role of digital technologies in the transition to a CE, 50% of the companies scored a 7 in the questionnaire, 16.67% of the companies gave a score of 9 in the questionnaire, another 16.67% of the companies gave a score of 2, and one company (16.67%) did not respond. Although a limited sample, the findings suggest digital technologies are generally seen of considerable significance in the transition to sustainability and CE.

Table 9.5 presents the digital technologies used by each of the companies that answered the questionnaire.

Table 9.5 indicates that social media and the Internet of Things are most used by the respondent companies, followed by Big Data/Analytics. Future intentions centred, in particular, on digital development/manufacturing and AI, followed by big data/analytics and robotics. Interestingly, none of the companies were familiar with blockchain, and there was a lack of knowledge and awareness across all technologies to some degree, which is confirmed by the high percentage of companies that did not respond.

9.4 Discussion

In this section, results are presented in a summarised form and emergent themes are discussed. Table 9.6 presents summarised findings regarding the sustainability-related actions and activities discussed in Sect. 9.3.1.

Table 9.6 Sustainability-related actions and activities: summary

Company/activity	Integration of sustainability into corporate culture/structure	Adoption of UN SDGs	Cross-supply chain transparency	Sustainability risk management	Supplier selection and influence	Emissions and chemicals reduction	Sustainable products and materials	Adopting sustainable business models
C1								
C2	x	x	x	x	x	x	x	x
C3								
C4								
C5							x	
C6								
C7								
C8	x		x	x	x	x	x	x
C9								
C10								
C11								
C12								
C13								
C14								
C15								
C16						x	x	
C17			x	x	x	x	x	x
C18	x	x	x	x	x	x	x	x
C19	x	x	x	x	x	x	x	x

(continued)

Table 9.6 (continued)

Company/activity	Integration of sustainability into corporate culture/structure	Adoption of UN SDGs	Cross-supply chain transparency	Sustainability risk management	Supplier selection and influence	Emissions and chemicals reduction	Sustainable products and materials	Adopting sustainable business models
C20	x	x	x	x	x	x	x	x
C21								
C22	x	x	x	x	x	x	x	x
C23								
C24	x		x		x			x
C25								
C26							x	
C27								
C28								
C29			x		x	x	x	
C30								
C31								
C32								
C33								
C34								
C35								

Source Chapter author

What is striking is that only twelve (34.29%) of the 35 surveyed companies register on this analysis. More specifically, eleven companies (31.43%) claim to offer sustainable products and/or materials; cross-supply chain transparency occurs at 25.71% of the companies; supplier selection and its influence features in 25.71% of companies; emissions and chemical reduction are mentioned in policies of 25.71% of the companies; adopting sustainable business models are observed in 22.86% of the studied enterprises; integration of sustainability in company culture and structure occurs in 20% of the companies; sustainability-related risk management is observed in 20% of the studied Bulgarian firms and adoption of the UN SDGs is observed in 14.29% of companies.

Five companies (C2, C18, C19, C20, C22) undertake activities related to all the sustainability-related measures. Of the twelve companies that are pursuing at least some sustainability activity, six are medium-sized enterprises, five are small enterprises and one is the large company (the only large company in the study). This suggests company size is of no great significance in determining sustainability activity, but rather the company management, irrespective of company size, determine to what extent and precisely which, sustainability-related actions and activities are pursued in the company's development activities. That said, it is noteworthy that none of the micro-sized companies appear to participate in sustainability activities.

Table 9.7 summarises the results regarding the activities related to the CE as discussed in Sect. 9.3.2. The key emergent themes were reduction of emissions and waste in operations; use of recycled materials; production of recyclable/biodegradable products; activities to extend product use phase; reverse logistics processes/take back schemes; reuse or recycling of materials or products and cooperation/collaboration with partners.

Of the 35 companies, eight (22.86%) were actively involved in one or more of these activities, and three companies (C18, C20 and C22) claimed to be pursuing all seven activities. As regards the activities themselves, six companies (17.14%) were pursuing a reduction of emissions and waste in operations, and the same number were involved in the use of recycled materials in the production of T&C products, and in activities related to the production of recyclable/biodegradable products. This was closely followed by activities to extend the use phase of the product, which were being undertaken by five companies (14.29%), as was cooperation with partners in the field of CE. Reuse or recycling of materials or products were observed at four companies (11.43%), and reverse logistics processes/take back schemes were underway at three (8.57%) companies.

Again, there was no obvious relationship between company size and the degree of take-up of CE actions and activities. The eight CE-active companies comprised four medium-sized companies, three small-sized companies and the one large-sized company in the study. As was the case with sustainability activities, however, no micro-sized companies appeared to carry out CE actions and activities.

Overall, Tables 9.6 and 9.7 suggest that the adoption of sustainability-related activities is somewhat ahead of the implementation of CE actions and practices. Five companies are pursuing all eight sustainability-related activities. Of these eight five are also implementing all seven CE practices. This points to a core of a few

Table 9.7 CE actions and activities: summary

Company/ CE activity	Reduction of emissions and waste in operations	Use of recycled materials	Production of recyclable/ biodegradable products	Activities to extend product use phase	Reverse logistics processes/ take back schemes	Reuse or recycling of materials or products	Cooperation/ collaboration with partners
C1							
C2	x			x			x
C3							
C4							
C5		x	x				
C6							
C7							
C8	x	x	x	x			x
C9							
C10							
C11							
C12							
C13							
C14							
C15							
C16							
C17							
C18	x	x	x	x	x	x	x
C19	x						
C20	x	x	x	x	x	x	x
C21							
C22	x	x	x	x	x	x	x
C23							
C24							
C25							
C26							
C27							
C28							
C29		x	x			x	
C30							
C31							
C32							
C33							

(continued)

Table 9.7 (continued)

Company/ CE activity	Reduction of emissions and waste in operations	Use of recycled materials	Production of recyclable/ biodegradable products	Activities to extend product use phase	Reverse logistics processes/ take back schemes	Reuse or recycling of materials or products	Cooperation/ collaboration with partners
C34							
C35							

Source Chapter author

companies who are leading the move towards sustainability, whilst the majority are lagging behind, with little involvement in either sustainability or CE operations.

From the analysis of the role of digital technologies in the transition to sustainability and CE, a general picture emerged from the six respondent companies. The majority (83.30%) of the surveyed companies are aware of the need to use digital technologies in their business activities, which is of great importance for the future development of companies in the T&C industry. Further, four of the six companies saw digital technologies as significant in transitioning to sustainability, and in the implementation of CE practices. As regards specific technologies, there was an awareness of the role of the major technologies (with the exception of blockchain and augmented reality), and a perception that they would play some part in future developments. Nevertheless, evidence from this small sample suggests that a small percentage of the companies are leading, with the majority as followers or laggards. Certainly, a greater and more widespread knowledge and awareness of the digital technologies and their potential role will be required for an effective transitioning to sustainable and CE practices in the industry (Alves et al., 2022; Goel et al., 2023; Kistamah, 2024; Kumar et al., 2023; Pereira et al., 2022; Špiler et al., 2023; Tripathi et al., 2021).

9.5 Conclusions

This chapter attempts to provide a comprehensive understanding of the transition to sustainability and CE in the Bulgarian T&C industry, and the role that digital technologies play in this transition. Despite the limitations of the research, related to the small number of studied companies and the lack of sufficiently detailed information regarding some aspects, the author believes the presented analysis provides the basis for assessment of the current situation in the sector, and development of future activities and policies in the T&C industry in Bulgaria.

Only twelve out of 35 surveyed companies develop sustainability-related actions and activities, with five of these companies pursuing all the eight listed sustainability-related activities. Eight out of 35 companies actively participate in one or more of the

CE-related activities, and three companies perform all seven activities. The majority of companies are aware of the benefits of using digital technologies and appreciate the importance of digital technologies in the processes related to sustainability and CE. Social media and the Internet of Things were the two digital technologies most used by companies surveyed.

In conclusion, the findings indicate that a significant part of the studied Bulgarian companies lag behind in adopting policies, strategies and activities related to sustainability, CE and digital technology deployment. Sustainability activities are better developed in a larger part of Bulgarian companies compared to activities related to the CE. Some of the digital technologies are currently used in the context of sustainability and CE, others are planned to be used in the future, and others are unknown to companies. It is vital and necessary to understand, develop and implement the symbiosis between sustainability, CE and digital technologies in the T&C industry in Bulgaria in the context of European policies and trends in a wider industrial context. This will require mutual efforts and cooperation between companies, State entities, industry and consumer organisations, scientific organisations and European institutions.

References

Akhtar, W. H., Watanabe, C., Tou, Y., & Neittaanmäki, P. (2022). A new perspective on the textile and apparel industry in the digital transformation era. *Textiles, 2*, 633–656. https://doi.org/10.3390/textiles2040037

Alves, L., Cruz, E. F., Lopes, S. I., Faria, P. M., & Cruz, A. M. R. (2022). Towards CE in the textiles and clothing value chain through blockchain technology and IoT: A review. *Waste Management & Research, 40*(1), 3–23. https://doi.org/10.1177/0734242X211052858

Angelova, Y. P. (2022). Textile industry in Bulgaria and the Covid-19 pandemic: Challenges and trends. In *CORTEP 2022, IASI-RO* (pp. 121–125). https://doi.org/10.2478/9788367405133-019

Bulgarian Association Circular Textile. (2022). Environmental savings from implementing a CE in the textile sector in Bulgaria. *Bulgarian Association Circular Textile (BACT)*, 3–31. https://bact.bg/wp-content/uploads/2022/12/BACT_report.pdf

Euratex. (2022a). Seizing growth opportunities in Bulgarian TCLF industries. *Euratex*. https://euratex.eu/news/seizing-growth-opportunities-in-bulgarian-tclf-industries/

Euratex. (2022b). Facts & key figures 2022 of the European T&C industry. *Euratex, Economic and Statistics*, 3–32. https://euratex.eu/wp-content/uploads/EURATEX_FactsKey_Figures_2022rev-1.pdf

European Commission. (2022). *EU strategy for sustainable and circular textiles*. European Commission. https://environment.ec.europa.eu/document/download/74126c90-5cbf-46d0-ab6b-60878644b395_en?filename=COM_2022_141_1_EN_ACT_part1_v8.pdf

European Court of Auditors. (2023). CE. Slow transition by Member States despite the actions of the EU. *European Union*, 2–67.

European Environment Agency. (2020). Sustainable development goals and the environment in Europe: A cross-country analysis and 39 country profiles. *European Environment Agency, 21*, 2–106. ISBN 978-92-9480-284-2. https://doi.org/10.2800/044724. file:///D:/Downloads/SDGs%20and%20the%20Environment_A%20cross-country%20analysis.pdf

Goel, P., Mahadevan, K., & Punjani, K. K. (2023). Augmented and virtual reality in apparel industry: A bibliometric review and future research agenda. *Foresight, 25*(2), 167–184. https://doi.org/10.1108/FS-10-2021-0202

IBIS World. (2024). *Clothing manufacturing in Bulgaria—Market size, industry analysis, trends and forecasts (2024–2029).* https://www.ibisworld.com/bulgaria/industry-statistics/clothing-manufacturing/805/

International Labour Organization. (2023). *Child labour and forced labour in Burkina Faso.* Food and Agriculture Organization of the United Nations. https://webapps.ilo.org/wcmsp5/groups/public/---ed_norm/---ipec/documents/publication/wcms_888736.pdf

ISO 14001:2015. *Environmental management systems—Requirements with guidance for use.*

Ivanov, M. (2021). Against the current: The Bulgarian textile industry 1800–1912. *CIELA,* 15–73. ISBN 978-954-28-3749-7.

Kim, J. K. (2023). The impact of digitalization on the textile industry: Mini review. *Trends in Textile Engineering & Fashion Technology (TTEFT), 8*(2), 000682. https://doi.org/10.31031/TTEFT.2023.08.000682

Kistamah, N. (2024). The applications of artificial intelligence in the textile industry. In T. P. Fowdur, S. Rosunee, R. T. F. Ah King, P. Jeetah, & M. Gooroochurn (Eds.), *Artificial intelligence, engineering systems and sustainable development* (pp. 257–269). Emerald Publishing Limited. https://doi.org/10.1108/978-1-83753-540-820241020

Kumar, A., Arora, M., Bhalerao, K., & Chhabra, M. (2023). Role of blockchain for sustainability and CE. In S. Dhar, D. T. Do, S. N. Sur, H. C. M. Liu (Eds.), *Advances in communication, devices and networking. Lecture notes in electrical engineering* (Vol. 902, pp. 413–425). Springer. https://doi.org/10.1007/978-981-19-2004-2_37

Marinova, V. (2022). Digitalization in packaging. *Izvestia Journal of the Union of Scientists—Varna. Economic Sciences Series, Varna, Union of Scientists—Varna, 11*(1), 47–57. ISSN(печатно) 1314–7390. ISSN(онлайн) 2603–4085.

Medarov, G., Nikolova, M., & Musiolek, B. (2019). Profile of Bulgaria country. *Friedrich Ebert Foundation*, 1–11. https://bulgaria.fes.de/fileadmin/user_upload/images/publications/CCC-CountryProfile-Bulgaria_BG.pdf

Ministry of Transport and Communications. (2020). *Digital transformation of Bulgaria for the period 2020–2030, Sofia* (pp. 1–31). https://www.mtc.government.bg/sites/default/files/digitaltransformationofbulgariafortheperiod2020-2030f.pdf

Minkov, I. (2009). Management of company culture. *Science and Economics* (Varna), *137*.

Minkov, I. (2023). Problems of company culture in Bulgarian industry. *Iko-consult* (Varna), *188*. (Bibl. Ph.D. protected doctoral dissertations.) ISBN (print) 978-954-8235-21-1.

National Development Program of Bulgaria 2030. (2020). *Circular and low-carbon economy* (pp. 29–32). National Development Program of BULGARIA 2030. https://www.mtc.government.bg/sites/default/files/bulgaria_2030_5_0.pdf

Pal, R., & Jayarathne, A. (2022). Chapter 15—Digitalization in the textiles and clothing sector. *The Digital Supply Chain*, 255–271. https://doi.org/10.1016/B978-0-323-91614-1.00015-0

Patarchanov, P. (2016). Situation and prospects in the development of consumer industry in Bulgaria. *Annual of Sofia University "St. Kliment Ohridski" Faculty of Geology and Geography, Book 2—Geography, 109,* 161–175.

Pereira, F., Carvalho, V., Vasconcelos, R., & Soares, F. (2022). A review in the use of artificial intelligence in textile industry. In J. Machado, F. Soares, J. Trojanowska, & S. Yildirim (Eds.), *Innovations in mechatronics engineering. icieng 2021. Lecture notes in mechanical engineering* (pp. 377–392). Springer. https://doi.org/10.1007/978-3-030-79168-1_34

Popova, D. (2019). The textile industry in Varna region (1878–1929). *Journal of the Union of Scientists—Varna. The Cultural Heritage of Varna and the Black Sea Region*, 93–101. ISSN 2738-7372 (Print). ISSN 2738-7380 (Online). https://doi.org/10.14748/isuvknv.v0i0.6571

Rusev, I. (1995). More about the First Factory in the Bulgarian Lands (More about the 1st Factory in the Bulgarian Lands + Sliven in the 1830s). *Bulgarian Historical Review, 2*, 69–82.

Rusev, I. (1998). About some new and unknown moments from the history of the first Bulgarian factory. *An Annual Book of Free University of Varna, Varna, Year IV, I*, 462–478.
Rusev, I. (2016). How many and which were the first factories in the Bulgarian lands before the liberation in (1878)? An attempt at analysis with the support of sources and historiography. *Izvestiya of the Center for Economic-Historical Research*. T. I. The urban economy in the Bulgarian lands over the centuries, Varna (pp. 33–55). ISSN (print) 25349244.
Špiler, M., Milošević, D., Miškić, M., Gostimirović, L., Beslać, M., & Jevtić, B. (2023). Investments in digital technology advances in textiles. *Industria Textile, 74*(1), 90–97. https://doi.org/10.35530/IT.074.01.202287
Statista. (2023a). *Number of enterprises in the manufacture of wearing apparel industry in Bulgaria from 2011 to 2020*. Statista Research Department. https://www.statista.com/statistics/369921/number-of-enterprises-in-the-manufacturing-of-wearing-apparel-sector-in-bulgaria/
Statista. (2023b). *Number of enterprises in the manufacture of textiles industry in Bulgaria from 2011 to 2020*. Statista Research Department. https://www.statista.com/statistics/369672/number-of-enterprises-in-the-manufacturing-of-textiles-sector-in-bulgaria/
Stoyanova, A. (2020). Assessment of compliance with the regulatory requirements when building environmental management systems. In *Law and Business in Contemporary Society, Collection of reports from the 3rd National Scientific Conference, Science and Economics*, Varna (pp. 85–93). https://conference.ue-varna.bg/pravni-nauki/ConfProceedings/LBCS2020/85-93.pdf
Texman. (2022). The T&C Industry of Bulgaria. *International Business Pages*. https://www.newclothmarketonline.com/international-business-pages/the-textile-and-clothing-industry-of-bulgaria/
Tripathi, G., Tripathi Nautiyal, V., Ahad, M. A., & Feroz, N. (2021). Blockchain technology and fashion industry-opportunities and challenges. In S. K. Panda, A. K. Jena, S. K. Swain, & S. C. Satapathy (Eds.), *Blockchain technology: Applications and challenges. Intelligent systems reference library* (Vol. 203). Springer. https://doi.org/10.1007/978-3-030-69395-4_12
Unicef. (2020). *Children's rights in the garment and footwear supply chain* (pp. 2–48). United Nations Children's Fund (UNICEF). https://www.unicef.org/media/70121/file/Childrens-rights-in-the-garment-and-footwear-supply-chain-2020.pdf
United Nations Global Compact. (n.d.). *The ten principles of the UN Global Compact*. UN Global Compact. https://unglobalcompact.org/what-is-gc/mission/principles
World Bank. (2022). *Bulgaria BG: GDP: % of manufacturing: Textiles and clothing*. World Bank. https://www.ceicdata.com/en/bulgaria/gross-domestic-product-share-of-gdp/bg-gdp--of-manufacturing-textiles-and-clothing
Wiegand, T., & Wynn, M. (2023). Sustainability, the CE and digitalisation in the German T&C industry. *Sustainability, 15*(11), 9111. https://doi.org/10.3390/su15119111. file:///D:/Downloads/Sustainability_the_Circular_Economy_and_Digitalisa.pdf
WITS World Integrated Trade Solution. (n.d.). *Bulgaria 2021 export partner share*. https://wits.worldbank.org/CountryProfile/en/Country/BGR/Year/LTST/TradeFlow/Export/Partner/by-country/Product/50-63_TextCloth
Wynn, M., & Felser, K. (2023). Digitalisation and change in the management of IT. *Computers, 12*, 251. https://doi.org/10.3390/computers12120251. file:///D:/Downloads/Digitalisation_and_Change_in_the_Management_of_IT.pdf
Zhelev, P. (2021). Competitiveness of the T&C industry in Bulgaria after the EU accession and in times of covid-19 pandemic. *Trakia Journal of Sciences, 19*(1), 89–98. ISSN 1313-3551 (online). https://doi.org/10.15547/tjs.2021.s.01.013

Radoslav Radev (r.radev@ue-varna.bg) is Chief Assistant Professor and holds a Ph.D. in economics at the Department of Commodity Science at the University of Economics, Varna, Bulgaria. He has over 15 years teaching experience, with his main disciplines including Commodity Science of Textile and Footwear, Commodity Science of Household Goods, and Fundamentals of Industrial Commodity Science. Radoslav's research interests are in the areas of sustainability, circular economy of textile goods; quality, consumer awareness of textile goods; edible coatings and their use in food technology; and food additives, food quality and safety. He has over 60 scientific publications and participated in five major research projects. He is a member of the International Society of Commodity Science and Technology, Austria (IGWT) and other scientific organisations. In addition, he is a member of the Academic Council of UE-Varna and a member of the Expert Council of the Center for Quality of Goods and Consumer Protection at the Research Institute of UE—Varna.

Chapter 10
Germany

Tina Wiegand and Martin Wynn

Abstract This chapter investigates the German textile and clothing (T&C) industry's involvement in sustainability and its transition towards the Circular Economy (CE). In emphasising the complex relationship between these two concepts, this study also explores the potential role of digital technologies in facilitating this transition. While digital technologies are expected to bridge the gap between the theory and practice of the CE, existing research predominantly consists of conceptual and review studies. Here, however, this chapter focuses on German SMEs in the T&C sector, and takes a pragmatic approach, drawing on academic papers, corporate sustainability reports, and expert interviews to examine the multiple aspects of this topic. An extended action matrix and critical success factors (CSFs) emerge from the findings, offering guidance for SMEs as they confront the challenges of transitioning to sustainable and circular practices. Despite the limited current use of digital technologies, the study suggests they will have an increasing impact on T&C sector competitiveness, especially for SMEs. As sustainability initiatives align with digitalisation, companies will face growing pressure to adopt circular business practices, both from consumer expectations and impending legislative requirements, such as the EU Strategy for Sustainable and Circular Textiles. The study provides a foundation for future case studies and a basis for more in-depth research of related aspects, including market influences, consumer awareness, and strategies for sustainable industry transformation.

Keywords Sustainability · Circular economy · Digitalisation · CE · German industry · Textile and clothing industry · Action list · Critical success factors

T. Wiegand
Hof University of Applied Sciences, Hof, Germany
e-mail: tina.wiegand@hof-university.de

M. Wynn (✉)
University of Gloucestershire, Cheltenham, UK
e-mail: mwynn@glos.ac.uk

M. Wynn and T. Wiegand (eds.), *Sustainability, the Circular Economy and Digitalisation in the European Textile and Clothing Industry*, SDGs and Textiles,
https://doi.org/10.1007/978-981-97-9116-3_10

10.1 Introduction and Background

The German textile and clothing (T&C) industry has a long history dating back to the Middle Ages, and it played a significant role in Germany's industrialisation during the eighteenth and nineteenth centuries. Particularly in the German uplands, textile clusters developed, but their subsequent evolution varied after the World Wars and the division of Germany into East and West (Breitenacher, 1989). In West Germany, the T&C industry experienced a boom in the 1950s and 1960s, while in East Germany, it was nationalised and restructured into socialist combines (Breitenacher, 1989; Breitenacher et al., 1991). From the 1970s onwards, a structural transformation began in the West German T&C industry, driven by international competition with Asian and later also African developing and emerging countries (Schneider, 2003). As a result, the clothing industry increasingly shifted its labour-intensive production abroad, while the textile industry, which is more capital-intensive and technology-dependent, was better able to adapt, partly due to German expertise in textile machinery manufacturing (Dispan, 2009; Marschall et al., 2020). The main driver of growth in this transformation was the technical textiles sector, which supplied a range of other industries, such as automotive, medicine, and lightweight construction. The East German textile industry also underwent an impressive transformation after reunification, and is now a leading hub for the development and production of technical textiles (Marschall et al., 2020).

However, in terms of gross value added (adjusted for inflation), clothing production in Germany decreased by around 90% between 1980 and 2020 (Fashion Council Germany e.V. & Oxford Economics, 2022). The remaining clothing industry in Germany is now primarily focused on functions such as product development, design, purchasing, marketing, distribution, logistics, and administration. There are often sewing units for samples or special orders and small batches (Dispan, 2009), while bulk production primarily takes place in countries of the so-called global south, such as China, Bangladesh, India, or Turkey (Worldbank, 2019). Consequently, most of the clothing consumption in Germany is provided by imports. With an import value of over €47 billion in 2023, clothing imports have almost doubled since 2008. However, a variety of textile products continue to be manufactured domestically, while the import and export of these products remain relatively low, as shown in Fig. 10.1 (Statistisches Bundesamt, 2024b).

This development is reflected in the current T&C industry structure, with a total of 879 companies in Germany employing approximately 83,000 people in 2023. Of these companies, 689 are in the textile sector (WZ08-13: manufacture of textiles) and 190 in the clothing sector (WZ08-14: manufacture of clothing). About 95% of these companies are small and medium-sized enterprises (SMEs) with up to 249 employees (Statistisches Bundesamt, 2024a).

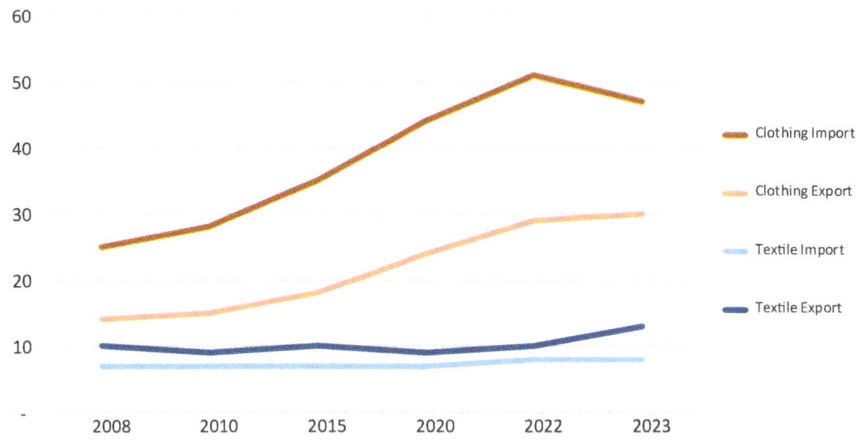

Fig. 10.1 Import and export of clothing and textiles (excluding shoes) in Germany 2008–2023 (€ billion). *Source* Statistisches Bundesamt (2024b)

The German T&C industry today is focused around regional clusters in

- North Rhine-Westphalia (224 companies)
- Bavaria (190 companies)
- Baden-Württemberg (158 companies)
- Saxony (120 companies)

To achieve a CE for textile products, extensive measures are planned by the EU. These range from regulating chemical use and design requirements to introducing a digital product passport (DPP) and expanding producer responsibility for the disposal and recycling of textile products, as well as increasing sustainability reporting (European Commission, 2022). EU regulations apply directly to German companies or are transferred into national standards and laws, showing that the political pressure on German T&C companies is increasing. This underlines the need for a holistic transformation of the industry, shifting from the linear "take-make-dispose" model towards closed-loop textile systems. Considering the highly globalised textile value chain, this shift urgently requires the integration of digital technologies. Their role as a driver, enabler, and facilitator for various circular strategies and business models (Neligan et al., 2022), for supply chain collaboration, as well as enabling transparency in material flows and environmental impacts (Bressanelli et al., 2022; Gebhardt et al., 2022; Hussain & Malik, 2020; Ramesohl et al., 2022), is undeniable. However, the T&C industry in Europe falls within sectors with relatively limited interconnectivity and collaboration, coupled with a relatively low level of digitalisation across the value chain (Diebel et al., 2022; Saccani et al., 2023; Schöggl et al., 2023).

According to one of the few existing studies on the state of the CE in the German T&C by Arnold et al. (2023), there are very few individual companies that have already adopted or introduced circular business models such as upcycling and recycling, product-as-a-service, repair and reuse, or companies with closed-loop supply chains. Studies focusing on the implementation of sustainability, the CE, and digitalisation in the T&C industry are currently limited, and existing research has primarily concentrated on analysing the driving forces, enabling factors, and challenges related to the adoption of CE practices in general (Brydges, 2021; Dissanayake & Weerasinghe, 2022; Franco, 2017; Jia et al., 2020).

Regarding the T&C industry in Europe, the few available empirical studies have mainly revolved around surveys and case studies conducted in Italy, the Netherlands, and Nordic countries. These studies have primarily explored business models (Abbate et al., 2023; Salmi & Kaipia, 2022; Stål & Corvellec, 2018), product development (Ræbild & Bang, 2017), operational practices (Ciccullo et al., 2023; Fischer & Pascucci, 2017), or the necessary capabilities and organisational re-configurations (Colucci & Vecchi, 2021; Sandberg & Hultberg, 2021; Vellesalu et al., 2023). The links between the CE and the transition to a sustainable future are complex, with several sources suggesting that digital technologies may play a key role in this transition (Okorie et al., 2018; Wynn & Jones, 2022). The World Economic Forum (2021, para. 1) claimed "we must accelerate the transformation to a circular economy in order to meet global climate goals by 2050", and that "this can only be achieved through focused and responsible digitalization", while Kottmeyer (2021, p. 17) argued that "digital technologies have the potential to close the realisation gap between theory and practice of the circular economy concept". Ranta et al. (2021, para. 1) point out the limitations of the relevant research to date, claiming it mainly concerns "conceptual and review studies" and that there remains "a lack of understanding of how digital technologies enable individual firms in real-life settings to improve their resource flows and value creation and capture, and thereby enable business model innovation to emerge". The role and deployment of digital technologies to facilitate and support a sustainable and circular T&C industry has, however, received limited empirical attention to date. Only one study by Wiegand and Wynn (2023) has addressed this topic so far, primarily focusing on only one portion of the T&C value chain due to their research design.

This chapter builds upon the work of Wiegand and Wynn (2023, 2024), focusing on the T&C manufacturing processes (Fig. 10.2) in the German T&C industry, to research how these companies are approaching sustainability, and the measures they are taking to move towards the CE.

In this context, this chapter addresses the following research questions (RQs):

- RQ1 How are German T&C companies addressing sustainability in their corporate strategies and activities?
- RQ2 What strategies and activities relating to the CE are being pursued in the German T&C industry?
- RQ3 What role are digital technologies playing in the transition to sustainability and the CE in the German T&C industry?

The chapter has five main sections. Following this introduction, the research method and a top-line categorisation matrix are set out. This is followed by the main research results and a discussion of the key emergent themes, in which an action

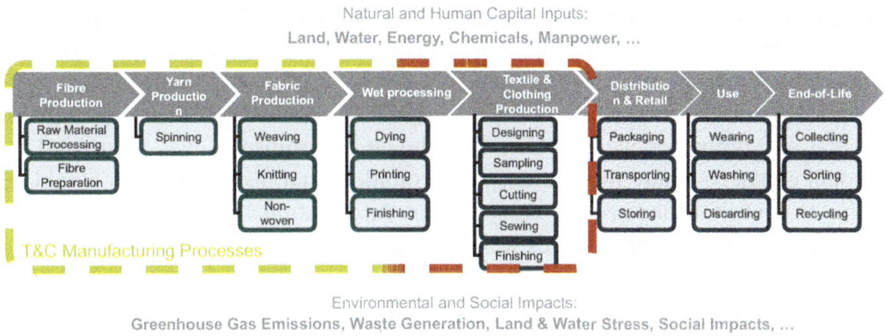

Fig. 10.2 Manufacturing processes in the T&C value chain. *Source* Based on Gözet and Wilts (2022), Reichel et al. (2014), Schumacher and Forster (2022)

list and top-line critical success factors (CSFs) for initiating the transition to the CE for SMEs in the T&C industry in Germany are presented. The concluding section provides a research summary and offers some thoughts on possible areas for future research in this field.

10.2 Research Method

The research methodology adopts a qualitative and inductive approach, emphasising the importance of practical implications and real-world applications, while acknowledging the complex nature of the T&C industry. The study aims to generate a comprehensive understanding of the transition towards sustainability and circularity within the main sectors of the German T&C manufacturing industry, with a particular focus on the role digital technologies play in this transition. The research is based on five case examples from different stages of the German T&C value chain (Fig. 10.3).

The operational framework and action list developed by Wiegand and Wynn (2023) for the transition to sustainable strategies and circular practices serves as

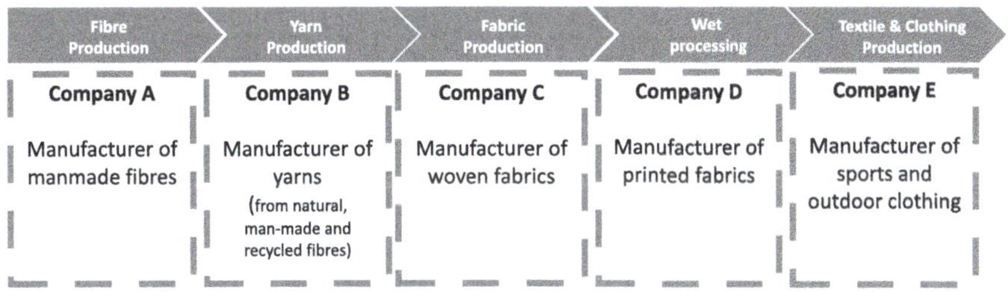

Fig. 10.3 Companies used as case examples in the study

the basis for this research, with the list being validated and expanded through empirical data. This study thus builds upon the authors' previous research and can be seen as part of a wider research programme (Fig. 10.4). In selecting companies as case examples, the researchers ensured that all stages of the value chain were represented, and that all companies were SMEs. The choice of companies was influenced by personal contacts, where the authors were confident that the participants would be willing to contribute and provide the necessary information. Anonymity for both the organisations and the interviewees was guaranteed.

Data collection was conducted using publicly available information related to sustainability and CE practices from the companies' websites and semi-structured interviews with the sustainability managers of each company. For Companies A and E, corporate sustainability reports were available as PDF documents on their websites, but sustainability-related information was more generally available on the websites of all companies. The interviews were conducted between June and August 2024. Each interview lasted between 50 and 70 minutes and was recorded, transcribed, and coded with the consent of the respondents. Interviews were conducted with one management representative responsible for sustainability at each company (Table 10.1).

The data analysis was conducted using Mayring's (2014) approach for qualitative content analysis, specifically employing the deductive category application method. The category system was based on the operational framework (Table 10.2)

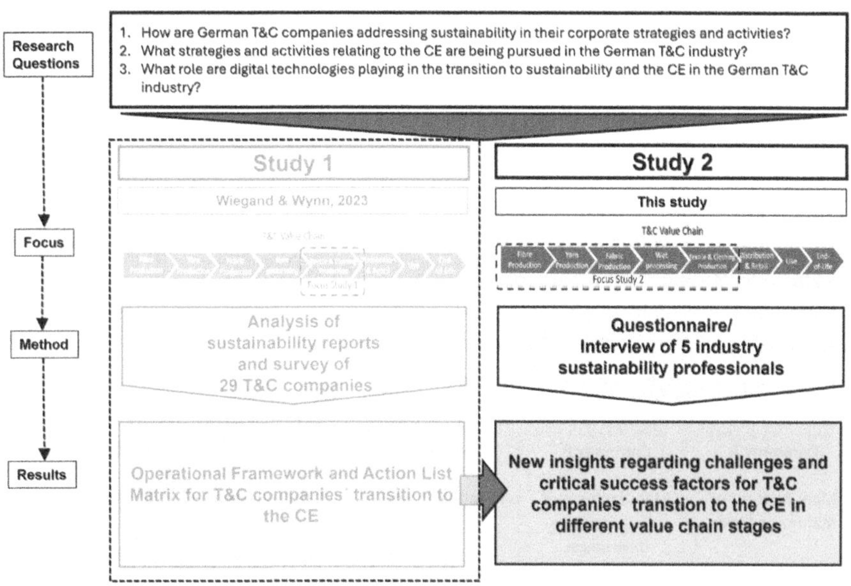

Fig. 10.4 The integration of this study into the wider research programme

Table 10.1 Sources of information utilised for the research

Company	Published sustainability report	Sustainability related website	Interview with a management representative
A	x	x	x
B		x	x
C		x	x
D		x	x
E	x	x	x

Table 10.2 Categorisation matrix for deductive content analysis to answer the research questions

	Dimension/ Action Area	ORGANISATION	PRODUCT	PROCESS
RQ 1	Strategy			
RQ 2	CE Practice/ Operations			
RQ 3	Digital Technology deployment			

Based on Wiegand and Wynn (2023, pp. 24–25)

proposed by Wiegand and Wynn (2023, pp. 24–25), with data structured and categorised accordingly. For elements that did not align with the predefined categories, new categories were developed inductively from the dataset.

10.3 Findings

This section addresses the three RQs using the methodological approach presented above. Information on the company websites was largely available in English. The interviews were conducted in German due to the nationality of the interviewees and to facilitate the discussion. The interviews were transcribed and translated into English by the author and checked by both AI translation tools and the co-author. Quotes from interviewees are therefore inserted as direct quotes, even though they may have originally been spoken in the German language.

10.3.1 RQ1. How Are German T&C Companies Addressing Sustainability in Their Corporate Strategies and Activities?

The analysis of the web sources and interviews revealed eight main themes, offering insights into how these companies approach and implement sustainability (Table 10.3). However, the specific emphasis and content within each category vary notably between the companies examined.

At the **organisational level**, sustainability is strategically and structurally embedded in all SMEs along the T&C value chain examined in this study, reflecting a long-term, generational commitment. Family-owned or owner-managed companies prioritise sustainable practices, with leadership actively integrating environmental and social responsibility into strategy and operations. This is evident in the structural and cultural integration of sustainability management and extensive stakeholder engagement, even though challenges remain in exploring alternative business models.

Integration of Sustainability into Company Culture and Structure

Except for Company A, all companies are family-owned or owner-managed, with leadership often passed down through generations, reflecting deep regional roots. Except for Company E, all of the companies maintain their production within Germany, avoiding outsourcing or offshoring. This long-term perspective is evident in the extended average tenure of employees, highlighting a commitment of the companies to the region and its people. For instance, interviewee B illustrates this generational thinking: "I am now the third generation of my family to work for [Company B]. Naturally, I hope that everything will continue as before. Therefore, we should not just think about profit. It is important for our society that people have jobs".

This mindset underscores a commitment to sustainability through long-term planning and intergenerational responsibility. Company D further exemplifies this by

Table 10.3 Sustainability-related strategies and activities on organisational, product, and process level identified in the case study companies

Dimensions	Sustainability-related sub-dimensions	Company				
		A	B	C	D	E
Organisation	Integration of sustainability into corporate culture/structure	x	x	x	x	x
	Adoption of UN SDGs	x				
	Adopting sustainable business models	x				x
Product	Sustainable products and materials	x	x	x	x	x
Process	Cross-supply chain transparency	x	x			x
	Sustainability risk management	x	x	x	x	x
	Emissions and chemical reduction	x	x	x	x	x

prioritising environmental and resource conservation in its investment strategies and future planning. Additionally, all companies, except for Company D, have dedicated sustainability departments that centralise all relevant sustainability issues and inquiries, but the responsibility for sustainability strategies and goals are managed at the top level, with a top-down approach to implementation.

Interestingly, none of these companies implement sustainability solely in response to recent or upcoming regulatory requirements. All have long prioritised sustainable resource utilisation and compliance with national and international regulations, because it is directly linked to financial success and ensures their survival in Germany. For example, interviewee D explained: "That's why I can tell you that as a small company, we're doing as much as we can, partly out of our own initiative and partly because we're forced to for cost reasons. So, I think we're already pretty far along; otherwise, we wouldn't still be around". At the same time, nearly all companies expressed a desire for more market regulation to ensure they remain competitive against less sustainability-focused manufacturers. As noted by Interviewee B, "We are competing with the entire world, and when I see that in China they simply decide to subsidise certain industries, which can then wipe out entire sectors in Europe, we have to fight against all of that".

These companies also place a strong emphasis on stakeholder engagement in their sustainability efforts, extensively involving various interest groups. Employees are a crucial stakeholder group and a significant lever in achieving the company's sustainability performance. They need to be informed about the goals, purpose, and measures of sustainability initiatives to adapt their actions and actively contribute. Company B, therefore, holds regular training and informational events for employees about corporate development, strategy, and goals, passing on the responsibility for sustainability to everyone. Similarly, all employees at Company A undergo relevant training. Interviewee E highlights the importance of this inclusion: "I notice internally that while I am deeply involved in the topic, if I go just one office over, others may not understand it as well. The challenge is getting everyone behind these stories. I think we're doing well, it's improving, and we're learning".

Alignment of Corporate Sustainability Goals with SDGs

A comprehensive sustainability strategy based on a systematic analysis of company impacts and alignment with the SDGs is evident only in Company A. Both Companies A and E have established specific indicators for monitoring and achieving their companies' sustainability goals. These objectives extend well beyond broad corporate targets, such as climate neutrality, and are broken down into more specific, internal goals. For example, Company E has set ambitious internal targets related to sustainable products, with incremental goals leading up to 2030 allowing for precise monitoring of progress towards the company's sustainability goals.

Adopting Sustainable Business Models

All the companies examined primarily generate value by processing raw materials and components into higher-value products for sale. Financially, they remain dependent on product sales, focusing their sustainability efforts on reducing raw materials and energy consumption, and minimising the negative impacts of their activities and products on people and the environment. However, they have yet to develop alternative revenue streams.

Only Company E, a brand manufacturer in direct contact with retail customers and consumers, has begun exploring new value creation methods, but is facing a number of challenges: "We've considered and tried different things [...], but it's all more of a side issue. We're more focused on the Second Life area now, like how we can improve repairs because nothing else works well. Even when we ask competitors, it's never a revenue driver. It's more effort than it's worth, and that's the challenge we face".

At **product level**, all the companies reported actively implementing measures to minimise the environmental impact of their products, particularly through the use of more sustainable materials such as organic cotton, plant-based fibres, or recycled materials. Although the demand for these products is rising, the proportion of sustainable products in the overall market still seems low, as suggested by Interviewee B: "We are noticing among buyers that sustainability is playing an increasingly important role. However, we still have to say that the share of organic yarns is very small compared to what we usually produce, although it is steadily growing". In this sense, companies also set themselves targets to increase the proportion of sustainable products in the product range.

Innovation also plays a significant role across all companies. Each company reports adhering to high-quality standards for their products and continuously striving for improvements. In developing new products and solutions, they increasingly collaborate with various external partners, ranging from machinery manufacturers and research institutions to downstream partners within their supply chains and even artists. As Interviewee A noted, "We are exploring how our fibres can contribute to making textile products more circular in their applications. This includes replacing synthetic fibres with natural ones or designing functionalised fibres and products that are biodegradable or recyclable. [...] We are also collaborating closely with our customers and the entire value chain in this effort".

At **process level** three sub-categories were identified, aiming to reduce the companies' social and environmental impact.

Sustainability-Related Risk Management

All companies studied have implemented sustainability-related risk management practices to mitigate potential negative impacts along their supply chains. These practices align with the OECD Due Diligence Guidance for Responsible Supply Chains in the Garment and Footwear Sector, which outlines key risks specific to the textile industry, such as human rights violations, hazardous working conditions, and environmental damage. This is particularly relevant when companies source globally for raw materials or components. All the companies examined reported that they use a Code of Conduct as the foundation for responsible procurement, and conduct regular audits or obtain external certifications to ensure compliance with social and environmental standards in upstream processes.

Cross-Supply Chain Transparency

Transparency is directly linked to risk management, as it allows the identification of possible risks and negative impacts, and thereby enables meaningful action. However, initiatives to achieve transparency, by collecting and sharing data on material origins, ingredients, and process impacts, vary among the companies studied. External pressures, particularly from customers demanding proof of sustainable practices, play a significant role in driving these transparency activities. For example, interviewee A observed: "We have focused a lot on the topic of traceability, because the requirements of brands in the textile industry are constantly increasing. The need to be able to trace where the products come from and where they might go to". Companies A, B, and E have taken steps to obtain certifications from suppliers that verify the origin of materials and compliance with environmental and social standards. They have also implemented measures to share this information with downstream partners, often using traceability platforms to securely track product information, while Company C and D provide no information on that.

Emissions and Chemical Reduction

Reducing harmful emissions is a top priority for all the companies analysed. This focus encompasses both air and water emissions, with comprehensive efforts aimed at controlling direct sources of harmful emissions (Scope 1 and Scope 2), such as by transitioning to renewable energy sources. However, efforts to mitigate emissions that are not directly controlled by the companies (Scope 3), which occur along the supply chain, are significantly less proactive. Interviewee B offered the following perspective: "Since last year, we've had a solar power system on our roof, and now we're planning to add an outdoor system as well, which will allow us to cover almost a third of our electricity needs ourselves. For the rest, we're aiming to purchase 100% green energy".

Companies also invest in energy-efficient production machinery and water recycling systems. These practices are not necessarily new, as some companies have long implemented innovative and efficient processes for energy management and resource conservation. The statements of Interviewee D exemplify the efforts to reduce water and energy consumption:

> We reuse our water up to four times. We collect it with special pumps and tanks [...] the machine is cooled from 98 degrees to 70 degrees with a jacket cooling. We also collect this water and use it as the first cooking bath. This way, we only need to heat the water from 60 to 98 degrees and do not have to use any fresh water. We have been doing it this way for over 20 years, mainly for energy-saving reasons. Nobody forced us to do it; we are business people and know how to calculate.

Overall, the findings suggest that the T&C sector is not only aware of sustainability challenges but is actively engaged in addressing them through strategic, cultural, and operational measures. Except for one representative from the clothing production stage, all the companies examined operate exclusively within their manufacturing sites in Germany, and are deeply embedded in their regional communities. This regional heritage brings with it a commitment to preserving local ecosystems and a long-term orientation in corporate management, which are central to organisational identity. Sustainability is a key consideration for all companies in their production processes, and in the selection of suppliers and materials. Across the board, measures have been implemented to reduce the use of chemicals and lower the carbon footprint, while the adoption of the UN SDGs and the development of sustainable business models are less widespread.

10.3.2 RQ2. What Strategies and Activities Relating to the CE Are Being Pursued in the German T&C Industry?

The CE is explicitly referred to by all the companies‘ sustainability-related websites. Their strategies and activities to reduce both resource consumption and waste were intensively discussed during the interviews. The findings suggest that the CE serves as a core component of the sustainability strategies in all companies. Although all companies are adopting some measures associated with the CE, they are presented as being part of a general move towards sustainable practices, rather than specifically in the context of pursuing a CE based strategy. Seven sets of activities relating to the CE can be identified in the companies studied (Table 10.4).

At the **organisational level**, CE principles form a core component of sustainability efforts. Many of the sustainability measures described under the first research question align with integral CE strategies, primarily focusing on reducing resource consumption and waste generation. A strong commitment to investing in renewable energy sources, such as solar panels, and advanced water or waste treatment facilities, exemplifies this approach. Furthermore, two main groups of CE-relevant activities are evident, discussed below.

Table 10.4 CE-related actions and activities at organisational, product, and process levels in the case study companies

Dimensions	CE-related sub-dimensions	A	B	C	D	E
Organisation	Cooperation/collaboration with partners	x	x	x		x
	Adopting a circular business model					x
Product	Procurement/use of recycled materials	x	x	x		x
	Production and sale of easily recyclable/ biodegradable products	x	x	x		x
	Activities to extend product use phase	x	x	x	x	x
	Reuse or recycling of materials/products	x	x	x		x
Process	Reduction of waste and/or emissions in operations	x	x	x	x	x
	Reverse processes					x
	CE-related marketing and communication		x	x		x

Cooperation and Collaboration with Partners

Nearly all the companies examined are involved in various projects with regional or national partners aimed at developing closed-loop textile cycles. The strong regional foundations of these companies facilitates networking and collaboration within their local contexts, enabling them to develop and pursue shared initiatives. For instance, Interviewee B reported, "We have had a conversation with [a customer], who has about 75 tonnes of cutting waste a year, which should be broken down to be reprocessed into yarn". In this context, Interviewee E's statement about collaborations among brand manufacturers is particularly revealing: "I would argue that no one really works together in this area; we're more focused on achieving exclusivity to be early adopters or to stay ahead. That's the challenge—if I can sell [my sustainability story], I wouldn't share it with my competitors". This highlights the competitive tension that hinders cooperation, even when collaboration could advance industry-wide goals.

Adopting a Circular Business Model

According to Geissdoerfer et al. (2020), a circular business model entails value creation through dematerialisation (e.g., substituting products with service and software solutions), intensifying product use, extending product lifespans, or implementing recycling measures. Among the companies studied, such practices are mainly observed at the brand manufacturer E. While fibre and yarn manufacturers, being in the early stages of the value chain, are engaged in developing recycling processes to eventually produce fibres and yarns from textile waste, these efforts are still in an early phase of development. The high-quality materials and intermediate products generated at all value chain stages may contribute to extended product lifespans, but the companies themselves have limited influence over this outcome. As Interviewee B pointed out, "We do not have influence on the downstream processes, such as how often a T-shirt is worn or whether it is reused or not".

At product level, the companies studied are actively incorporating circularity approaches into their product strategies by using recycled materials, though challenges such as higher costs and maintaining quality standards persist. Efforts are also being made to produce easily recyclable or biodegradable products, but progress is limited by conflicting requirements for product quality and recyclability. Additionally, all companies focus on extending product lifespans through high-quality standards and services.

Procurement and Use of Recycled Materials

All companies pursue the use of recycled materials or the reuse of production waste, though approaches vary significantly. Company D, for example, uses recycled polyester in its sportswear and repurposes production of leftovers or textile promotional materials into new products. However, this practice is less common in the earlier stages of the value chain, where, as noted by interviewee D, "the quantity is too small, and it's too expensive. The customers ask about it, and I say, you can have it, but it costs more".

Additionally, maintaining specific quality standards when using secondary raw materials is challenging, leading to situations where "the customer demands a certain quality from us, and if we can't deliver it, they just won't buy" (Interviewee B). Nonetheless, companies in the early stages of the value chain are engaged in various projects and research collaborations to develop high-quality recycled materials and products.

Production and Sale of Easily Recyclable or Biodegradable Products

Except for Company D, all companies are committed to producing materials and products that are as easily recyclable as possible, and a wide range of measures is being implemented. On the one hand, these include the development of mono-material products based on synthetic fibres to facilitate recycling; and on the other hand, replacement of synthetic fibres with natural ones helps to make products compostable. However, progress in this area is still limited, as certain product properties or high-quality standards often conflict with recyclability. For example, Interviewee C explained, "We have sourced a new yarn that comes from a fibre-to-fibre recycling process that we are currently testing. However, it does not meet our high standards, so we will probably have to mix it with a synthetic fibre". This mixing of different fibre types, however, can negatively impact the recyclability of the final product.

Activities to Extend the Product Use Phase

All companies reported that they adhere to the highest quality standards, thereby contributing to the longevity of materials and textile end products. For instance, Company C designs all its woven goods to be "washable and easy to care for", and offers training and care service for customers and consumers to ensure that products remain functional for as long as possible. Company E also extends product life by offering a repair service for its sports and outdoor clothing, in addition to maintaining high-quality standards.

At process level the companies examined prioritise energy efficiency, chemical reduction, and quality management to minimise waste and emissions, supported by relevant certifications. Reverse processes, such as recycling and reuse of materials, are being explored, with some companies involved in closed-loop supply chain projects. Marketing and communication efforts related to CE practices are more prominent among brand manufacturers, though upstream suppliers are also providing consumer-facing sustainability information.

Reduction of Waste and/or Emissions in Operations

All the studied companies prioritise energy and resource efficiency, which is supported by certifications such as ISO 50001, 14001, or EMAS. These certifications not only ensure reduced energy consumption but also facilitate the transition to renewable energy sources, thereby lowering greenhouse gas emissions. Quality management is maintained through ISO 9001 certification, aiming for continuous improvement and minimal waste. In wet processes, like printing and dyeing, the use of chemicals is reduced, leading to less water and soil contamination. As noted by interviewee D, "What we've eliminated—and what used to be a huge mess—is the printing ink that we used to have to dispose of. Now, it's almost nothing because there's hardly any ink used anymore. I no longer have those large amounts of leftover ink like before when I used to dump 30% into the wastewater. This change came about because we've eliminated rotary printing and switched entirely to digital printing". Additionally, all companies reported that their compliance with international standards like ZDHC, REACH, or BREF ensures that materials and products are free from harmful chemicals.

Reverse Processes

The companies examined in this study are adopting various strategies and initiatives to manage and recycle production waste, highlighting their efforts in reverse logistics and material reuse. Company A has implemented modern facilities that allow for the utilisation of production waste, some of which is incinerated for energy generation. Company E has developed a recyclable collection where clothing can be returned at the end of its lifecycle for recycling, in partnership with a German recycling company. For quantity reasons, production offcuts are also supposed to be recycled there, "but transportation is, of course, an issue since the offcuts are located in [production plants] in Asia. However, we are already working on ideas, such as placing them in the container with the finished goods". This illustrates the logistical challenges of textile recycling in a globally distributed supply chain.

However, such reverse processes seem to be less relevant for companies further down the supply chain. Still, some are involved in research projects aimed at creating (regional) closed-loop supply chains to consolidate and recycle their textile waste streams. Companies C and D have not yet undertaken such activities but interviewee D expresses interest in having "someone to collect our fabric waste and repurpose it, so they can create value from our waste without us having to finance it".

In contrast, all companies undertake activities aimed at the internal reuse of materials, such as repurposing packaging or production scraps.

Even if companies report on collaborations and developments to close textile loops, these are at best at the project stage, and do not yet indicate a comprehensive textile material loop. Nevertheless, they point to corresponding future trends and developments that could lead to large-scale reverse supply chains involving various actors.

CE-Related Marketing and Communication

There is a clear trend in that activities targeting end consumers decrease as one moves up the supply chain, which is also reflected in marketing and communication efforts. Brand manufacturer E is highly active on various online platforms and in retail, informing customers about their products and related sustainability efforts and CE activities. For example, they acknowledge challenges in communicating the complexities of their recyclable clothing line, which "is not perfectly explained yet; we're still working on it because it's a very complex topic. We're trying to simplify it now". Effective communication and simplified processes are critical for the success of circular business models that require consumer participation.

However, CE-related marketing and communication efforts are not completely limited to the brand manufacturer. Yarn and fabric manufacturers B and C also provide consumer-facing information to highlight their innovations and the sustainable qualities of their materials in various end products. For example, Interviewee C mentioned offering hangtags on end products with care instructions and QR codes for videos to promote the longevity of their materials, and "when consumers buy a [product] made with our fabric, we offer a service hotline where they can ask how to properly wash and care for the fabric. For example, they can ask what to do about pressure marks or if something like marker or nail polish gets on it. We offer this service to help the fabric last as long as possible". Similarly, Company B offers hangtags with information on sustainable yarns, including material origin, properties, and certifications.

Overall, the findings show that the companies studied integrate CE principles at various levels, emphasising energy efficiency, chemical reduction, and the use of recycled materials to minimise waste and emissions. Collaboration with partners is key, particularly in efforts to develop closed-loop supply chains and advance recycling processes, although challenges persist, particularly in logistics and maintaining quality standards. Companies are also making strides in extending product lifespans through high-quality standards and consumer communication efforts related to CE practices. However, while significant progress has been made, many CE initiatives are still in the early stages of development.

10.3.3 RQ3. What Role Are Digital Technologies Playing in the Transition to Sustainability and the CE in the German Textile and Clothing Industry?

As digital technologies are not mentioned in any detail in the public reporting and communication of the companies studied, notably on their corporate websites, the interviews focused extensively on the potential connections between digital technologies and the companies' sustainability and CE initiatives. As noted in Sect. 10.3.2, the CE is an integral part of the sustainability strategies for all companies involved. However, the findings on the role of digital technologies suggest that while these technologies do support various CE activities, they are more frequently associated with general optimisation efforts rather than with strategic sustainability or CE goals. From the analysis of interview data, eight key themes emerged regarding the role of digital technologies in the transition towards sustainability and the CE (see Table 10.5).

At **organisational level**, the companies examined use similar strategies for digital technology deployment with a strong focus on traditional applications, and they face similar challenges related to digitalisation.

Table 10.5 Digital technology deployment for sustainability and CE in the case study companies

Dimensions	Sub-dimensions related to digital technology deployment	A	B	C	D	E
Organisation	Adoption and experimenting with new technologies	x	x		x	x
	Data management and integration	x	x		x	x
	Organisational challenges		x	x		x
Product	Product development and innovation				x	
	Transparency and traceability	x	x			x
Process	Process optimisation and efficiency	x	x	x	x	x
	Resource management and impact assessment	x	x		x	

Adoption and Experimentation with Digital Technologies

The findings indicate that various technologies are utilised across different stages of the value chain, offering potential to support circular practices. However, there seems to be limited enthusiasm for experimenting with new technologies. The most commonly used technologies among all companies are typical administrative software systems like Enterprise Resources Planning (ERP) or Product Data Management (PDM) systems. Two of the companies studied have expressed the need to transition to new ERP solutions due to outdated systems.

Regarding emerging technologies like AI and blockchain, the industry appears to be in the early stages of adoption. While experimentation indicates their clear potential, widespread acceptance is hindered by factors such as high costs and integration challenges. Furthermore, there is a lack of knowledge about profitable applications or industry-specific trained models to drive such projects forward.

For example, Interviewee A stated "We wanted to integrate AI into product development, such as market research or trend screening, but other projects were prioritised". In this context, Interviewee C summarised:

> It's always easy to say, 'Let's implement AI somewhere'. Larger companies have an advantage here, because they produce with less variability and have much more money available for such systems. What sets us apart, though, is our complexity—producing custom items for each client in runs of only 50 or 100 meters. Large companies would laugh at that—they produce tens of thousands of meters. For them, it's much more worthwhile to use AI, for example, in a fabric inspection machine where it's about detecting defects. [...] AI can learn and correctly identify errors there. But when the fabric changes every 30 to 40 meters, how is AI supposed to learn?

In contrast, Company D has shown how AI, when properly implemented, can open new horizons in product development for small companies, even leading to cost savings, as discussed in the product development section below.

Blockchain technology is scarcely used among the companies studied, with only Company A utilising a blockchain-based traceability platform tailored to the T&C

industry. However, this platform is not widely adopted by higher-value chain partners, likely due to the lack of standardised industry-wide solutions, making it challenging for companies to select a platform without having to supply data to multiple systems simultaneously. This highlights the need for industry-wide agreements and collaboration to enhance traceability and data consistency.

Similarly, digital manufacturing technologies like 3D techniques are scarcely used, with the exception of Company D, which utilises digital printing. This has the additional benefit of making all product and printing data available digitally, creating a form of digital twin. Robotics, on the other hand, plays no role in the T&C production of the companies studied. Interviewee C again cites the high product variability as a reason for low automation: "There is no standard process; each product goes through the company in a different way, with different production steps [...], and a lot is still handled manually". This suggests that maintaining a high level of flexibility in production impedes automation and digitalisation, which resonates with related debates concerning the adoption of standard processes prior to technology implementation (Wynn et al., 2024).

Social media is widely used across all companies for traditional marketing and communication, especially in human resources for employer branding. However, the potential of these technologies to play more strategic roles in driving sustainability and innovation is not yet being realised.

Data Management and Integration Challenges

Companies are increasingly collecting large amounts of diverse data from various sources, such as material origins, lifecycle assessment (LCA) related data, and machine data in the production process. Effective management involves ensuring that this data is collected automatically, securely stored, and easily accessible, accurately reflecting the processes and materials used. Proper infrastructure is necessary to achieve this, as highlighted by Interviewee D: "We buy €50,000 worth of storage every year, not machines, but storage to manage our data". Additionally, information exchange between value chain stages is often not standardised or automated, but manually requested and recorded. Brand manufacturer E expressed a desire for "more data when we purchase materials. Today, we manually research for and enter this data to calculate CO_2 or water consumption". The willingness to be transparent and the ability to automate and efficiently capture, analyse, and share data is increasingly becoming a competitive advantage. However, Company C faces difficulties in this regard: "Data is in our systems, but it's not easily accessible. We can analyse data retrospectively, but we lack real-time data to inform us about the current state, such as error rates". This suggests there may be a role for data analytics tools in such companies to extract, analyse and report on data from disparate sources.

Organisational Challenges

One challenge identified in various interviews is resistance to new technologies among long-term employees. This, along with the high costs of implementation, are important organisational challenges in embracing digital transformation. Company culture and employee mindset play a significant role.

At **product level**, product development and innovation through advancements in AI tools could reshape industry practices and drive sustainability. While most companies examined consider tools to be important in product development, they have so far hardly been used. However, the example of company D shows how AI can be leveraged to revolutionise processes such as print design, eliminating traditional sampling and travel costs. Concurrently, efforts are underway to enhance the recyclability of products through precise material data and transparent supply chain information, with digital technologies playing a crucial role in improving transparency and traceability.

Product Development and Innovation

In product development, the use of AI-based systems is currently limited to a few applications. However, company D has significantly transformed its entire development process by using generative AI for developing print designs: "This is the latest ground-breaking thing where real money is being saved, and it is a sign of where things may be heading". The digital development and printing process has made it possible to dematerialise the entire sampling process, completely replacing the need to attend creative fairs for sourcing print designs, along with the associated travel and investments, by using AI design software. Interviewee D further sees potential in connecting these AI systems with printing machine systems to simulate and optimise design developments in terms of material and energy consumption.

Efforts to develop recyclable mono-material products and improve recycling processes are being advanced across different value chain stages. In this context, digital tools for better material data and lifecycle assessments are seen as critical for achieving these goals. Brand manufacturer E, for example, is developing clothing products from mono-materials intended to be entirely recyclable. To ensure this, much more precise information about the materials' contents and compositions is necessary. For instance:

> Polyester is currently only listed as an ingredient, but there are various types of polyester that can significantly impact later recycling processes. Due to a lack of data and accuracy, extensive tests are conducted during product development in collaboration with recycling partners to ensure recyclability.

The more precise and transparent the data on materials and contents along the supply chain, the simpler and more efficient recycling processes can be.

Transparency and Traceability

Transparency will be of particular relevance in the context of the EU Strategy for Sustainable and Circular Textiles (European Commission, 2022), which envisages the introduction of a DPP, which should contain key product information along the supply chain. Thus, transparent collection and publication of data on the origin and composition of products and materials are becoming increasingly important for T&C companies due to anticipated legal requirements like those required by the DPP. Technologies such as blockchain, RFID, and other tracing systems offer solutions for tracking material authenticity and supporting transparency. However, these technologies are implemented inconsistently in the companies studied. Company A produces “fibres with a physical tracer”, while Company E plans to “use QR codes by early next year, allowing end consumers to scan a code and access information”.

At **process level**, the companies primarily use digital technologies for driving process optimisation and efficiency. Nevertheless, real-time data capture and IoT systems are also enhancing resource management and impact assessment. Meanwhile, the integration of ERP systems and mobile apps is refining operational workflows, though challenges remain in adapting these systems to meet evolving sustainability and data management needs. The focus is increasingly on leveraging digital tools to support CE principles and foster industry-wide collaboration for sustainable practices.

Process Efficiency and Optimisation

The ability to capture, monitor, and analyse data in real-time through IoT systems enhances process efficiency and resource management. Yarn manufacturer B states, “We have online monitoring systems on almost all machines, where the yarn thickness is continuously measured, and we have automatic foreign fibre detection, which is a big plus”. This reduces error rates and contributes to waste reduction.

Mobile apps support existing processes, such as sales presentations and project management, leading to improved efficiency and reduced costs. However, their role is mainly supplementary rather than transformative. Process integration within companies through ERP systems is reported by all companies studied. These systems primarily serve to manage operational processes efficiently. However, increasing demands for integration with external systems and partners to capture and process sustainability and CE-related data are also affecting these systems, requiring them to be expanded or replaced by more advanced systems (Wynn et al., 2024). In this context, interviewee C commented: “Our data is not easily accessible, which is why I’m not a big fan of our [current system]. Therefore, we are now exploring new systems”.

Resource Management and Impact Assessment

Digital technologies also play an important role in resource management and impact assessment within company and value chain processes. For instance, yarn manufacturer B measures and manages all material flows:

> Input, output—in other words, what comes into the company, the products, the energy, operating materials, and what goes out again, including our own generated electricity and waste. We’ve been consistently recording this internally since 1997, enabling us to create key figures and stay up to date using a self-developed management system.

Brand manufacturer E, on the other hand, evaluates its products using “internal calculations, meaning we have our own assessments in our PDM and ERP systems, to calculate the impact of our products”. This is built around an “internal points system based on factors such as materials used or dyeing processes. For example, depending on the type of cotton used, water consumption and CO_2 emissions may be higher, resulting in fewer points”. The aim is to objectively assess the sustainability of products and identify measures to increase it.

Overall, there seems to be a need for cooperation and innovation forums to promote the integration of digital technologies with sustainability and CE practices. Social media is seen as an instrument for promoting cooperation, which is crucial for sustainable practices, but it is not used accordingly. Efforts to integrate digital technologies into the CE, such as the use of QR codes for returns, suggest that there is an increasing focus on the principles of the CE, although widespread application is still in development. Although AI, big data, and analytics have the potential to reduce resource consumption and waste, the industry struggles with data sharing and comprehensive analysis due to system silos, the continuation with manual processes, and the lack of an appropriate change culture.

10.4 Discussion

The findings outlined above raise several issues that merit further discussion. Firstly, the CE has emerged as a strategic priority across all stages of the T&C value chain. This shift is likely driven by the increasing politicisation of the issue and the expected impact on businesses, as the EU is implementing multiple regulatory requirements for the T&C industry to promote sustainability, the CE, and digitalisation. A key aspect of this is the Extended Producer Responsibility (EPR) system (European Commission, 2022), which will hold companies accountable for the entire lifecycle of their products, thereby encouraging recycling and waste prevention. Furthermore, the Eco-design for Sustainable Products Regulation (ESPR) (European Commission, 2024) introduces eco-design requirements to enhance recyclability and sustainability, and includes a DPP to ensure transparency regarding the sustainability attributes of products (European Commission, 2021).

Secondly, sustainability awareness is deeply embedded in the strategic goals of companies along the textile value chain in Germany. CE is explicitly addressed in sustainability strategies, with CE principles, such as resource utilisation and waste reduction, guiding many of the company's environmental objectives. However, the CE concept is not yet fully integrated, with only specific aspects being incorporated into existing linear value chains and business models. For instance, the reduction of harmful emissions is a significant focus across all companies studied. However, the emphasis is primarily on direct emissions caused by the companies themselves (Scope 1 and 2), leaving a substantial portion of emissions from purchased products largely unaddressed.

Thirdly, the business models, especially in companies further down the supply chain, have been minimally impacted by CE measures. Additionally, digital technologies do not yet play a widespread role in implementing CE initiatives. This disconnect between the two concepts is evident in a recent survey of global CEOs by Software AG (2023), which states that "84% of companies view digital transformation and sustainability transformation as separate initiatives" (para. 9). More specifically, Arnold et al. (2023), in their quantitative study of German textile manufacturers, concluded that "digitalisation opportunities are recognized, but a stringent method of implementation for this is not yet available nationwide" (p. 17). A recent survey by the European Commission (Izsak & Moreno, 2024) among T&C SMEs also reveals a lower level of engagement in digital transformation compared to the entire European T&C industry. This underscores the urgent need for the development of best practice examples and industry-wide success factors through the involvement of research and industry partners from all stages of the value chain.

Fourthly, reverse material flows have been implemented only to a very limited extent. While some of the companies studied, particularly those in the lower value chain levels like fibre and yarn manufacturers, are involved in pilot projects with partners to process textile waste into new secondary materials, these initiatives are still in their early stages, and do not yet offer scalable, industry-wide solutions. The existing concepts for reuse and recycling considered by the companies currently focus mainly on packaging materials and only in rare cases on the textile products themselves. The foundations for a product's recyclability are laid during the product development phase, where the material composition ensures that products are manufactured without hazardous chemicals, used for as long as possible, repaired if necessary, and recycled at the end of the lifecycle as required by the guidelines of the EU Strategy for Sustainable and Circular Textiles (European Commission, 2022). However, other principles, such as extending the lifespan through repair, or rental and leasing models, require cooperation and corresponding processes. The development of these processes and infrastructure is, however, only sporadically evident. Intensive collaboration among actors at all stages of the value chain, as well as joint innovations, are necessary. The study by Izsak and Moreno (2024) shows that the number of start-ups with sustainable and digital business models is rapidly increasing, with solutions ranging from sustainable materials to platform solutions, sharing business models, and digital solutions for sorting and recycling. The competition for established companies to meet the new demands and challenges is thus increasing.

The CE transition will therefore require appropriate partnerships and collaborative investment. Without knowledge of the entire supply chain, the goals set out by the EU will not be achieved. In particular, the requirement for a DPP, which is to be introduced in line with EU strategy, will force companies to further address the digital transparency and traceability of their entire supply chains. However, size and resource constraints make it difficult for SMEs to develop and implement closed-loop supply chains on their own. On the other hand, as the T&C industry in Germany is locally very concentrated, collaboration between various actors in the value chain and the inclusion of new market players (start-ups) to form local hubs could be an approach to quickly develop joint innovations and effective processes to extend product lifespans and create functioning recycling processes. Arnold et al. (2023, p. 18) also call for "further investigations on new forms of collaboration between manufacturers, retailers, and consumers ensuring materials are kept in use for as long as possible". The expected introduction of producer responsibility in line with the EU's Strategy for Sustainable and Circular Textiles is, therefore, not yet matched by any functioning concepts or capacities on the part of the SMEs considered.

Fifthly, to address these issues requires significant change within organisations, involving people who are motivated and skilled to face new challenges and adapt. This, however, seems to be a central challenge, leading to (further) competitive disadvantages for the domestic industry compared to new market players and international competition. On the one hand, engaged employees with the right skills and competencies are needed to implement new technologies and processes, and on the other hand, all stakeholders must contribute and be involved. Company culture and employee mindset play a significant role and act as a critical success factor, as the following two statements related to the utilisation of advanced digital technology in product development illustrate. Company D asserts that "our company culture is always positive towards these things because we have the right people". Company C, however, points up a key issue regarding staff reskilling: "a new employee in product development is experimenting with AI design tools, but long-term employees struggle with new technologies because they've been doing it the same way for 20 or 30 years. But we're trying to push this more".

Sixthly, for SMEs operating in the T&C sector, there are less exemplars of best practices in companies of this size, with little guidance available on how to progress the CE transition. Although innovative ideas may surface, implementing them is often challenging due to limited resources. As Company C noted: "We are ultimately a small player, so it's difficult to make many things feasible with our manpower without costs skyrocketing. There are many cool ideas and approaches, but whether they are truly scalable and realistic is always a question for me". On the other hand, new technologies have the potential to address these challenges. However, as Schöggl et al. (2023) highlight, the adoption of digital technologies for sustainability and CE practices in the T&C manufacturing sector is still relatively low compared to other industries. They emphasise the importance of developing business cases to demonstrate how digital technologies can better facilitate sustainability and circularity management.

This research aims to contribute to filling this gap in the literature through the expanded action list, distilled from the research findings, for transitioning to the CE presented in Table 10.6.

In addition, a further review of the interview feedback and literature sources suggests a number of top-line CSFs for the successful implementation of CE practices within T&C organisations. These CSFs are briefly set out below and in Table 10.7, and are grouped within the three change dimensions of organisation, product, and process.

In the **organisation domain**, an open and innovative culture is paramount. Such a culture fosters creativity and adaptability, encouraging employees to embrace new ideas and drive sustainable innovations that are crucial for the CE. Knowledgeable and skilled employees across all organisational levels are also needed to ensure sustainability strategies are integrated effectively into daily operations, which drives progress towards achieving circularity. Additionally, access to cooperation and collaboration with external partners, as well as securing funding, plays a significant role in advancing CE initiatives. Partnerships allow companies to share resources and best practices, thereby accelerating the implementation of circular strategies.

Table 10.6 Extended action list matrix for T&C companies transitioning to the CE

Change dimension/action area	Organisation	Product	Process
Strategy	• Set the circular economy as a core component of corporate sustainability objectives • Integrate responsibility for sustainability/CE into organisational structure and culture • Adopt SDGs into corporate strategy (business strategy, annual reporting, CSR) • Create an open and flexible corporate culture and empower and encourage employees to get involved and experiment with new technologies	• Develop and implement sustainable product design and development strategies (e.g., design for durability, repairability, recyclability) • Initiate cross-supply chain transparency of product lifecycle • Set targets for product carbon footprints	• Establish process owners for sustainability/CE across company • Review, adapt and adopt sustainable and circular business models (rental, leasing, and complementary services, e.g., repair) • Introduce auditing and certification procedures along the supply chain • Implement supply chain risk assessment to identify potential negative social and environmental impacts • Collaborate with business partners in end-of-life treatment concepts
Practice/Operations	• Establish sustainability/CE function and structure • Cascade sustainability and CE objectives throughout the organisation • Include triple bottom line accounting in financial reporting	• Implement product development and manufacture changes to reduce waste (e.g., use recycled and recyclable materials) • Reduce packaging and increase use of sustainable packaging materials • Introduce certification by sustainability labels as appropriate • Implement product and material traceability and tracking	• Publish audit related information about suppliers and production sites • Consider vertically integrated suppliers (e.g., weaving, dying, printing, etc.) • Implement reduction programmes for emissions, water, and chemicals • Introduce cross-company waste reduction and recycling objectives and practices • Establish and implement take-back systems and reverse logistics processes

(continued)

Table 10.6 (continued)

Change dimension/action area	Organisation	Product	Process
Technology	• Develop digitalisation/IT strategy to incorporate sustainability/CE drivers and benefits • Transition IT function to championing sustainability/CE ("Green IT") • Use social media and other digital media technologies to support stakeholder communication	• Use digital technologies to support product development and production efficiencies (e.g., IoT, 3D technologies) • Implement a digital product passport • Deploy digital technologies to reduce waste and enhance recycling	• Trial digital technologies in support of sustainability and CE objectives • Incorporate technology use cases into sustainability/CE processes (e.g., mobile apps, blockchain) • Digitalise administrative processes to reduce resources and materials consumption and emissions • Implement digital tracing tools to provide cross-supply chain transparency

Based on Wiegand and Wynn (2023)

Table 10.7 Critical success factors (CSFs) for T&C companies transitioning to the CE

	Organisation	Product	Process
Critical success factors (CSFs)	• Open and innovative culture • Knowledgeable and skilled employees at all organisational levels • Access to cooperation and collaboration with external partners and funding	• Sustainability and CE as the core and starting point for product development • Systematic collection and publication of product-related data	• Real-time data monitoring and analysis • Seamless integration of internal and external data and tools • Cross-functional collaboration across internal departments and external partners

In the **product sphere**, another key factor is the integration of sustainability and CE principles at the very beginning of product development. This approach ensures that products are designed with their entire lifecycle in mind, minimising environmental impact and maximising resource efficiency. Moreover, the systematic collection and publication of product-related data enhance transparency and accountability, enabling stakeholders to track sustainability performance and make informed decisions.

As regards the **process dimension**, real-time monitoring and analysis of data are essential for optimising processes and ensuring compliance with sustainability goals. This capability allows companies to quickly address inefficiencies and adjust operations to reduce waste and emissions. Furthermore, the seamless integration of data and tools across internal and external systems is critical for achieving a comprehensive view of the supply chain, ensuring that all stakeholders are aligned in working towards common sustainability goals. Cross-functional collaboration, both within the organisation and with external partners, is another critical factor. This collaboration facilitates the sharing of knowledge and resources, helping to overcome challenges and drive innovation in CE practices.

10.5 Conclusions

This chapter has examined how manufacturing SMEs in the German T&C industry are approaching the issues of sustainability and the CE, and has assessed the role of digital technologies in the transition to sustainability and the CE. The results suggest that these companies consider sustainability and the CE to be key strategic issues and are aligning their business activities to reduce resource use and waste in order to gain competitive advantage over international competitors, and meet legal requirements. In addition to providing insights into the relationship between digitalisation and CE, other important contributions of the chapter are the action list (Table 10.6) and success factors (Table 10.7) derived from the research results. The chapter thus makes a small contribution to practice and adds to the existing literature on digitalisation and CE in the T&C industry by complementing the perspectives put forward by other authors in different geographical contexts (Koszewska, 2018; Pal & Jayarathne, 2022; Schumacher & Forster, 2022).

Although T&C manufacturers in Germany currently make little use of digital technologies to support their sustainability efforts, the results suggest that digitalisation is increasingly influencing the competitiveness of companies in the T&C industry, especially SMEs. At the same time, the pressure on companies to make their processes and products more sustainable is increasing. So far, initiatives to introduce more sustainable and circular processes, products, and business models have not been primarily driven by digitalisation, which does not yet appear to have acted as a catalyst for such a transition. However, feedback from interviews with five companies at different levels in the T&C value chain suggests that many companies recognise the need for a transition to CE practices facilitated by digitalisation, in order to effectively meet compliance requirements and changing customer expectations, while ensuring cost and human resource efficiency. However, lack of knowledge, skills, and financial resources is impeding adoption.

Furthermore, companies will face increasing pressure from upcoming legislative requirements to adapt their products, business models, and supply chains to the demands of the CE. The EU Strategy for Sustainable and Circular Textiles (European Commission, 2022), for example, envisions the implementation of the DPP which the Commission views as a crucial tool for a climate-friendly and resource-efficient economy. Among its other objectives, the DPP is intended to contain information such as the product's origin, composition, repairability, and disassembly methods, as well as how the product should be handled at the end of its lifecycle. A key element of the DPP is the standardised exchange of data that operators, waste management companies, or public authorities can access. Without transparent processes supported by digital technologies, companies will struggle to provide the necessary data and information and will likely face regulatory penalties.

The introduction of a mandatory DPP is not an end in itself but primarily aims at supporting a transition to sustainable and circular products. It is essential for companies to build up relevant knowledge, develop capabilities, and take the necessary measures to reduce the environmental impact of their products. This includes not only the materials used but also the production processes and chemical usage, as well as durability, repairability, and recyclability. Transparency through intensive exchange of data and information provides the basis for this (Götz et al., 2022), and the exchange of detailed data on resource and material usage and consumption throughout the entire supply chain can also support process optimisation and efficiency gains, while at the same time reducing errors and waste.

This study has limitations in that it is based on publicly available information and interview data from 5 SMEs, each representing a different T&C value chain stage. The analysis of this small number of companies provides an incomplete picture and needs to be considered with caution when drawing conclusions at the industry level. The findings are based on the T&C industry in Germany and are not necessarily applicable in other geographical environments or business sectors. Future work based on case studies is needed to investigate best practices and evaluate the CSFs put forward here.

Further research could also focus on examining external market influences such as the expected laws and regulations under the EU Strategy for Sustainable and Circular Textiles, and how SMEs can most effectively align their sustainability strategies and actions. In addition, other research initiatives could profitably examine how growing customer awareness of the environmental impact of the T&C industry in different geographical locations is affecting the industry. As Claudio (2007) noted almost twenty years ago: "the biggest impacts for increasing sustainability in the clothing industry rests with the consumer" and "consumer awareness about the fate of clothing through its lifecycle may be the best hope for [future] sustainability" (p. 454). Indeed, this may be the key factor that finally pushes the industry into adopting the CE in its practices and operations.

References

Abbate, S., Centobelli, P., Cerchione, R., Nadeem, S. P., & Riccio, E. (2023). Sustainability trends and gaps in the textile, apparel and fashion industries. *Environment, Development and Sustainability*, 1–28. https://doi.org/10.1007/s10668-022-02887-2

Arnold, M. G., Pfaff, C., & Pfaff, T. (2023). Circular business model strategies progressing sustainability in the German textile manufacturing industry. *Sustainability, 15*(5), 4595. https://doi.org/10.3390/su15054595

Breitenacher, M. (1989). Textilindustrie im Wandel [In English: Textile industry in transition]. In Gesamttextil (Ed.), *Schriften zur Textilpolitik, Heft 6* [In English: Letters on textile policy, Issue 6].

Breitenacher, M., Adler, U., Brander, S., & Haase, D. (1991). *Die Textil- und Bekleidungsindustrie der neuen Bundesländer im Umbruch* [In English: The textile and clothing industry in Eastern Germany]. *Ifo Studien zur Industriewirtschaft: Vol. 41*. Ifo-Institut für Wirtschaftsforschung.

Bressanelli, G., Adrodegari, F., Pigosso, D. C. A., & Parida, V. (2022). Towards the smart circular economy paradigm: A definition, conceptualization, and research agenda. *Sustainability, 14*(9), 4960. https://doi.org/10.3390/su14094960

Brydges, T. (2021). Closing the loop on take, make, waste: Investigating circular economy practices in the Swedish fashion industry. *Journal of Cleaner Production, 293*, 126245. https://doi.org/10.1016/j.jclepro.2021.126245

Cicculo, F., Pero, M., & Patrucco, A. S. (2023). Designing circular supply chains in start-up companies: Evidence from Italian fashion and construction start-ups. *The International Journal of Logistics Management, 34*(3), 553–581. https://doi.org/10.1108/IJLM-04-2022-0158

Claudio, L. (2007). Waste couture: Environmental impact of the clothing industry. *Environmental Health Perspectives, 115*(9), A448–A454. https://doi.org/10.1289/ehp.115-a449

Colucci, M., & Vecchi, A. (2021). Close the loop: Evidence on the implementation of the circular economy from the Italian fashion industry. *Business Strategy and the Environment, 30*(2), 856–873. https://doi.org/10.1002/bse.2658

Diebel, J., Eckert, J., Gloy, I.-S., & Schlichter, S. (2022). *Kreislaufwirtschaft: Textile Kreisläufe schaffen, Zukunf gestalten* [In English: Circular economy: Creating textile loops, shaping the future]. Forschungskuratorium Textil e. V. https://textil-mode.de/de/forschung/blog-und-medien/

Dispan, J. (2009). *Bekleidungswirtschaft: Branchenanalyse 2009; Strukturwandel - Entwicklungstrends - Herausforderungen - arbeitsorientierte Handlungsfelder. Informationsdienst des IMU-Instituts: Vol. 2009, 2*. IMU-Inst. http://www.imu-institut.de/papers/publication.2009-06-10.0930855862/2009-06%20IMU-Infodienst%20Bekleidung.pdf

Dissanayake, D., & Weerasinghe, D. (2022). Towards circular economy in fashion: Review of strategies, barriers and enablers. *Circular Economy and Sustainability, 2*(1), 25–45. https://doi.org/10.1007/s43615-021-00090-5
European Commission. (2021). *Sustainability strategy for textiles*. https://ec.europa.eu/growth/industry/sustainability/strategy-textiles_en
European Commission. (2022). *EU Strategy for Sustainable and Circular Textiles: COM(2022) 141 final*. https://environment.ec.europa.eu/strategy/textiles-strategy_en
European Commission. (2024). *Ecodesign for sustainable products regulation—Making sustainable products in the EU the norm*. https://commission.europa.eu/energy-climate-change-environment/standards-tools-and-labels/products-labelling-rules-and-requirements/sustainable-products/ecodesign-sustainable-products-regulation_en
European Environment Agency. (2019). *Textiles and the environment in a circular economy: Eionet report—ETC/WMGE 2019/6*. https://www.eea.europa.eu/publications/textiles-and-the-environment-the
Fashion Council Germany e.V., & Oxford Economics. (2022). *The status of German Fashion 2021*. https://md-master.htw-berlin.de/files/Stg/MD/Bachelor/0._Startseite/Bilder_zu_Aktuelles/2021_Status_of_German_Fashion_FCG_vEN.pdf
Fischer, A., & Pascucci, S. (2017). Institutional incentives in circular economy transition: The case of material use in the Dutch textile industry. *Journal of Cleaner Production, 155*, 17–32. https://doi.org/10.1016/j.jclepro.2016.12.038
Franco, M. A. (2017). Circular economy at the micro level: A dynamic view of incumbents' struggles and challenges in the textile industry. *Journal of Cleaner Production, 168*, 833–845. https://doi.org/10.1016/j.jclepro.2017.09.056
Fu, B., Shu, Z., & Liu, X. (2018). Blockchain enhanced emission trading framework in fashion apparel manufacturing industry. *Sustainability, 10*(4), 1105. https://doi.org/10.3390/su10041105
Gebhardt, M., Kopyto, M., Birkel, H., & Hartmann, E. (2022). Industry 4.0 technologies as enablers of collaboration in circular supply chains: A systematic literature review. *International Journal of Production Research, 60*(23), 6967–6995. https://doi.org/10.1080/00207543.2021.1999521
Geissdoerfer, M., Pieroni, M. P., Pigosso, D. C., & Soufani, K. (2020). Circular business models: A review. *Journal of Cleaner Production, 277*, 123741. https://doi.org/10.1016/j.jclepro.2020.123741
Götz, T., Berg, H., Jansen, M., Adisorn, T., Cembrero, D., Markkanen, S., & Chowdhury, T. (2022). *Digital product passport: The ticket to achieving a climate neutral and circular European economy?* University of Cambridge Institute for Sustainability Leadership. http://nbn-resolving.de/urn/resolver.pl?urn:nbn:de:bsz:wup4-opus-80497
Gözet, B., & Wilts, H. (2022). The circular economy as a new narrative for the textile industry value chain with a view to Germany's opportunities for a circular economy transformation. Zukunftsimpuls, 23. Wuppertal. Available online: https://epub.wupperinst.org/frontdoor/deliver/index/docId/8108/file/ZI23_Textile_Industry.pdf
Hussain, M., & Malik, M. (2020). Organizational enablers for circular economy in the context of sustainable supply chain management. *Journal of Cleaner Production, 256*, 120375. https://doi.org/10.1016/j.jclepro.2020.120375
Izsak, K., & Moreno, C. (2024). *Monitoring the twin transition of industrial ecosystems: Textiles: Analytical report*. https://op.europa.eu/en/publication-detail/-/publication/06780edf-0911-11ef-a251-01aa75ed71a1/language-en/format-PDF/source-343263872
Jia, F., Yin, S., Chen, L., & Chen, X. (2020). The circular economy in the textile and apparel industry: A systematic literature review. *Journal of Cleaner Production, 259*, 120728. https://doi.org/10.1016/j.jclepro.2020.120728
Koszewska, M. (2018). Circular economy—Challenges for the textile and clothing industry. *Autex Research Journal, 18*(4), 337–347. https://doi.org/10.1515/aut-2018-0023
Kottmeyer, B. (2021). Digitisation and sustainable development: The opportunities and risks of using digital technologies for the implementation of a circular economy. *Journal of Entrepreneurship*

and Innovation in Emerging Economies, 7(1), 17–23. https://doi.org/10.1177/239395752096 7799

Marschall, P., Peters, R., Jordan, J., Fromhold-Eisebith, M., Thomes, P., & Kemper, M. (2020). *Wisotex 4.0 -Wirtschaftsräumliche Implikationen von Industrie 4.0 in relationaler Perspektive: Innovation, Evolution, Organisation und Interaktion, untersucht am Beispiel der deutschen Textilindustrie* [In English: Wisotex 4.0—Economic-spatial implications of Industry 4.0 in relational perspective: Innovation, evolution, organization and interaction, examined on the example of the German textile industry]. https://doi.org/10.13140/RG.2.2.33453.92647

Mayring, P. (2014). *Qualitative content analysis: Theoretical foundation, basic procedures and software solution.*

Neligan, A., Baumgartner, R. J., Geissdoerfer, M., & Schöggl, J.-P. (2022). Circular disruption: Digitalisation as a driver of circular economy business models. *Business Strategy and the Environment*, Article bse.3100. Advance online publication. https://doi.org/10.1002/bse.3100

Okorie, O., Salonitis, K., Charnley, F., Moreno, M., Turner, C., & Tiwari, A. (2018). Digitisation and the circular economy: A review of current research and future trends. *Energies, 11*(11), 3009. https://doi.org/10.3390/en11113009

Pal, R., & Jayarathne, A. (2022). Digitalization in the textiles and clothing sector. In B. L. MacCarthy & D. Ivanov (Eds.), *The digital supply chain* (pp. 255–271). Elsevier. https://doi.org/10.1016/B978-0-323-91614-1.00015-0

Ræbild, U., & Bang, A. L. (2017). Rethinking the fashion collection as a design strategic tool in a circular economy. *The Design Journal, 20*(sup1), S589–S599. https://doi.org/10.1080/14606925.2017.1353007

Ramesohl, S., Berg, H., & Wirtz, J. (2022). *The circular economy and digitalisation: Strategies for a digital-ecological industry transformation.* https://doi.org/10.48506/opus-7900

Ranta, V., Aarikka-Stenroos, L., & Väisänen, J.-M. (2021). Digital technologies catalyzing business model innovation for circular economy—Multiple case study. *Resources, Conservation and Recycling, 164*, 105155. https://doi.org/10.1016/j.resconrec.2020.105155

Reichel, A., Mortensen, L.F., Asquith, M., & Bogdanovic, J. (2014). Environmental indicator report 2014: Environmental impacts of production-consumption systems in Europe. EUR-OP: Luxembourg. ISBN 978-92-9213-487-7

Roos, S., Jönsson, C., Posner, S., Arvidsson, R., & Svanström, M. (2019). An inventory framework for inclusion of textile chemicals in life cycle assessment. *The International Journal of Life Cycle Assessment, 24*(5), 838–847. https://doi.org/10.1007/s11367-018-1537-6

Saccani, N., Bressanelli, G., & Visintin, F. (2023). Circular supply chain orchestration to overcome circular economy challenges: An empirical investigation in the textile and fashion industries. *Sustainable Production and Consumption, 35*, 469–482. https://doi.org/10.1016/j.spc.2022.11.020

Salmi, A., & Kaipia, R. (2022). Implementing circular business models in the textile and clothing industry. *Journal of Cleaner Production, 378*, 134492. https://doi.org/10.1016/j.jclepro.2022.134492

Sandberg, E., & Hultberg, E. (2021). Dynamic capabilities for the scaling of circular business model initiatives in the fashion industry. *Journal of Cleaner Production, 320*, 128831. https://doi.org/10.1016/j.jclepro.2021.128831

Schneider, A. (2003). *Internationalisierungsstrategien in der deutschen Textil- und Bekleidungsindustrie : eine empirische Untersuchung* [In English: Internationalisation strategies in the German textil and clothing industry: An empirical investigation]. RWTH Aachen. https://publications.rwth-aachen.de/search?p=id:%22629%22

Schöggl, J.-P., Rusch, M., Stumpf, L., & Baumgartner, R. J. (2023). Implementation of digital technologies for a circular economy and sustainability management in the manufacturing sector. *Sustainable Production and Consumption, 35*, 401–420. https://doi.org/10.1016/j.spc.2022.11.012

Schumacher, K. A., & Forster, A. L. (2022). Textiles in a circular economy: An assessment of the current landscape, challenges, and opportunities in the United States. *Frontiers in Sustainability, 3*, Article 1038323. https://doi.org/10.3389/frsus.2022.1038323

Software AG. (2023). *How to invest in sustainability by investing in technology—The "Genius of the And". White Paper.* https://www.softwareag.com/en_corporate/resources/software-ag/wp/sustainability-digital-transformation.html?utm_source=google&utm_medium=cpc&utm_campaign=swag-brand_umbrella&utm_region=hq&utm_subcampaign=stg-1&utm_content=stg-1_whitepaper_how-to-invest-in-sustain-by-technology&gclid=CjwKCAjwq-WgBhBMEiwAzKSH6GYqWqGSReIiYDXyqEyz5efb2vAH-AmWtM5Tc2pK3gtmB8ntaaKe_RoCd0kQAvD_BwE

Stål, H. I., & Corvellec, H. (2018). A decoupling perspective on circular business model implementation: Illustrations from Swedish apparel. *Journal of Cleaner Production, 171*, 630–643. https://doi.org/10.1016/j.jclepro.2017.09.249

Stamm, A., Altenburg, T., Müngersdorff, M., Stoffel, T., & Vrolijk, K. (2019). *Soziale und ökologische Herausforderungen der globalen Textilwirtschaft* [In English: Social and ecological challenges of the global textile industry]. https://doi.org/10.23661/rep1.2019

Statistisches Bundesamt. (2024a). *Anzahl der Betriebe in der deutschen Textil- und Bekleidungsindustrie in den Jahren 2021 bis 2023* [In English: Number of companies in the German textile and clothing industry between 2021 and 2023]. https://de.statista.com/statistik/daten/studie/209701/umfrage/unternehmen-in-der-deutschen-bekleidungsindustrie-nach-beschaeftigtenzahl/

Statistisches Bundesamt. (2024b). *Import und Export von Textilien und Bekleidung in bzw. aus Deutschland* [In English: Import and export of textiles and clothing to and from Germany]. https://www-genesis.destatis.de/genesis/online

UN Environmental Programme. (2020). *Sustainability and circularity in the textile value chain—Global stocktaking.* https://wedocs.unep.org/handle/20.500.11822/34184;jsessionid=63D96871DBFEC9815E6F2E743A706A7C

Vellesalu, A., Chkanikova, O., Hjelmgren, D., & Salomonson, N. (2023). Institutional reconfiguration and value co-creation in circular product development: A service ecosystem perspective in the textile and apparel industry. *Journal of Cleaner Production, 414*, 137682. https://doi.org/10.1016/j.jclepro.2023.137682

Wiegand, T., & Wynn, M. (2023). Sustainability, the circular economy and digitalisation in the German textile and clothing industry. *Sustainability, 15*(11), 9111. https://doi.org/10.3390/su15119111

Wiegand, T., & Wynn, M. (2024). Circularity and Digitalisation in German Textile Manufacturing: Towards a Blueprint for Strategy Development and Implementation. *Processes, 12*(12), 2697. https://doi.org/10.3390/pr12122697

World Economic Forum. (2021). *Why digitalization is critical to creating a global circular economy.* https://www.weforum.org/agenda/2021/08/digitalization-critical-creating-global-circular-cconomy

Worldbank. (2019). *European Union textiles and clothing imports by country in US$ thousand 2019.* https://wits.worldbank.org/CountryProfile/en/Country/EUN/Year/2019/TradeFlow/Import/Partner/by-country/Product/50-63_TextCloth/Show/Partner%20Name;MPRT-TRD-VL;MPRT-PRDCT-SHR;AHS-WGHTD-AVRG;MFN-WGHTD-AVRG;/Sort/MPRT-TRD-VL/Chart/top10

Wynn, M., Ilyas, J., Isleyen, O. F., Brüntrup, H., & Metin, B. (2024). Reassessing critical success factors for ERP implementation in the digital era. *Digital Technologies Research and Applications, 3*(2), 140–154. https://doi.org/10.54963/dtra.v3i2.297

Wynn, M., & Jones, P. (2022). Digital technology deployment and the circular economy. *Sustainability, 14*(15), 9077. https://doi.org/10.3390/su14159077

Tina Wiegand (tina.wiegand@hof-university.de) is a lecturer at Hof University of Applied Sciences, Germany. She completed a dual degree in Business Administration with a focus on supply chain management and a Master's degree with a focus on project management. She worked in the textile industry for over ten years in project and process management, with managerial responsibilities in operational and strategic purchasing for different corporate brands. Additionally, she gained experience in significant systems projects. Currently, she is pursuing a Ph.D. degree at the University of Gloucestershire, UK, researching digitalisation and circularity in the textile supply chain.

Martin Wynn (mwynn@glos.ac.uk) is Associate Professor in Information Technology in the Business, Computing and Social Sciences at the University of Gloucestershire, UK and holds a Ph.D. from Nottingham Trent University. He was appointed Research Fellow at East London University, and he spent 20 years in industry at Glaxo Pharmaceuticals and HP Bulmer Drinks. His research interests include digitalisation, information systems, sustainability, project management, and urban planning. His latest book, Handbook of Research on Digital Transformation, Industry Use Cases, and the Impact of Disruptive Technologies, was published in 2022.

Chapter 11
Conclusion

Tina Wiegand and Martin Wynn

Abstract This concluding chapter identifies four key issues for each of the three change dimensions discussed in this book (sustainability, the CE and digital technology deployment), plus two further issues that apply to all three change dimensions—the impact of regulation and governance, and the significance of stakeholder influence. This provides the basis for the development of a maturity model specifying four levels of sustainability and circular practice, similar classifications for digital technology deployment, and illustrative profile definitions for companies in the T&C industry. These models may be of value in subsequent research studies that aim to advance the transition to circularity in the T&C industry.

This book offers a broad range of perspectives on the T&C industry across a number of European countries. It assesses and analyses various aspects of the industry, including domestic manufacturing companies (Denmark, Bulgaria, Portugal), large international textile companies and fashion brands (UK, Türkiye), small and medium-sized enterprises (Germany, France), the unique characteristics of specific T&C industrial regions (Italy), and the sustainability communication of major international brands with their consumers (Romania).

One overarching theme that emerges is the recent transformation of the European T&C industry in response to economic, environmental, and regulatory pressures. This transformation includes significant structural change, the shift from mass production and outsourcing towards more sustainable practices, and the adoption of new strategies and digital technologies. These changes are driven by the need to adapt to global challenges, such as environmental impact and the demand for ethical practices, and

T. Wiegand
Hof University of Applied Sciences, Hof, Germany
e-mail: tina.wiegand@hof-university.de

M. Wynn (✉)
University of Gloucestershire, Cheltenham, UK
e-mail: mwynn@glos.ac.uk

M. Wynn and T. Wiegand (eds.), *Sustainability, the Circular Economy and Digitalisation in the European Textile and Clothing Industry*, SDGs and Textiles,
https://doi.org/10.1007/978-981-97-9116-3_11

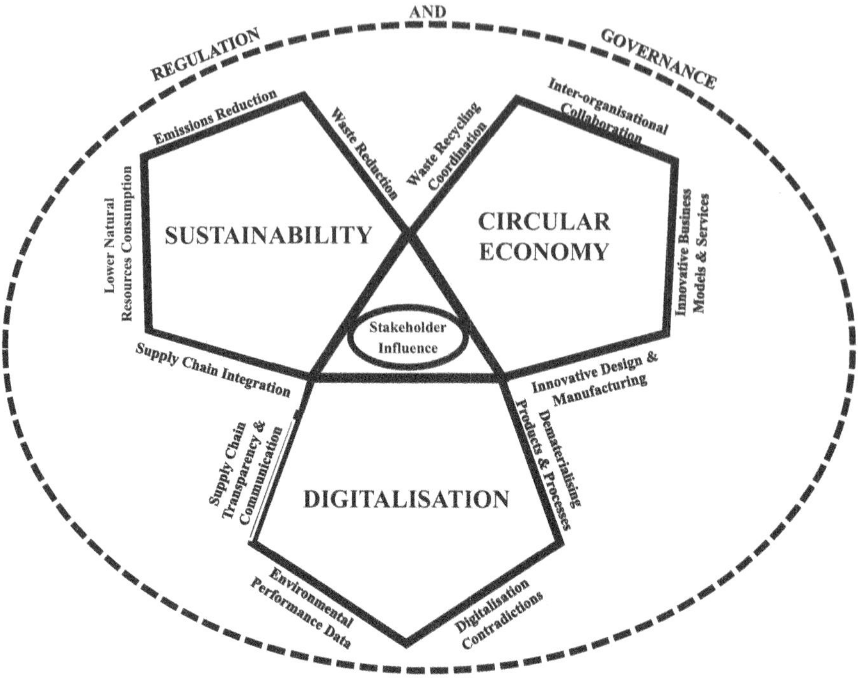

Fig. 11.1 Sustainability, circular economy, and digitalisation: key themes evidenced in country analyses

are in some cases facilitated by emerging opportunities, like the development of eco-industrial districts and the implementation of holistic, sustainable approaches within the industry. Each chapter focuses on a particular European country, addressing issues concerning sustainability and the circular economy (CE), as well as the role of digital technologies in this transition.

This conclusion provides a summary analysis and overview of sustainability strategies (Sect. 11.1), CE practices (Sect. 11.2), and digital technology deployment (Sect. 11.3) in these European countries, discussing similarities and differences, with reference to existing literature and research studies. For each of these three main themes (sustainability, the CE, and digitalisation), there were four key issues that were recurrent in the majority of the chapters. In addition, two further key issues—the impact of regulation and governance, and the significance of stakeholder influence—emerged as being of relevance to all three themes. This is illustrated in Fig. 11.1 and discussed in more detail below.

Based on this analysis, a framework is put forward (Sect. 11.4) for classifying and categorising organisations in the T&C industry, as regards their sustainability and CE practices, and level of digitalisation. Finally, Sect. 11.5 provides some concluding remarks, pointing out the limitations of the study and possible future research agendas.

11.1 The Transition to Sustainability

There are only a few years left to achieve the ambitious 17 UN Sustainable Development Goals. However, progress—especially concerning environmental objectives within the EU—has been limited, requiring increased effort and more stringent regulatory measures in the future if these targets are to be met (European Union, 2023).

Although the majority of the companies in the study indicated their awareness of the negative impacts of the T&C industry on the environment, climate, and natural resources, the development and implementation of appropriate strategies and practices vary significantly among companies and between countries. For example, relatively few companies explicitly reference the SDGs in their sustainability initiatives. This suggests a low level of strategic integration and alignment of corporate sustainability with wider policies and objectives.

Notably, large fast fashion brands, despite being identified by the EU as unsustainable due to their enormous resource consumption and waste production, largely continue to adhere to their current business models and practices. Some even oppose existing sustainability regulations instead of leveraging their size and influence to lead the industry towards a sustainable future (see the UK chapter, for example). Conversely, some small businesses and start-ups are capitalising on the growing niche of responsible consumers by offering innovative approaches, such as local supply chains, on-demand production, maintenance and repair services, as well as digital and transparent product information (see examples from Italy, Germany, Denmark, and France). These companies are thus already ahead of foreseeable regulatory measures in some regards. Overall, four key themes relating to sustainability were in evidence across various chapters and merit further discussion here.

Firstly, *the reduction of emissions* in the face of climate change is cited as the main focus of sustainable activities in nearly all the studied companies. However, the scope and strategic direction of measures and approaches differ considerably between companies. For instance, companies in France, the UK, Germany, and Denmark report on extensive efforts to measure and reduce emissions, often focusing on Scope 1 and Scope 2 emissions, which are directly controlled by the companies themselves through their own activities and energy consumption. Measures such as renewable energy, electric vehicles, alternative fuels, and reduced business travel are highlighted by companies of all sizes. However, fewer companies report actions to reduce Scope 3 emissions, which are outside their direct control, such as those from suppliers or logistics third-parties. Yet, these emissions constitute the majority of total emissions, as shown by various international studies (Sadowski et al., 2021; Stridsland et al., 2023). In contrast, companies in the Prato district in Italy and a quarter of the studied Turkish companies report conducting Life-Cycle Assessments (LCAs) to capture all emissions associated with the company or its products and to develop targeted reduction measures. This approach is significantly different from that pursued by the companies noted above. Bulgarian companies, however, appear to focus more on technical modernisation aimed at cost reductions, efficiency gains, and productivity increases. While these changes undoubtedly have sustainability benefits, they are more of a welcome side effect rather than the primary objective of these measures, suggesting that sustainability is not a strategic priority.

Secondly, nearly all companies aim to *lower the consumption of natural resources*. Key measures include reducing water consumption in production processes through more efficient dyeing techniques, and water treatment and reuse. At product level, regardless of company size or location, there is a push towards using recycled materials or alternatives to oil-based materials to achieve more sustainable products. A noteworthy example is the Prato industrial area in Italy, where wastewater treatment is collaboratively managed by and for all local companies in a shared treatment plant, also producing additional sustainability benefits such as energy and other resource savings. Given the regional clustering of the T&C industry in Germany, the development of similar districts, where circular textile clusters could emerge, could enhance the industry's competitiveness on a global scale.

Thirdly, *waste reduction* is a common theme across all the companies studied. The range of measures is broad. Some companies highlight compliance with legal requirements concerning hazardous chemicals (as mentioned in the Türkiye and Bulgaria chapters), while Central European manufacturing companies pursue innovative approaches, such as reusable packaging, on-demand production through digital printing and knitting (as mentioned in the Denmark, Germany and Italy chapters), substituting chemical dyes with natural ones (noted in the France, Germany and Denmark chapters), or recycling and reusing textile waste (as mentioned in the Italy chapter).

Fourthly, *the integration of sustainability measures into supply chains* is addressed by nearly all companies, via a wide range of activities and initiatives. Whether for large clothing brands or small and medium-sized manufacturing companies, suppliers play a crucial role in ensuring and maintaining sustainability measures. This requires knowledge and transparency regarding the actions of upstream suppliers (as mentioned in the Germany and France chapters), and a strategic risk management approach to support informed and appropriate decision-making (as observed in the Germany and Türkiye chapters). An awareness of downstream retail operations (as illustrated in the UK and Romania chapters) should also contribute to the overall sustainability strategy.

The types of measures implemented across the extended supply chain can be categorised as environmental and/or social. Environmental measures include sourcing sustainable materials and raw materials, as evidenced by certifications (e.g., GOTS, FSC, and others), as well as efficient water use and purity assessment (as noted in the UK, Italy, Germany, Denmark, Portugal, and Bulgaria chapters). Social measures include concern for responsible procurement, which is discussed in all chapters. A Code of Conduct, which outlines minimum requirements such as human and labour rights as a contractual basis for business relationships with suppliers, is frequently raised as a requirement, regardless of company size or location. A notable observation is the difference between the expectations placed on suppliers—such as adherence to human rights, fair working conditions, payment of living wages, and compensated overtime—and the working conditions within the company headquarters in Europe, which aim to provide comfortable working environments, sports facilities, flexible working hours, and similar benefits for their employees. This reflects the reality that many companies aim to maintain lower labour costs in their supply chain, as compared with their domestic manufacturing operations.

Just using local supply chains is an interesting aspect of this debate. Examples from small companies in France, Italy, and Denmark show that focusing on local value chains can mitigate many of the discussed risks and sustainability challenges noted above. The vertical integration of multiple production stages within a single company and local sourcing allows for production on demand and quick response to change, reduced transport needs, and adherence to environmental and labour standards, that often results in high-quality products, which, however, tend to be more expensive than mass-produced goods from low-wage countries. As a result, these companies often serve a niche market or operate in the luxury segment.

11.2 Circular Economy Strategies and Activities

The CE is seen as a "system which necessitates a paradigm shift to replace the 'end of life' concept with reusing, recycling and recovering materials throughout the supply chain" (Kirchherr et al., 2023, p. 4). Despite increasing efforts and regulations regarding the reduction of resource consumption and waste generation within and outside the EU, this theme plays a minor role in the sustainability reporting of many companies (as outlined in France, Denmark, Bulgaria, UK, Türkiye, and Portugal chapters).

The CE is based on the relatively simple principle that products and materials should be maintained at their highest value and in the best condition for as long as possible to reduce the extraction and processing of new resources, and avoid the negative impact of waste treatment. Principles aimed at reducing, intensifying, and prolonging product use (Refuse, Reduce, Repair) are therefore prioritised over strategies for reusing (Remanufacture, Repurpose) and recycling materials.

According to Uhrenholt et al. (2022), companies can be divided into two types concerning CE activities: those that were founded on circular principles and those that were not, and thus must undergo a transition to the adoption of circular practices. The authors refer to these as "born-circular" companies and "legacy organisations", each with different conditions and transformation paths. This distinction can be observed among the companies studied in this book. There are examples from Italy, Denmark, and France of start-ups that have built their entire organisation, products, services, and processes around circular and sustainable practices, and these are now just as much a part of the current T&C landscape as incumbents of various sizes. Many of these older legacy enterprises are simultaneously driving changes and innovations in the dual and interrelated transition towards digitalisation and sustainability, while maintaining and improving their core business (see examples from Germany, France, the UK, Türkiye, and Denmark). This simultaneous pursuit of exploration and exploitation strategies is discussed in the literature under the concept of "ambidexterity", which requires specific skills, management capabilities, and resources due to parallel structures and processes (Santa-Maria et al., 2022). Overall, this constitutes a fragmented landscape concerning the pursuit of CE principles in European companies. However, four overarching themes concerning the CE, discussed in the individual chapters, can be identified and are summarised here.

Firstly, *innovations in product design and manufacturing processes* are being advanced across the board, though there are significant differences in the rate of progress. For instance, a focus on high-quality materials and fibres "that age with beauty" (as one Danish company states) should provide product durability and longevity. Even large fast fashion brands mention durability and longevity in the context of their sustainability reporting. However, it remains unclear how these concepts align with the fast fashion business model. Moreover, there is no direct connection made between these claims and concrete actions such as resale, repair, or take-back initiatives. This disconnect suggests that these statements may be more of a marketing exercise than a genuine commitment to sustainability. At the same time, in other countries such as Türkiye, there is no mention of durability approaches.

Examples of innovative materials include water-dissolvable sewing threads (UK), recycled ocean plastics for swimwear (France), and recycled raw materials from other industries' waste streams (Denmark). Notably, the use of recycled materials for products varies widely in how and in what form they appear. Some of the older legacy organisations maintain conventional product lines alongside circular lines that partially use recycled materials (see examples from the UK, Germany, Bulgaria, Türkiye, and France), while examples from born-circular companies from France and Italy show a more predominant use of recycled materials in their products. While secondary raw materials may have a better environmental footprint compared to primary materials (Liu et al., 2020; Wiedemann et al., 2022), the circular impact of using these materials remains limited if they are not integrated into a broader context that includes product design (e.g., recyclability, durability) and corresponding take-back and recycling initiatives (Laitala et al., 2018). As findings from a Scandinavian study demonstrate, the development and use of new materials, in conjunction with appropriate design and the collaboration of stakeholders along the supply chain, are key components in the creation of closed-loop textile systems (Sandvik & Stubbs, 2019).

The use of mono-materials—being easier to recycle than products made from multiple materials because they can be sorted into their constituent parts—and certifications such as GOTS, Oeko-Tex, Cradle-to-Cradle, RCS, and GRS for transparency and communication, are evident in Germany, Denmark, Romania, Portugal and Bulgaria. The European Textile and Apparel Confederation (EURATEX) also observed a steady increase in the issuance of certifications for environmentally and socially responsible raw materials, products, and companies (EURATEX, 2024). In this context, a relatively small study conducted among Czech companies (Krause, 2018) revealed that certified companies, on average, achieved a higher return on investment compared to non-certified ones, suggesting that sustainable and circular practices could add to profitability and constitute a competitive advantage.

In terms of packaging, there are efforts to reduce plastics and incentivise recycling, with some companies offering discounts for returning packaging, notably in Italy. The shift towards natural, cellulose- or protein-based materials instead of oil-based ones is also observed in Denmark, Germany, Italy, France, and the UK. In production and manufacturing, various companies are adopting innovative and sustainable practices that reflect differing levels of commitment to CE principles. In Denmark, for instance, companies are employing traditional hand-dyeing techniques using natural, plant-based dyes, emphasising a return to more environmentally friendly methods. Similarly, in Italy, companies are utilising patented processes to eliminate chemicals and dyes in wool spinning, further reducing the environmental impact of their production processes. A German example of shifting to digital printing over conventional rotary printing illustrates a concerted effort to minimise chemical usage and waste. Additionally, there is a notable trend towards manufacturing in moderation, where companies prioritise handcraft processes that focus on quality rather than mass production. This approach not only enhances the longevity and durability of products but also aligns with the broader objectives of the CE by reducing overproduction and waste. However, the situation in Bulgaria highlights the challenges faced by some companies in fully embracing CE strategies. Despite strict testing of materials for harmful substances, the continued production of mixed fibres and fabrics indicates a lack of awareness of CE issues and the absence of a holistic strategy. This contrast underscores the varying degrees of progress within the industry and the need for more comprehensive approaches to adopting circular practices and achieving sustainability in general.

Secondly, *innovative business models and services* such as resale, rental, and repair are gaining traction. In the UK, a large company (Marks & Spencer) joined a kidswear peer-to-peer marketplace, operating alongside conventional sales models, though profitability remains unclear. Similarly, in Germany, a rental model for sports goods has been reported, though it has yet to achieve profitability. In Romania, where there is high poverty risk and a lack of separate collection systems, there is significant potential for new business models based on take-back, resale, and repair. However, large retailers seem to be slow to offer these services, with only H&M providing vouchers for used clothing. Loyalty programs aimed at promoting CE activities are underutilised, though they hold great potential.

The degree to which the product use phase is extended through repair services varies between countries. In the Romanian chapter, the inclusion of repair kits or spare parts with products is reported. Although this is a widely known practice, without a clear CE strategy, this often leads to additional waste rather than a genuine extension of product life. Luxury brands like Jimmy Choo (UK) offer repair services for up to five years, though it is unclear how widely these services are used. In contrast, lifelong repair services are available in Germany and France, and free repair and recycling options are offered in Italy. This illustrates how integrating circular product design with corresponding processes can enhance consumer value and create a competitive advantage. By ensuring that products are designed for durability and repairability, and by providing the necessary infrastructure to support these processes, companies can distinguish themselves in a market increasingly focused on sustainability.

However, repair, care, and maintenance guidance for consumers seems mainly relevant for the marketing of brands, with manufacturers generally overlooking the potential benefits. This represents an untapped opportunity for manufacturers to engage directly with the market by sharing material knowledge and repair skills or collaborating with repair communities to supply spare parts. Studies have shown that consumers prefer to purchase new items rather than repair existing ones, with the primary barriers to repair in developed societies being a lack of skills and knowledge among consumers (Fachbach et al., 2022; Laitala et al., 2021). The absence of additional support mechanisms, such as online communities, instructional videos, and dedicated repair services, further diminishes the likelihood of consumers choosing to repair their products. Equally, there is a notable lack of information from companies regarding initiatives that could encourage repair practices, indicating a missed opportunity to foster a more repair-friendly culture and contribute to the CE. Companies could invest in educating consumers, allowing them to participate and contribute more meaningfully in reducing environmental impacts, as was concluded in the chapter on Portugal. An example from Denmark, where small embroidery shops offer repair services, demonstrates the significant potential for collaboration between brands and locally distributed small providers, which could lead to a more robust and accessible repair infrastructure.

In manufacturing companies, there are contrasting approaches to extending the use phase of technology and machinery, presenting a contradictory picture. On the one hand, there is a focus in some companies on maintaining existing machinery instead of replacing it, emphasising longevity and sustainability in equipment use (for example in Denmark). On the other hand, some companies favour modernising or digitalising machinery to enhance efficiency and productivity (for example, in Bulgaria and Germany). This contrast reflects the different ways in which companies manage their technology and machine resources, in striking the right balance between extending the life of current equipment and investing in new technologies for improved performance.

Thirdly, the *coordination of pre- and post-consumer waste recycling* (closing the loop through managed recycling of waste materials) is in evidence, but exhibits significant regional variation. Some companies are embracing circular thinking by designing and producing mono-materials or easily disassembled products and providing comprehensive take-back schemes, some of which are blockchain-based for traceability (as seen in Denmark and Italy). These companies are actively integrating circular strategies into their operations. Conversely, other companies offer collection or take-back bins without further detailing any subsequent recycling activities, reflecting a more superficial commitment to circularity (as observed in examples from the UK and Romania). A notable example is an Italian denim fabric producer that has initiated its own collection and recycling activities for old textiles, demonstrating how companies can vertically integrate their supply chains to secure a consistent supply of high-quality secondary raw materials.

Fourthly, *collaboration among various actors within and outside of existing networks* is essential for advancing CE strategies. Recent studies highlight the positive impact of inter-organisational cooperation and collaboration on the design and success of circular products and processes (e.g., Schöggl et al., 2024; Suchek & Franco, 2024). For example, in Italy, an industrial district pools the strengths of various actors to jointly process waste, purify water, and engage in collaborative projects, such as working with art institutes to produce easy-to-repair garments and educational programs for future textile professionals and consumers. Gucci's open-innovation platform for CE transformation is a potential best practice that could be scaled by larger brands to have a broader impact. In Portugal, 56 business and non-business entities collaborate in a consortium to seek traceability and transparency, eco-design, eco-engineering, and the development of new sustainability metrics. Meanwhile, in Germany and Türkiye, joint projects with NGOs in textile-to-textile recycling are leading to larger quantities and faster scaling, and in Türkiye, partnerships for the resale of used jeans demonstrate real-world examples for the potential of collaborative efforts in the CE domain.

11.3 Digitalisation in Support of Sustainability and the CE

Digital technology deployment is increasingly seen as a critical enabler and supporter in the context of CE transformation (Piscicelli, 2023), as it offers significant potential for enhancing sustainability and circularity at various levels of industrial processes. As Luoma et al. (2022, p. 13) conclude from their study with European, U.S., and South African T&C experts, "the more ambitious the circularity objectives are, the more vital the role of data becomes".

Digital technologies allow products and materials to be monitored across multiple life cycles, thereby enhancing data availability and quality to engender improved product and supply chain management and sustainability decisions. Simultaneously, well-managed and effective digital technology deployment can help to build trust and transparency along value chains, and support new sustainable business models like product-service systems (Bressanelli et al., 2022; Wynn & Jones, 2022). A recent study by various European institutions conceptualises the integration of circular design approaches and digital technologies like body scans, e-libraries, and digital twins, demonstrating that a fully dematerialised product development for clothing is possible (Donmezer et al., 2023).

However, the reporting of the companies examined in this study provides little insight into the success or failure of the use of digital technologies in the support of sustainability and the CE. It may be, however, that companies are indeed utilising digital technologies but do not directly link them to their sustainability and circularity initiatives, or they may simply perceive no benefit in mentioning these technologies in their public communications. A recent study by Tolentino-Zondervan and DiVito (2024), who analysed the published sustainability information of 94 Dutch T&C companies, reached a similar conclusion, noting the minimal reporting on the use of digital technologies and offering comparable reflections on why these technologies are not prominently highlighted in sustainability reporting.

The findings on the degree and purpose of digitalisation among the companies covered in this study present an incomplete picture, but nevertheless, four overarching themes emerge from the cases studied here.

Firstly, *digital technologies are increasingly being utilised for environmental performance data* (generation, collection, processing, analysis, and dissemination). For instance, companies are using IoT and analytics technologies for gathering data related to their corporate carbon footprint (ccf) or product carbon footprint (pcf), which serves as the basis for further actions. Knowing exactly where and what negative impacts occur is an essential starting point for deriving goals, implementing effective measures, and tracking progress. The Denmark chapter, for example, points to a case where smart meters are being employed to enhance data accuracy.

This is also of relevance in the context of forthcoming legislative requirements, such as those concerning the Digital Product Passport (DPP). The EU Strategy for Sustainable and Circular Textiles (European Commission, 2022b) has introduced the DPP as a key instrument for tracking and sharing information between stakeholders. The DPP, linked to individual textile products, manages information required by regulations like the Eco-design Regulation for Sustainable Products (ESPR) and Extended Producer Responsibility (EPR) (European Commission, 2022b). These requirements encompass product identification, compliance documentation, and recycling information. As the DPP is set to become mandatory for all textile products sold in the EU, it plays a pivotal role in the EU's mission for a sustainable textile industry, placing significant demands on companies to provide comprehensive digital product information. For example, a Danish company has implemented administrative software to ensure extensive and accurate product reporting. In Türkiye, a company utilises a "Green-Map-Tool" for preliminary analysis in collaboration with manufacturers and suppliers during product development. Similarly, an Italian company has partnered with "BCome" (a sustainability platform) to integrate traceability information and transparent data regarding the environmental impact of its best-selling products via QR codes on product labels.

Secondly, *digital technologies are being employed to enhance supply chain transparency and communication.* Activities range from simple tools like digital meetings to reduce business travel (as seen in Germany and Türkiye), to software tools that facilitate real-time information sharing on quality issues (Germany, Italy), and to more complex traceability platforms that track product, material, or chemical data across the value chain (Germany, UK, Italy, and Türkiye). These platforms help ensure compliance with environmental and social standards and prepare for the requirements of the DPP. Of relevance here is the case study of an Austrian fibre manufacturer, reported elsewhere (Ahmed & MacCarthy, 2021), which details the immense complexity and challenges associated with making entire supply chains transparent.

Blockchain technology will likely play a major role in providing secure and transparent data. Many companies in this study have embarked upon initiatives to increase supply chain transparency and traceability. In this context, digital libraries or material databases like the Mindful Materials or the Higg Index are mentioned, which comprehensively include product and material information across value chain stages, although this remains a distant goal for many companies (as noted, for example, in the Denmark and Germany chapters).

Social media and other interactive tools also play a significant role in cross-supply chain communication and information exchange. There is a generational issue here—for example, a study among Italian students revealed that younger generations are increasingly attentive to sustainability in textile consumption (Gazzola et al., 2020). Social media and apps offer relatively simple but effective means to inform consumers and involve them in circular practices like textile care, repair, sustainable consumption, resale, or return, as evidenced in a Portuguese research project (Cruz & Da Rosado Cruz, 2023). However, examples from the Romanian chapter show that major fashion brands rarely use these channels for such purposes, suggesting a lack of awareness or genuine interest in transitioning from the linear model.

Thirdly, *digital technologies are instrumental in dematerialising products and processes and enhancing overall efficiencies.* Examples include digital printing as an alternative to conventional wet printing processes (see Germany and Denmark chapters) and laser-based denim bleaching and engraving as alternatives to chemical-based methods (as in Türkiye, for example). These technologies replace water- and chemical-intensive processes, contributing to more sustainable production methods.

Additionally, 3D visualisation and prototyping (as evident in Germany) or fully digitalised samples (as in Türkiye) demonstrate how digital tools can reduce time, waste, and resources while providing advantages in creating, visualising, and evaluating products during the early stages of product development. These benefits align closely with CE principles (Spahiu et al., 2021).

Textile recycling processes have also benefitted from automated sorting systems based on IoT and AI, as described in the Italian and Danish chapters, significantly enhancing sorting quality and, consequently, the recycling process and its output. A recent McKinsey study on textile recycling in Europe concluded that "increasing collection rates and improving the quality and efficiency of sorting and pre-processing will likely be key to scaling textile recycling in Europe" (McKinsey, 2022, p. 36).

Fourthly, *there are clear contradictions in how digitalisation is impacting sustainability.* For example, the increase in online shopping presents a double-edged sword: while it leads to resource savings through reduced inventory, logistics, and storage space, it also contributes to negative environmental impacts through increased energy consumption (as highlighted in the UK chapter). Similarly, the shift to digital production processes results in higher energy demands (as discussed in the Danish chapter), which should be considered within comprehensive sustainability strategies by also prioritising renewable energy sources. Additionally, the dependence on a few tech companies due to increased digitalisation, and the need to ensure data and system security, are identified as potential risks (as examined in the UK chapter). A recent incident where a Microsoft update led to global software outages and disruptions confirms the relevance of these critical perspectives (Nelson & Kaye, 2024), and the significance of corporate digital responsibility and addressing cybersecurity issues will only increase as digitalisation impacts more widely and more deeply in the corporate environment (Wynn & Jones, 2023).

Finally, two key issues for the future that will affect all three main themes discussed above are the role of stakeholders (and particularly the end consumer), and the growing impact of regulation and corporate governance. Firstly, consumer influence is scarcely discussed in most chapters, despite being inherently linked to the textile supply chain and its sustainable transformation. However, as noted in the UK chapter, the tensions between sustainable and economic development highlight the growing significance of consumer knowledge and consumer choice. Only when consumers have easy access to a wider range of sustainable products and associated information are these products likely to gain real traction in the market. The findings from Romania, where the communication of sustainability from a consumer perspective was examined, reach a similar conclusion. All the large fashion and sports brands studied focus on selling new products, but sustainable alternatives, sustainable consumption, usage, and disposal concepts are either missing or not being communicated effectively to consumers; and a study conducted among Polish consumers (Popowska & Sinkiewicz, 2021) found that sustainability is rarely considered in clothing purchase decisions. In contrast, a regularly conducted survey of German consumers (Greenpeace, 2022) concluded that awareness of sustainable fashion practices has significantly increased in recent years. This underlines the diverse nature of consumer knowledge and awareness in Europe and highlights the importance of effectively communicating with consumers as regards the impact of their consumption habits and the availability of alternative, more sustainable options.

In addition to consumers, other stakeholders also play a key role, as illustrated in the Italian, Portuguese, and German chapters. Existing and future employees must be able and willing to support and shape the changes required for sustainable transformation. This demands appropriate knowledge and new skills. As shown in the Italy chapter, companies can actively contribute to creating and disseminating this knowledge and the necessary skills by collaborating with schools, universities, training institutes, and the broader public to define and implement new standards. An open and innovative corporate culture, along with a willingness to experiment, are derived as critical success factors for a successful sustainable and digital transition in the German chapter. A recent study on the relationship between digitalisation and the CE (Neri et al., 2024) reaches a similar conclusion, emphasising that stakeholder involvement, corporate culture, education, training, and management commitment strongly influence the success of the digital and sustainable transition.

Secondly, as regards regulation and corporate governance, the EU is pushing forward with a comprehensive set of legislative proposals under the Green Deal (European Commission, 2019), specifically targeting the T&C industry to foster sustainability, circularity, and digitalisation. Central to these efforts are the EU Strategy for Sustainable and Circular Textiles (European Commission, 2022b) and the European Industrial Strategy (European Commission, 2020) among others. These will introduce various laws and regulations, which cannot all be exhaustively covered here. However, some key initiatives will be briefly discussed below.

The revised Waste Framework Directive (European Commission, 2023g) requires EU member states to implement a separate collection of textiles by 2025. This directive emphasises developing new capacities and infrastructure for textile collection, sorting, reuse, and recycling, with a focus on advancing sorting and recycling technologies. A significant component of this initiative is the EPR scheme, which will obligate textile companies to take responsibility for the entire lifecycle of their products. The EPR scheme aims to reduce textile waste by promoting resale and recycling over disposal, incineration, or export. A few EU member states have already introduced EPR schemes, such as France (European Environment Agency, 2024). By holding manufacturers accountable for the environmental impacts of their products, this regulation encourages a shift towards more sustainable and circular practices throughout the textile value chain.

Another key piece of legislation is the ESPR (European Commission, 2024c), which will set eco-design requirements for various product groups to enhance their recyclability, energy efficiency, and overall environmental sustainability. This regulation also includes the introduction of a DPP, which will provide consumers with accessible information about a product's sustainability attributes, such as durability, reparability, recycled content, and availability of spare parts. The DPP will apply not only to manufacturers but also to importers, distributors, and retailers, aiming to empower consumers to make informed decisions and increase transparency regarding the environmental impact of products throughout their lifecycle.

In addition, the EU Taxonomy Regulation (European Commission, 2023c) sets criteria to facilitate sustainable investments by integrating environmental objectives into financial risk management. This regulation is designed to make sustainability a central criterion in the financial sector, guiding investment decisions towards more sustainable projects. Furthermore, the Corporate Sustainability Reporting Directive (CSRD) (European Commission, 2023a) will mandate sustainability reporting for a wide range of companies, and the Corporate Sustainability Due Diligence Directive (CSDDD) (European Commission, 2024a) will enhance due diligence requirements throughout supply chains to ensure ethical and sustainable practices, whilst the Regulation on Deforestation-Free Supply Chains (European Commission, 2023f) will require companies to confirm that their products do not contribute to deforestation or forest degradation worldwide. All these regulations aim to increase transparency and accountability regarding companies' environmental and social impacts.

At the same time, new laws and regulations emerge from the EU's new Industrial Strategy (European Commission, 2020) and Data Strategy (European Commission, 2023e), aimed at supporting the green and digital transition (the "twin transition") of EU industries, which will also impact the T&C industry. Examples include the Regulation on European Data and Governance (Data Governance Act and Data Act) (European Commission, 2023d, 2024b), which regulates access to and user rights for data, while the General Data Protection Regulation (GDPR) (European Commission, 2022a) aims at ensuring the protection of personal data and the EU AI Act aims to regulate the use of artificial intelligence (European Commission, 2023b). These legislative measures highlight the EU's commitment to fostering sustainable development and responsible business practices across the T&C industry, establishing rigorous standards, accountability mechanisms, and data regulations that companies must adhere to in the coming years.

11.4 Towards a Framework for Company and Country Profiling

Even though the case studies and country analyses offered in this book reveal regional particularities and differences in individual company strategies, they do suggest, as noted above, general trends regarding the adoption of sustainable and circular practices as well as digital technology deployment. In assessing the industry's evolution and current status, a number of categories can be identified based on a matrix, with the adoption of sustainability strategy and CE practices forming one axis and digital technology deployment on the other.

Building upon evidence from the country accounts, and using the Organisation-Process–Product dimensions of the model for T&C transitioning referenced in the Introduction chapter (as well as in Chaps. 9 and 10), levels of maturity along both these axes can be discerned. As regards sustainability and CE adoption, four levels of adoption are detailed in Table 11.1, these being Static, Basic, Intermediate, and Advanced. Other labels could have been used but these adequately describe the general profiles, and also build upon other maturity model labelling (Aguiar et al., 2018; Irizar & Wynn, 2018).

In terms of digital technology deployment, four levels of maturity can be identified—Rudimentary, Operational, Integrated, and Transformed (Table 11.2). These classifications build upon those found in other maturity models (e.g., Costa et al., 2022), and apply to digitalisation in the industry in general.

Within this matrix of sustainability and CE activity on the one axis, and digital technology adoption on the other, four company profiles can be identified. These can be illustrated with case examples drawn from the country chapters (Fig. 11.2).

Laggards neither pursue advanced and holistic sustainability strategies and CE practices nor exhibit any significant degree of digitalisation. These companies tend to provide general, sometimes opaque or unverifiable information in their sustainability reports or refer to future-oriented objectives. Their digitalisation efforts suggest isolated measures aimed more at intra-organisational efficiency than inter-organisational cooperation and collaboration or the pursuit of sustainability strategies. These companies are lagging behind general developments, remain stuck in their linear business models, and are inadequately prepared for upcoming regulatory requirements like the DPP or reporting obligations under CSRD and CSDDD, as well as changing consumer patterns. This group is exemplified by some case examples from **Bulgaria, such as Company C18** where sustainability hardly seems to be established as a strategic business field and the implementation of digital technologies primarily serves to increase operational efficiency and productivity.

In contrast, at the other end of the scale, are the **Transformers**, companies that have clear sustainability strategies aligned with widely recognised sustainability goals, such as the SDGs, and implement comprehensive measures and practices to minimise resource use and waste generation in line with CE principles. They also collaborate with relevant stakeholders and report transparently and verifiably on their progress towards these goals. Additionally, they pursue a digitalisation strategy

Table 11.1 Levels of sustainability strategy and CE practice adoption

Company level of sustainability and CE adoption	Definition		
	Organisation	Product	Process
Static	Lack of overarching awareness of sustainability and CE principles. Sustainability and CE are not integrated into the strategy or structure No targeted development of sustainability and CE skills in the organisation The business model remains predominantly linear Efforts to reduce resource consumption or waste/emission generation are driven largely by cost-saving and marketing motives rather than genuine environmental concerns	Materials and products are not designed with sustainability or circularity in mind No transparency or traceability of product-related environmental and social impacts No product or material-related repair or recycling activities	The focus is primarily on compliance with environmental and social regulations, with no significant efforts to reduce energy, water, materials, or chemicals consumption within internal and/or supply chain processes No CE-related cooperation or collaboration with external partners

(continued)

Table 11.1 (continued)

Company level of sustainability and CE adoption	Definition		
	Organisation	Product	Process
Basic	Emerging awareness of sustainability and CE principles Simple sustainability and CE activities are initiated sporadically and without a strategic framework. E.g., certain SDGs are pursued, but without linking to corporate objectives Organisational unit for sustainability/CE being developed with the relevant knowledge and skills. Efforts to reduce emissions and waste are evident, but they operate within the confines of a linear business model Data and fact-based communication of sustainability and CE activities to strengthen the company's image	Pilot projects to explore alternative products or services. But activities are limited and often isolated, e.g., the use of more sustainable materials and packaging or sustainability certification for individual materials, products, or product lines The emphasis remains on incremental improvements rather than a shift towards full circularity	The reduction of material, energy, water, and chemicals use in internal processes (e.g., through ISO management certifications) and the minimisation of scope 1 and 2 emissions are being pursued but no CE-related cooperation or collaboration with external partners

(continued)

Table 11.1 (continued)

Company level of sustainability and CE adoption	Definition		
	Organisation	Product	Process
Intermediate	Full awareness of sustainability and CE principles with a strategic alignment and specific objectives based on the SDGs, although implementation is still in its infancy Exploratory monitoring of sustainability and CE indicators and derivation of measures Organisational unit for sustainability/CE established, building knowledge and skills across the organisation Efforts to reduce emissions and waste are evident, but they operate within the confines of a linear business model Data and fact-based public communication of sustainability and CE activities to strengthen the company's image	Materials and products are designed with a focus on product and material durability, upgradability, and eco-design. Products, materials, and services are developed with the intent of closing, slowing, or regenerating resource loops. Product and materials traceability is implemented and transparent life cycle assessments or certifications for all materials, products, and services are provided (e.g., GRS, Cradle2Cradle, GOTS, Green Bottom)	The reduction of material, energy, water, and chemicals consumption in supply chain processes and the minimisation of scope 1, 2, and 3 emissions are actively pursued. Joint CE-related projects and innovations with external partners (e.g., for reverse logistics, recycling, or repair/reuse activities)

(continued)

Table 11.1 (continued)

Company level of sustainability and CE adoption	Definition		
	Organisation	Product	Process
Advanced	Sustainability and CE principles are fully embedded in both organisational strategy and operations. Systematic monitoring of sustainability and CE indicators and derivation of measures The business model is oriented towards value creation through maximising resource use and minimising waste (e.g., through product-service-systems or sharing/renting models) Sustainability and CE knowledge and skills are rooted throughout the company and are constantly being developed strategically through training and education for both internal and external partners Data- and fact-based, transparent public communication of sustainability and CE activities, as well as targeted information and involvement of stakeholders (e.g., consumers) on life cycle extension and recovery	Materials and products are fully designed for longevity, multiple uses, and regenerative recycling Fully transparency and traceability of product-related environmental and social impacts through LCA or certifications for all materials, products, and services are provided Product or material-related repair, refurbish, or remanufacture activities	All processes along the value chain are aligned with CE principles and focus on minimising resources and waste Production on demand is being pursued Optimisation and localisation of value and logistics processes in cooperation with all relevant partners, with the aim of closed material cycles Joint CE-related innovation projects with external partners for integrated, efficient, and customer-centred product and process development

Table 11.2 Levels of digital technology adoption

Company level of digital technology adoption	Definition
Rudimentary	Digitalisation occurs on an ad hoc basis, without strategic objectives. They are used internally within the company or individual processes. There is no overarching digital strategy. Technologies deployed would typically be the early digital techs—social media and mobile apps, with limited use of the Cloud to host some of the company's systems
Operational	Digital technologies are deployed function by function for process improvements and efficiency gains. Change is incremental and the business model is not radically altered. IoT is typically used on the shop floor and elsewhere for data capture, and analytics applications are used for data analysis and reporting
Integrated	The use of digital technologies is aligned with overall business strategy goals. Digitalisation is addressed at corporate level either as part of the IT strategy or as a separate digital strategy. Integration of digital technologies with existing information systems is prioritised. Data quality and cybersecurity are recognised as key corporate issues allied to digitalisation. AI, Robotics, digital twin, and digital fabrication may all be used in various process areas
Transformed	Digital technologies are seen as a key lever to enable significant business change, possibly involving transitioning to a new business model. Digital technologies may now be deployed within the company's products and services, as well as supporting internal processes. Digitalisation follows a cross-company and cross-process approach, and supports inter-company transactions and communication across the extended supply chain. Advanced technology deployment, including blockchain and digital fabrication, allows participation in digital platforms and ecosystems

aligned with their sustainability objectives, using digital technologies purposefully as enablers and supporters of these goals. Among the companies analysed in this book, one of the **Italian cases, Company C** serves as a representative example of this group.

Two other groups can be identified. **Digital Leaders** have a strong focus on digitalisation, emphasising efficiency, and productivity gains. While sustainability benefits may occur as a by-product, these companies lack a comprehensive sustainability and CE strategy or, at best, have a highly fragmented one. It can be conjectured that this group is relatively small, because digitally advanced companies, at the enterprise stage, are unlikely not to have, at the very least, sustainability strategies in place. A study conducted as part of the European Monitor of Industrial Ecosystems (EMI) also finds that "SMEs in the textiles industrial ecosystem have been less engaged in digital technologies than in the green transition" (Izsak & Moreno, 2024, p. 8). Among the companies analysed in this book, one of the **German cases, the small company D** could serve as an example for this group.

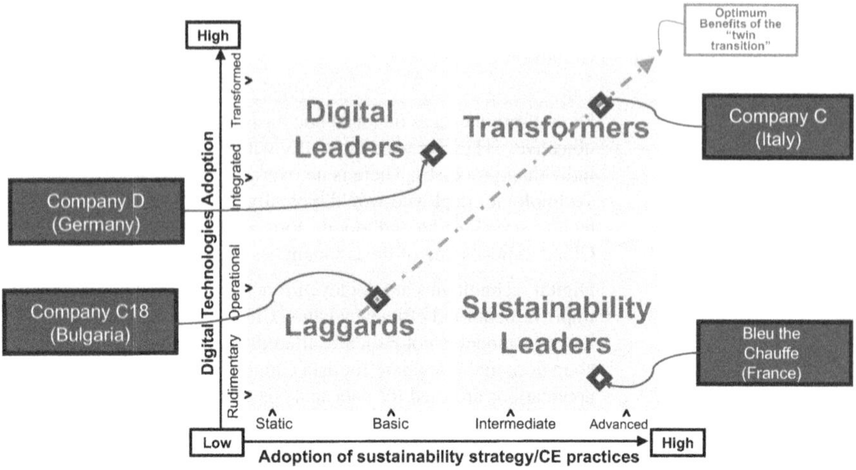

Fig. 11.2 DT/sustainability profiles for T&C companies: laggards, transformers, digital leaders and sustainability leaders

Sustainability Leaders are companies that have clear sustainability goals and extensive, holistic CE practices, which they report on comprehensively, transparently, and verifiably. However, in terms of digitalisation, they follow a strategy that deviates from the general trend by emphasising craftsmanship, individualisation, and direct customer interaction over innovation, automation, and collaboration. While they do use digital technologies, such as online shops, websites, or social media channels, these are more focused on communication rather than supporting operational processes. Among the companies analysed in this book, some micro companies, such as **Bleu the Chauffe** and others from **Denmark and France** are representative of this group.

Some cases from the Italy, Germany, and Türkiye chapters illustrate the significance of integrating both sustainability and digital strategies to achieve what the EU terms a "twin transition". This encompasses the combination and coordination of the green transition, driven by the need to reach the aims of climate neutrality and sustainability" and "the digital transition, [which] is an ongoing process of technology-driven change" (Muench et al., 2022, p. 9). These companies have clear sustainability strategies and use digital technologies with positive sustainability effects but the strategic integration of these two aspects is only partially evident, limiting progress towards **Transformer** profile and advanced twin transition.

11.5 Concluding Remarks

This study clearly has its limitations. Although the chapters in this book provide a considerable amount of detail, they cannot fully capture all the regional differences and specificities of the European T&C industry, not least because only nine European countries have been researched here. While general trends regarding the progress of sustainable and circular practices, as well as the adoption of digital technologies, can be discerned, the diversity in company size, value chain stages, and regional contexts limit the generalisability of the findings. Furthermore, the majority of analyses presented in this book are based on publicly available information, which inherently limits the depth and breadth of insights, particularly concerning digital technology deployment within European T&C companies. Nevertheless, the authors believe that this study provides a valuable piece of comparative work that can provide the basis for further research in a number of areas.

In-depth analyses of digital technology deployment in support of sustainability and circularity in the T&C industry would be of value, as little detailed information is publicly available on how companies are integrating digital technologies into their sustainability strategies and CE practices. This could provide insights into the optimal combination of digital technologies and CE practices in the industry's various processes. It is also evident that collaboration and partnerships are critical in engendering the transition to sustainability and the CE, and that digital technologies play a major role in building trust and ensuring real-time and transparent information flows. Future research could profitably focus on identifying the barriers and challenges to successful digitally-enabled and supported partnerships, as well as the key factors that contribute to their success. Then, the role of consumers, including their consumption patterns, knowledge, and skills related to sustainability and CE, and especially, the role of digital technologies in engaging consumers in sustainable and circular consumption patterns, represents another critical area for research. Understanding these factors could provide valuable insights into how to effectively engage consumers in the transition to a sustainable and circular T&C industry.

Such research initiatives could be complemented by comprehensive quantitative studies that could build upon studies included in this book to shed more light on the state of the European T&C industry, particularly in terms of regional differences and company sizes in the twin transition to sustainability and digitalisation.

References

Aguiar, J., Pereira, R., Braga Vasconcelos, J., & Bianchi, I. (2018). An overlapless incident management maturity model for multi-framework assessment (ITIL, COBIT, CMMI-SVC). *Interdisciplinary Journal of Information, Knowledge, and Management*, *13*, 137–163. https://doi.org/10.28945/4083

Ahmed, W. A. H., & MacCarthy, B. L. (2021). Blockchain-enabled supply chain traceability in the textile and apparel supply chain: A case study of the fiber producer, Lenzing. *Sustainability, 13*(19), 10496. https://doi.org/10.3390/su131910496
Bressanelli, G., Adrodegari, F., Pigosso, D. C. A., & Parida, V. (2022). Circular economy in the digital age. *Sustainability, 14*(9), 5565. https://doi.org/10.3390/su14095565
Costa, I., Riccotta, R., Montini, P., Stefani, E., de Souza Goes, R., Gaspar, M. A., Martins, F. S., Fernandes, A. A., Machado, C., Loçano, R., & Larieira, C. L. C. (2022). The degree of contribution of digital transformation technology on company sustainability areas. *Sustainability, 14*(1), 462. https://doi.org/10.3390/su14010462
Cruz, E. F., & Da Rosado Cruz, A. M. (2023). Digital solutions for engaging end-consumers in the circular economy of the textile and clothing value chain—A systematic review. *Cleaner and Responsible Consumption, 11*, 100138. https://doi.org/10.1016/j.clrc.2023.100138
Donmezer, S., Demircioglu, P., Bogrekci, I., Bas, G., & Durakbasa, M. N. (2023). Revolutionizing the garment industry 5.0: Embracing closed-loop design, E-libraries, and digital twins. *Sustainability, 15*(22), 15839. https://doi.org/10.3390/su152215839
European Union. (2023). *Sustainable development in the European Union: Monitoring report on progress towards the SDGS in an EU context: 2023 edition*. Publications Office of the European Union.
EURATEX. (2024). *Facts and key figures 2023 of the European textile and clothing industry*. https://euratex.eu/wp-content/uploads/EURATEX_FactsKey_Figures_2022rev-1.pdf
European Commission. (2019). *The European green deal: COM(2019) 640 final*. https://eur-lex.europa.eu/resource.html?uri=cellar:b828d165-1c22-11ea-8c1f-01aa75ed71a1.0002.02/DOC_1&format=PDF
European Commission. (2020). *COM (2020) 102 final: A new industrial strategy for Europe*. https://eur-lex.europa.eu/legal-content/EN/TXT/PDF/?uri=CELEX:52020DC0102&from=EN
European Commission. (2022a). *Data protection in the EU*. https://commission.europa.eu/law/law-topic/data-protection/data-protection-eu_en
European Commission. (2022b). *EU strategy for sustainable and circular textiles: COM(2022) 141 final*. https://environment.ec.europa.eu/strategy/textiles-strategy_en
European Commission. (2023a). *Corporate sustainability reporting*. https://finance.ec.europa.eu/capital-markets-union-and-financial-markets/company-reporting-and-auditing/company-reporting/corporate-sustainability-reporting_en
European Commission. (2023b). *EU AI Act: First regulation on artificial intelligence*. https://www.europarl.europa.eu/topics/en/article/20230601STO93804/eu-ai-act-first-regulation-on-artificial-intelligence
European Commission. (2023c). *EU taxonomy for sustainable activities*. https://finance.ec.europa.eu/sustainable-finance/tools-and-standards/eu-taxonomy-sustainable-activities_en#legislation
European Commission. (2023d). *European data governance act*. https://digital-strategy.ec.europa.eu/en/policies/data-governance-act
European Commission. (2023e). *European data strategy—Making the EU a role model for a society empowered by data*. https://commission.europa.eu/strategy-and-policy/priorities-2019-2024/europe-fit-digital-age/european-data-strategy_en
European Commission. (2023f). *Regulation on deforestation-free products*. https://environment.ec.europa.eu/topics/forests/deforestation/regulation-deforestation-free-products_en
European Commission. (2023g). *Waste framework directive*. https://environment.ec.europa.eu/topics/waste-and-recycling/waste-framework-directive_en
European Commission. (2024a). *Corporate sustainability due diligence*. https://commission.europa.eu/business-economy-euro/ng-business-eu/sustainability-due-diligence-responsible-business/corporate-sustainability-due-diligence_en
European Commission. (2024b). *Data act*. https://digital-strategy.ec.europa.eu/en/policies/data-act
European Commission. (2024c). *Ecodesign for sustainable products regulation—Making sustainable products in the EU the norm*. https://commission.europa.eu/energy-climate-change-enviro

nment/standards-tools-and-labels/products-labelling-rules-and-requirements/sustainable-products/ecodesign-sustainable-products-regulation_en
European Environment Agency. (2024). *Management of used and waste textiles in Europe's circular economy*. https://www.eea.europa.eu/publications/management-of-used-and-waste-textiles
Fachbach, I., Lechner, G., & Reimann, M. (2022). Drivers of the consumers' intention to use repair services, repair networks and to self-repair. *Journal of Cleaner Production, 346*, 130969. https://doi.org/10.1016/j.jclepro.2022.130969
Gazzola, P., Pavione, E., Pezzetti, R., & Grechi, D. (2020). Trends in the fashion industry. The perception of sustainability and circular economy: A gender/generation quantitative approach. *Sustainability, 12*(7), 2809. https://doi.org/10.3390/su12072809
Greenpeace. (2022). *Nachhaltigkeit ist tragbar* [In English:Sustainability is wearable]: *Repräsentative Greenpeace-Umfrage zu Kaufverhalten, Tragedauer und Nutzung der Alternativen zum Neukauf von Mode* [In English: Representative greenpeace survey on purchasing behavior, wearing time and use of alternatives to buying new fashion]. https://www.greenpeace.de/publikationen/nachhaltigkeit-tragbar
Irizar, J., & Wynn, M. G. (2018). A new maturity model for project risk management in the automotive industry. *International Journal of Risk and Contingency Management, 7*(3), 53–72. https://doi.org/10.4018/IJRCM.2018070103
Izsak, K., & Moreno, C. (2024). *Monitoring the twin transition of industrial ecosystems: Textiles: Analytical report*. https://op.europa.eu/en/publication-detail/-/publication/06780edf-0911-11ef-a251-01aa75ed71a1/language-en/format-PDF/source-343263872
Kirchherr, J., Yang, N.-H.N., Schulze-Spüntrup, F., Heerink, M. J., & Hartley, K. (2023). Conceptualizing the circular economy (revisited): An analysis of 221 definitions. *Resources, Conservation and Recycling, 194*, 107001. https://doi.org/10.1016/j.resconrec.2023.107001
Krause, J. (2018). Relationship between the voluntary instrument of CSR in the textile industry in the Czech Republic and Financial Performance. *Fibres and Textiles in Eastern Europe, 26*(5(131)), 8–12. https://doi.org/10.5604/01.3001.0012.2524
Laitala, K., Klepp, I. G., Haugrønning, V., Throne-Holst, H., & Strandbakken, P. (2021). Increasing repair of household appliances, mobile phones and clothing: Experiences from consumers and the repair industry. *Journal of Cleaner Production, 282*, 125349. https://doi.org/10.1016/j.jclepro.2020.125349
Laitala, K., Klepp, I., & Henry, B. (2018). Does use matter? Comparison of environmental impacts of clothing based on fiber type. *Sustainability, 10*(7), 2524. https://doi.org/10.3390/su10072524
Liu, Y., Huang, H., Zhu, L., Zhang, C., Ren, F., & Liu, Z. (2020). Could the recycled yarns substitute for the virgin cotton yarns: A comparative LCA. *The International Journal of Life Cycle Assessment, 25*(10), 2050–2062. https://doi.org/10.1007/s11367-020-01815-8
Luoma, P., Penttinen, E., Tapio, P., & Toppinen, A. (2022). Future images of data in circular economy for textiles. *Technological Forecasting and Social Change, 182*, 121859. https://doi.org/10.1016/j.techfore.2022.121859
McKinsey. (2022). *Scaling textile recycling in Europe—Turning waste into value*. https://www.mckinsey.com/industries/retail/our-insights/scaling-textile-recycling-in-europe-turning-waste-into-value
Muench, S., Stoermer, E., Jensen, K., Asikainen, T., Salvi, M., & Scapoo, F. (2022). Towards a green & digital future: Key requirements for successful twin transitions in the European Union (JRC science for policy report JRC129319). *Europäische Gemeinschaften*. https://doi.org/10.2760/977331
Nelson, E., & Kaye, D. (2024, July 19). What we know about the global Microsoft outage. *The New York Times*. https://www.nytimes.com/2024/07/19/technology/microsoft-crowdstrike-outage-what-happened.html
Neri, A., Cagno, E., Susur, E., Urueña, A., Nuur, C., Kumar, V., Franchi, S., & Sorrentino, C. (2024). The relationship between digital technologies and the circular economy: A systematic literature review and a research agenda. *R&D Management*, Article radm.12715. Advance online publication. https://doi.org/10.1111/radm.12715

Piscicelli, L. (2023). The sustainability impact of a digital circular economy. *Current Opinion in Environmental Sustainability, 61*, 101251. https://doi.org/10.1016/j.cosust.2022.101251
Popowska, M., & Sinkiewicz, A. (2021). Sustainable fashion in Poland—Too early or too late? *Sustainability, 13*(17), 9713. https://doi.org/10.3390/su13179713
Sadowski, M., Perkins, L., & McGarvey, E. (2021). Roadmap to net zero: Delivering science-based targets in the apparel sector. *World Resources Institute*. Advance online publication. https://doi.org/10.46830/wriwp.20.00004
Sandvik, I. M., & Stubbs, W. (2019). Circular fashion supply chain through textile-to-textile recycling. *Journal of Fashion Marketing and Management: An International Journal, 23*(3), 366–381. https://doi.org/10.1108/JFMM-04-2018-0058
Santa-Maria, T., Vermeulen, W. J. V., & Baumgartner, R. J. (2022). How do incumbent firms innovate their business models for the circular economy? Identifying micro-foundations of dynamic capabilities. *Business Strategy and the Environment, 31*(4), 1308–1333. https://doi.org/10.1002/bse.2956
Schöggl, J.-P., Stumpf, L., & Baumgartner, R. J. (2024). The role of interorganizational collaboration and digital technologies in the implementation of circular economy practices—Empirical evidence from manufacturing firms. *Business Strategy and the Environment, 33*(3), 2225–2249. https://doi.org/10.1002/bse.3593
Spahiu, T., Manavis, A., Kazlacheva, Z., Almeida, H., & Kyratsis, P. (2021). Industry 4.0 for fashion products—Case studies using 3D technology. *IOP Conference Series: Materials Science and Engineering, 1031*(1), 12039. https://doi.org/10.1088/1757-899X/1031/1/012039
Stridsland, T., Biørnstad, W., Vigen, K., Østergaard, K. L., & Sanderson, H. (2023). No-one left behind: An open access approach to estimating the carbon footprint of a Danish clothing company. *Journal of Cleaner Production, 426*, 139126. https://doi.org/10.1016/j.jclepro.2023.139126
Suchek, N., & Franco, M. (2024). Inter-organisational cooperation oriented towards sustainability involving SMEs: A systematic literature review. *Journal of the Knowledge Economy, 15*(1), 1952–1972. https://doi.org/10.1007/s13132-023-01196-x
Tolentino-Zondervan, F., & DiVito, L. (2024). Sustainability performance of Dutch firms and the role of digitalization: The case of textile and apparel industry. *Journal of Cleaner Production, 459*.
Uhrenholt, J. N., Kristensen, J. H., Rincón, M. C., Adamsen, S., Jensen, S. F., & Waehrens, B. V. (2022). Maturity model as a driver for circular economy transformation. *Sustainability, 14*(12), 7483. https://doi.org/10.3390/su14127483
Wiedemann, S. G., Biggs, L., Clarke, S. J., & Russell, S. J. (2022). Reducing the environmental impacts of garments through industrially scalable closed-loop recycling: Life cycle assessment of a recycled wool blend sweater. *Sustainability, 14*(3), 1081. https://doi.org/10.3390/su14031081
Wynn, M., & Jones, P. (2022). *Can digital technologies deliver the circular economy?* https://breakthrough.neliti.com/digital-technologies-circular-economy/
Wynn, M., & Jones, P. (2023). Corporate responsibility in the digital era. *Information, 14*(6), 324. https://doi.org/10.3390/info14060324

Tina Wiegand (tina.wiegand@hof-university.de) is a lecturer at Hof University of Applied Sciences, Germany. She completed a dual degree in Business Administration with a focus on supply chain management and a Master's degree with a focus on project management. She worked in the textile industry for over ten years in project and process management, with managerial responsibilities in operational and strategic purchasing for different corporate brands. Additionally, she gained experience in significant systems projects. Currently, she is pursuing a Ph.D. degree at the University of Gloucestershire, UK, researching digitalisation and circularity in the textile supply chain.

Martin Wynn (mwynn@glos.ac.uk) is Associate Professor in Information Technology in the School of Business, Computing and Social Sciences at the University of Gloucestershire, UK and holds a Ph.D. from Nottingham Trent University. He was appointed Research Fellow at East London University, and he spent 20 years in industry at Glaxo Pharmaceuticals and HP Bulmer Drinks. His research interests include digitalisation, information systems, sustainability, project management, and urban planning. His latest book, Handbook of Research on Digital Transformation, Industry Use Cases, and the Impact of Disruptive Technologies, was published in 2022.

The manufacturer's authorised representative in the EU is Springer Nature Customer Service Centre GmbH, Europaplatz 3, 69115 Heidelberg, Germany. If you have any concerns regarding our products, please contact ProductSafety@springernature.com

Printed and bound by CPI Group (UK) Ltd, Croydon, CR0 4YY
16/09/2026
02217780-0008